INDUCED MUTAGENESIS

MOLECULAR MECHANISMS AND THEIR
IMPLICATIONS FOR ENVIRONMENTAL PROTECTION

BASIC LIFE SCIENCES

Alexander Hollaender, General Editor

Associated Universities, Inc., Washington, D.C.

Recent volumes in the series:

A Continuation Order Plan is available for this series. A continuation order will bring delivery of each new volume immediately upon publication. Volumes are billed only upon actual shipment. For further information please contact the publisher.

INDUCED MUTAGENESIS

MOLECULAR MECHANISMS AND THEIR IMPLICATIONS FOR ENVIRONMENTAL PROTECTION

Edited by

Christopher W. Lawrence

School of Medicine and Dentistry
The University of Rochester
Rochester, New York

PLENUM PRESS • NEW YORK AND LONDON

Library of Congress Cataloging in Publication Data

Rochester International Conference on Environmental Toxicity (14th: 1981)
    Induced mutagenesis.

    (Basic life sciences; v. 23)
    Proceedings of the Fourteenth Rochester International Conference on Environmental
Toxicity, held June 1-3, 1981, at the University of Rochester, Rochester, New York.
    Includes bibliographical references and index.
    1. Mutagenesis—Congresses. 2. Chemicals—Environmental aspects—Congresses. I.
Lawrence, Christopher W. II. Title. III. Series. [DNLM: 1. Environmental exposure—
Congresses. 2. Mutagenicity tests—Congresses. 3. Mutagens—Adverse effects—Con-
gresses. W3 BA255 v.23/WA 754 R676 1981i]
QH465.A1R63 1981   574.87′3282   82-16706
ISBN 0-306-41163-6

Proceedings of the Fourteenth Rochester International Conference
on Environmental Toxicity, held June 1 – 3, 1981, at the University
of Rochester, Rochester, New York
Conference number CONF-810608

© 1983 Plenum Press, New York
A Division of Plenum Publishing Corporation
233 Spring Street, New York, N.Y. 10013

Printed in the United States of America

PREFACE

Concern is often expressed that our environment may include an increasingly large variety of mutagens, but the extent of the potential hazard they pose has yet to be fully evaluated. A variety of empirical procedures has been devised with which to estimate the mutagenic potency of suspect agents, and the relative merits of different tests are currently under debate. Although such tests are of great value, and are indeed indispensable, they are not, nevertheless, sufficient. In the long term, accurate estimation of hazard will also require a better understanding of the various mechanisms of mutagenesis, and in many instances these remain remarkably elusive. Our knowledge and appreciation of the problem has increased substantially over the last few years, but the precise way in which many mutagens cause mutations is not yet known.

The aims of this conference were therefore two-fold. The first was to survey present information about mutagenic mechanisms, drawing together data from work with various experimental approaches and organisms, in order to discern the principles governing the action of different mutagens. The second was to examine the implications of such principles for the execution and evaluation of test procedures, and critically assess the research areas that need further attention in order to improve the interpretation of test results.

Chris Lawrence

ACKNOWLEDGEMENT

We gratefully acknowledge the support provided for this Conference by the U.S. Department of Energy, The Foundation for Microbiology, Exxon Corporation and the University of Rochester.

We are especially indebted to Dorris Nash for efficient administrative and secretarial service in the preparation and conduct of the Conference, and to the Word Processing Center, Department of Radiation Biology and Biophysics, for typing the discussion sections.

The Conference Committee

Christopher W. Lawrence, Chairman
Louise Prakash
Fred Sherman

CONTENTS

SESSION I: MUTAGEN/DNA INTERACTIONS

SESSION II: INFIDELITY AND SPECIFICITY

SESSION III: GENETIC ANALYSIS OF MUTAGENESIS

SESSION IV: MAMMALIAN SYSTEMS

SESSION V: HUMAN SYSTEMS AND ENVIRONMENTAL PROTECTION

---

*Conference participant

SESSION I

MUTAGEN/DNA INTERACTIONS

MODERATOR: LOUISE PRAKASH

MUTAGENIC EFFECTS OF NUCLEIC ACID MODIFICATION

AND REPAIR ASSESSED BY IN VITRO TRANSCRIPTION

B. Singer

Department of Molecular Biology and Virus Laboratory
University of California
Berkeley, CA 94720

Mutagenesis is defined as a heritable change which can occur through indirect as well as direct change in the genetic message. This paper, however, will focus on the direct effects of chemical modification of nucleic acids. Since man is the species we are most concerned about, the effect of modification on mammalian cells or whole animals will be stressed whenever possible.

Many of the types of chemicals which have been assessed in terms of their mutagenicity in mammalian systems (1-4) and for which there are data on the mechanism of chemical modification of nucleic acids are shown in Figure 1. These are primarily directly acting aklyating agents or other simple point mutagens.

Chemical Reactions of Simple Alkylating Agents

A large group of mutagens are alkylating agents, although these differ greatly in their mutagenicity. For this reason, we have been working for a number of years to elucidate the chemical reactions of nucleic acids with the "good" mutagens (e.g., N-nitroso compounds), as compared to the "poor" mutagens (e.g., alkyl sulfates).

Single-stranded homopolyribonucleotides were first used as models for RNA in most of these studies. Significant differences were found between methylating agents and the analogous ethylating agents, suggesting that the greater mutagenicity of, for example, ethyl methanesulfonate (EtMS) compared to methyl methanesulfonate (MeMS) could be due to reaction with oxygens or exocyclic amino groups. However, in RNA the exocyclic amino groups were not

1

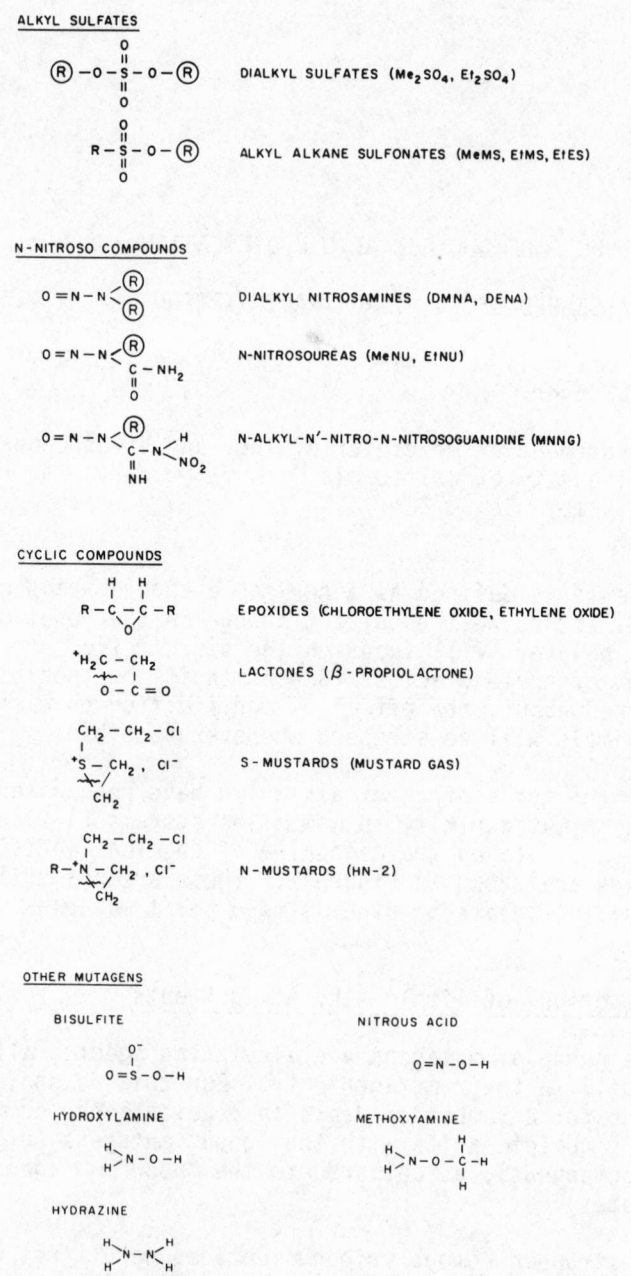

Figure 1. Structural formulas and abbreviations of various types of simple mutagens. With the exception of dialkyl nitrosamines, all are directly acting. The first three groups are alkylating agents.

measurably reactive, while all oxygens including phosphodiesters and ribose could be modified with ethylnitrosourea (Table I) (5). Both ethylnitrosourea (EtNU) and ethylnitrosoguanidine (ENNG) ethylated oxygens predominantly. Two new derivatives were found O^2-alkyl C and O^2-alkyl U, and one previously marginally detected, O^4-alkyl U, was also found to be a significant product.

Turning to double-stranded nucleic acids, all the same reaction products could be quantitated, and again EtNU (and ENNG) reacted to a high extent with the O^6 of G, the O^2 of C and T, the O^4 of T and phosphodiesters (5) (Table II). The specificity of eight reagents toward oxygens apparently correlates with the reported mutagenicities in various systems: EtNU (ENNG) > MeNU (MNNG) > EtMS > MeMs > Et_2SO_4 > Me_2SO_4. While the metabolically activated nitrosamines are not included in the Tables, data for their reaction in vivo indicates that they resemble the nitrosoureas in oxygen specificity.

There are several points in Tables I and II which should particularly be noted when discussing chemical reactivity. First, the Tables deal with percent of total alkylation, or proportion, rather than absolute amounts. Methylating agents are about 20 times more efficient in reacting with nucleic acids (in vitro or in vivo) than ethylating agents under the same reaction conditions. Thus, 6.3 percent O^6-MeG (MeNU) represents in moles about 15 times the amount of O^6-EtG (EtNU) (Table II). It cannot be said, therefore, that quantitatively methylation of oxygens occurs to a lower extent than ethylation. Yet the biological effects of EtNU are greater than with MeNU at similar dose levels (1).

Regarding phosphotriesters, which are clearly the major product of EtNU reaction, here too, quantitatively there are as many or more methyltriesters than ethyltriesters, as is also the case when comparing triesters formed by MeMS and EtMS. All these comparisons, of course, are only true when comparing equimolar reactions under identical conditions of time, temperature, pH, etc., and in vivo there are the additional factors of repair and errors in replication which may differ for methyl and ethyl substituents.

Another point to be noted is that both the N-1 of A and N-3 of C are hydrogen-bonded in double-stranded (ds) nucleic acids and are presumed to be less reactive than is single-stranded (ss) nucleic acids. However, a definite and unexpectedly high reactivity was observed for these sites (Table II). Having convinced ourselves that the ds nucleic acids we were studying were completely ds, we then found that alkylation of hydrogen-bonded sites, at and even below 37°C, occurred as a result of thermal denaturation (15). This finding, also described by others using different techniques (16-18), opens up many possibilities and must be considered when

Table I. Alkylation of Single-Stranded Nucleic Acids in Vitro[a]

Aklylation Site	% of Total Alkylation[b]					
	Me_2SO_4	MeMS	MeNU	Et_2SO_4	EtMS	EtNU
Adenine						
N-1	13.2	18	2.8	11	8	2
N-3	2.6	1.4	2.6	3	1	1.2
N-7	3.1	3.8	1.8	3	3	0.6
Guanine						
N-3	∿0.1	∿1	∿0.4	2	1	0.5
O^6	<0.2	nd	3	2	1	7
N-7	62	68	69	62	77	10
Uracil/Thymine						
O^2				2		6
N-3				nd		
O^4				1		4
Cytosine						
O^2				2		5.5
N-3	9.5	10	2.3	11	5	1.7
Diester	<2	2	∿10	6	10	65
Ribose						12

[a]Analyses were from experiments using DNA from M13 phage and RNA from TMV, yeast, HeLa cells, animal ribosomes and μ2 phage. Much of the data was compiled by Singer (6). Other references are Singer and Fraenkel-Conrat (7) and Singer (5).

[b]The absolute amount of alkylation varied greatly but the proportion of derivatives was not noticeably affected. "nd" indicates that the derivative was not detected.

Table II. Alkylation of Double-Stranded Nucleic Acids in Vitro[b]

Aklylation Site	% of Total Alkylation[b,c]					
	Me_2SO_4	MeMS	MeNU	Et_2SO_4	EtMS	EtNU
Adenine						
N-1	1.9	3.8	1.3	2	1.7	0.2
N-3	18	10.4	9	10	4.0	4.0
N-7	1.9	(1.8)	1.7	1.5	1.1	0.3
Guanine						
N-3	11	(0.6)	0.8	0.9	0.9	0.6
O^6	0.2	(0.3)	6.3	0.2	2.	7.8
N-7	74	83	67	67	65	11.5
Thymine						
O^2			0.11	nd	nd	7.4
N-3			0.3	nd	nd	0.8
O^4			0.4	nd	nd	2.5
Cytosine						
O^2	(nd)	(nd)	0.1	nd	nd	3.5
N-3	(<2)	(<1)	0.6	0.7	0.6	0.2
Diester		0.8	17	16	13	57

[a]Analyses were from experiments using DNA from salmon sperm, calf thymus, salmon testes, rat liver and brain, human fibroblasts, and HeLa cells. Some of the data were complied by Singer (6). Other key references are: Goth & Rajewsky (8); Lawley et al. (9); Sun & Singer (10); Singer (5,11); Shackleton et al. (12); Newbold et al. (13); Beranek et al. (14).

looking for products of nucleic acid-mutagen interaction. In a
later section, the potential mutagenic effect of modification of
the base-paired positions, the N-1 of A or N-3 of C or U/T, will be
discussed. Among the other positions on bases that are involved in
hydrogen-bonding in a Watson-Crick structure are the O^6 of G, O^4
of U/T and O^2 of C. These, however, have a free electron pair
as shown in Figure 2, allowing alkylation to take place. This is
well illustrated by the data for EtNU reaction with ss and ds
nucleic acids. Here, in contrast to the hydrogen-bonded nitrogens,
N-1 of A or N-3 of C, there is no significant difference in oxygen
reactivity as a result of strandedness.

Both ss and ds polynucleotides have proven very useful in the
identification of reaction products and in testing hypotheses, such
as reactivity of hydrogen-bonded sites. Much of our technical
expertise, such as separation methods, stability measurements,
chemical parameters, etc., have come from the use of synthetic
polynucleotides. It must be noted, however, that a number of
derivatives, clearly identified in polymer reactions with EtNU, are
found in RNA and DNA reactions. These are shown on the right-hand
side of Figure 3 and include N^6-alkyl A, N^4-alkyl C and
N^2-alkyl G. The latter alkylation site has been found only as
O^6, N^2-diethyl G after EtNU reaction, or as N^2-benzylguanosine,
reported for guanosine reaction with N-nitroso-N-benzylurea (19).
In no experiment could we identify a C-8 substituted purine.

There thus appears little overlap, regarding reactive sites,
between the products of many polycyclic aromatic hydrocarbons
with nucleic acids, and those formed by simple alkylating
agents (Fig. 3). An appropriate caution is that results with
synthetic polymers cannot always be extrapolated directly to
nucleic acids. However, no reaction has been found to occur in
nucleic acids that does not occur in model experiments.

If polymers are not the same as nucleic acids in terms of
reactivity, then are test tube experiments valid as predictors of
in vivo reactions? Data on the reactions of DNA, cells, and whole
animals, from our and other laboratories, are very similar to in vitro
data in Table II, although the extent of in vivo reaction is 2-3
orders of magnitude less than in many in vitro experiments. It has
not yet been mentioned but almost all the data in vivo (and usually
in vitro) are obtained by using radioactive alkylating agents. The
limitation of specific activity and dose when studying reactions in
mammalian cells or whole animals makes the results more variable
when trying to quantitate minor derivatives. Nevertheless, the
data indicate that nucleic acid reactions are independent of milieu
or extent of reaction.

We would conclude that the reactive sites in EtNU treated DNA
are: Phosphate diester >> N-7 G > O^2-T, O^6-G > N-3 A > O^2C,
O^4-T > and Other N-.

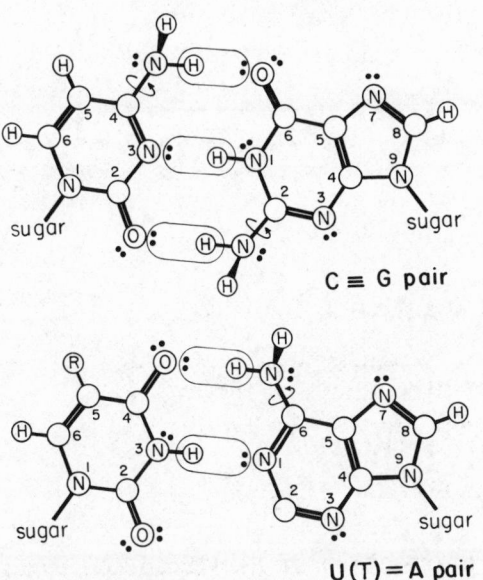

C ≡ G pair

U(T) = A pair

Figure 2. Participation of electron pairs in base pairing. The ovals represent hydrogen bonds in a Watson-Crick structure. Solid pairs of dots represent electron pairs. Electron pairs inside ovals, when involved in hydrogen-bonding, are unreactive. Others may react when in double-stranded structures, if physically accessible. Note that the oxygens all have a free electron pair, reactive both in single and double-stranded polymers (Tables 1, 2). The three exocyclic amino groups may rotate, as indicated by arrows, so that the two hydrogens are equivalent. A substituent may therefore interfere with normal H-bonding if rotated into the base pairing side.

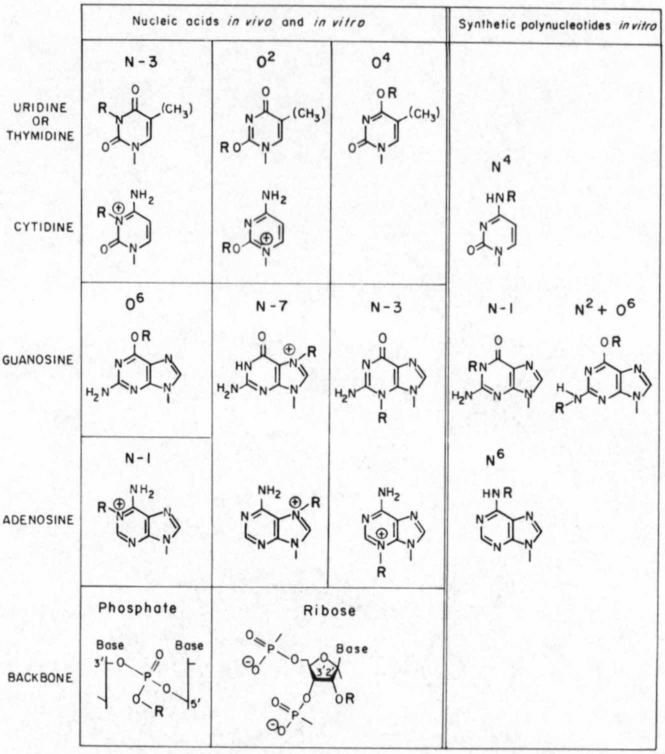

Figure 3. Sites of reaction of N-nitroso compounds with nucleic
acids or polynucleotides in neutral aqueous solution.
All derivatives shown are formed with ethyl and methyl
nitrosoureas and nitrosoguanidines. See Table 2 for
products formed by other alkylating agents. The
derivatives on the right side have been found only
after reaction with synthetic polynucleotides.

Chemical Reactions of Cyclic Alkylating Agents

These comprise a number of unrelated mutagens and carcinogens.
Representative identified nucleic acid derivatives are shown in
Figure 4. The alkylating agents studied earliest were the N- and
S-mustards, the former of which (HN-2) is also used
therapeutically. Brookes and Lawley (19-22), in a pioneering
series of studies, found that mustard gas reacts with the N-7 of G
and both a monoadduct and cross-linked adduct were formed. Later,
derivatives resulting from reaction of the N-1 of A and N-3 of C

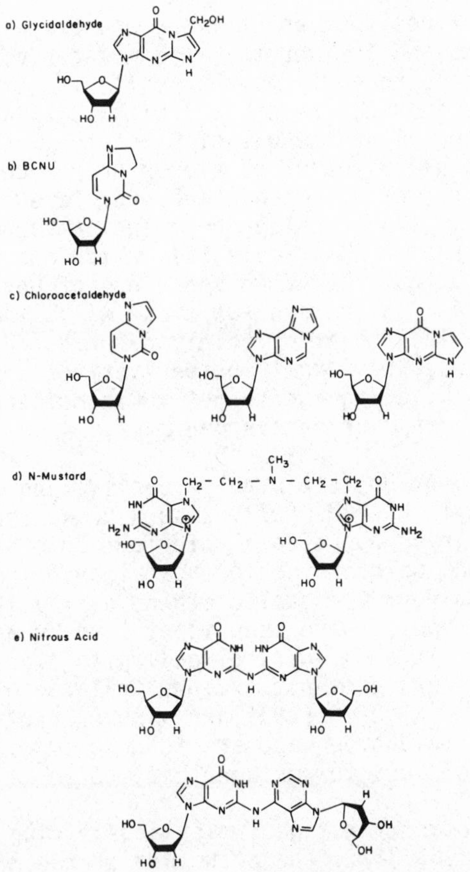

Figure 4. Examples of mutagenic reaction resulting in cyclic or
           cross-linked derivatives. a. The glycidaldehyde
           product with guanosine; b. 3,N^4-Ethano C from BCNU
           reaction with cytidine; c. Etheno derivatives from
           reaction of cytidine, adenosine and guanosine with
           chloracetaldehyde; d. Cross-linked N-7 guanines
           resulting from reaction with nitrogen mustard; e. Two
           types of cross-links, G-G or G-A, resulting from
           nitrous acid reaction.

were identified so that it now appears that S-mustard resembles
alkyl sulfates in its reactions except that, being bifunctional,
inter- and intrastrand cross-linking also occurs. N-mustard also
cross-links through the N-7 of G (23) (Fig. 4d).

   β-Propiolactone also reacts with the N-7 of G (24) and N-1 of
A (25). There is less information on other reactions although it
might be predicted that the N-3 of A could also be modified.

Aliphatic epoxides such as ethylene and propylene oxide, widely used as sterilants and in industry, are mutagenic and ethylene oxide has been added recently to the small list of human carcinogens. The epoxides resemble typical alkylating agents and react with DNA and RNA at the N-7 of G and the N-1 and N-3 of A (26), forming the hydroxyethyl or hydroxypropyl derivatives. No O^6-alkyl derivative was detected using propylene oxide. N-7 alkyl G and N-3 alkyl A are also the principal products from dimethylsulfate-treated DNA (Table II), which has also been termed a human carcinogen, but the known reactions of both dimethylsulfate and ethylene oxide with DNA are not those which would be likely to cause mutations. This point certainly deserves further investigation, since the identified derivatives are the same as those formed by alkylating agents not now considered to be significantly mutagenic or carcinogenic.

Chloroethylene oxide is a highly reactive and mutagenic metabolite of vinyl chloride (27), also a human carcinogen. Like all epoxides, the ring opens easily to form chloroacetaldehyde. Both chloroethylene oxide and chloroacetaldehyde react with polynucleotides to form the cyclic etheno derivatives, ϵA, ϵC, and ϵG (Fig. 4C). ϵA and ϵC have been identified by mass spectrometry as products in the liver of rats given vinyl chloride in drinking water (28). This work does not exclude that the other known derivatives, $1,N^2$-ϵG and $N^2,3$-ϵG (29), are also products. The bromine analogue, bromoacetaldehyde, appears to be chemically identical but is evidently of very low mutagenicity (30).

Another bifunctional mutagen which adds a ring to a nucleoside is glycidaldehyde (31,32). The structure of the modified guanine derivative from reaction with calf thymus DNA is shown in Figure 4A. This mutagen, like chloroacetaldehyde, should also form cyclic derivatives with other nucleosides.

The reactions of the widely used antineoplastic agent, 1,3-bis-(2-chloroethyl)-1-nitrosourea, (BCNU), have been studied in synthetic polynucleotides and three new derivatives identified: 3-(β-hydroxyethyl)C, 3, N^4-ethano C (Fig. 4b) and 7-(β-hydroxyethyl)G (33). Cross-links have been measured indirectly (34) and these could result from initial reaction at either the N^4 of C or the N-7 of G, which would make the entire molecule an alkylating agent. The fluorine analogue, FCNU, would be likely to react with nucleic acids in a similar manner but as yet, only the N-3 derivative of C has been reported (35).

Simple, Non-Alkylating Mutagens

These might be termed "classical" mutagens since their predominant effect is to change base-pairing by deamination

(nitrous acid, bisulfite), or addition to the N^4 of C
(hydroxylamine, methoxyamine, hydrazine), causing a shift in the
tautomeric equilibrium (chemistry reviewed by Brown (36)). The
major products of each of these mutagens is shown in Figure 5.
Figure 5a illustrates modification of the exocyclic amino group
while 5b, c, and d are deamination reactions. In addition,
bisulfite, hydroxylamine, methoxyamine, and hydrazine can add to
the 5,6 double-bond of C or (except for methoxyamine) to the 5,6
double-bond of U. Hydroxylamine and methoxyamine also react with
the N^6 of A (37,38) which is a potentially mutagenic reaction.
While nitrous acid primarily deaminates C→U and A→HX, deamination
of G→X and evidence for cross-links has also been reported (39).
Both these latter reactions are inactivating, rather than
mutagenic. The two types of cross-links identified in nitrous acid
treated DNA are shown in Figure 4d.

Effect of Nucleic Acid Structures on Reactivity

Conformation and secondary structure of nucleic acids play a
major role in chemical reactivity (36,40). That is, in general,
mononucleotides are more reactive than single-stranded
homopolynucleotides which are more reactive than single-stranded
nucleic acids, which in turn, are more reactive than
double-stranded nucleic acids. In particular, it has been found
that hydroxylamine, methoxyamine and bisulfite are poor reagents
for ds nucleic acids even though their preferred sites of
modification are not hydrogen-bonded. Bisulfite and formaldehyde
are unreactive toward the N^4 of C or the N^6 of A, respectively,
in ds polymers and this selectivity has been used as a test for the
extent of ds structure (17,41,42). Chloroacetaldehyde is much less
reactive in a ds DNA compared to denatured DNA but is nevertheless
a mutagen (43).

All DNA, when being transcribed or replicating is
single-stranded, and single-stranded regions are more reactive than
double-stranded regions. Therefore, it follows that reaction on ss
segments can be more extensive and on different sites (i.e., N-1 of
A, N-3 of C) than on ds segments. In fact, DNA treated with
alkylating agents in vivo is found to react at what are normally
base-paired sites.

Quantitation of Repair or Removal of Modified Nucleosides

There are many excellent reviews on the general subject of
DNA repair in eukaryotes (2,44-47) and only conclusions or data
not yet in review will be discussed. Most of the well-known and
well-characterized repair enzymes have been isolated from
procaryotes, notably E. coli (reviewed by Lindahl, (48)). These

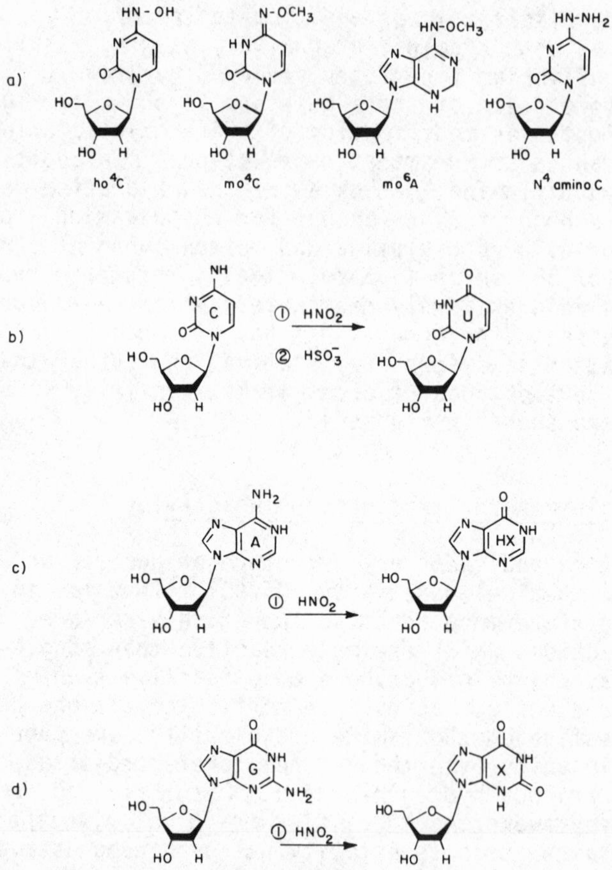

Figure 5.   Examples of reactions with mutagens producing shifts in
            the tautomeric equilibrium or deamination.
            a.   Reactions of hydroxylamine (ho⁴C), methoxyamine
            (mo⁴C, mo⁶A) or hydrazine (N⁴-amino C).
            b.   Deamination of C to U by nitrous acid or bisulfite.
            c.   Deamination of A to hypoxathine by nitrous acid.
            d.   Deamination of G to xanthine by nitrous acid.

include enzymes repairing UV dimers and apurinic/apyrimidinic
sites; glycosylases acting on 3-alkyl A, ring-opened 7-methyl G,
deaminated A and C; and an enzyme dealkylating O^6-alkyl G by
transfer of the alkyl group to an amino acid.

In vivo many nucleic acid derivatives discussed in this paper
are found to be removed/repaired at rates varying according to the
chemical stability of the derivative, amount of derivative, the
cell or animal species and the specific organ or cell studied.
These factors affect the $t_{1/2}$ of derivatives and for this reason

specific data are not given in this paper.

Chemical evidence for disappearance of mutagen products from DNA has been shown for all ethyl and methyl base alkylated derivatives identified in mammalian cells or animals treated with alkylating agents. (Examples are Frei et al. (49); Bodell et al. (50)). Two laboratories have found that rat liver or human lymphoblasts contain a glycosylase for 7-MeG (52,53). The lymphoblast extract also depurinates alkylated ds DNA containing 7-EtG and other N-3 and N-7 methyl and ethyl purines, although the ethyl derivatives are poorer substrates than the methyl derivatives (52). Rat liver homogenates (54) and a factor in rat liver chromatin (55) dealkylate O^6-alkyl G by the same mechanism as that reported for the E. coli enzyme (56-58).

Three other O-alkyl nucleosides are also enzymatically lost, particularly in EtNU treated rat liver (51) and human fibroblasts (50). There are no data on the enzymes responsible other than that they probably differ from the O^6-alkyl G enzyme on the basis that Renard and Verly reported that rat liver chromatin dealkylated O^6-alkyl G but not O^2-EtT (55).

Repair of alkylphosphotriesters, which represent a high proportion of the reaction of N-nitroso alkylating with nucleic acids, is questionable. Shooter and Slade (59), using an indirect method of quantitation (alkali strand breaks), conclude that methylphosphotriesters are decreased at a reasonable rate in vivo but ethylphosphodiesters apparently not. Our in vivo data (51) suggest that ethylphosphotriesters persist for much longer times than do base-ethylated products. Whether loss of triesters is due to dealkylation, which reforms normal diesters, or to an excision mechanism is still unclear.

To summarize, most derivatives formed by mutagens reacting with DNA appear to be removed/repaired by mammalian enzymes, although such enzymes have not been purified. The extent or rate of loss of derivatives is strongly dependent on the initial dose of mutagen since there is much evidence that repair systems can be saturated. Regardless of how efficiently repair occurs, it is seldom complete and certainly some potentially mutagenic derivatives are present during replication.

Transcription of Modified Nucleotides in Nucleic Acids and Polynucleotides

A modification of DNA which avoids repair may be mutagenic in at least three ways: it may directly mispair (60,61); it may cause frameshifts by stabilizing extrahelical forms, as proposed by

Streisinger (62); it may cause induction of a misrepair system:
SOS repair (63).

The mutagenicity of large aromatic derivatives appears to be
partly a function of their conformation in the DNA. Termination of
synthesis or termination with reinitiation (gaps) are most
frequently found (64-66). Secondary effects, such as the
production of polymerases with decreased fidelity have also been
seen (67). In polymers modified with the simple alkylating agent,
β-propiolactone, polynucleotide replication reveals increased
noncomplementary nucleotide incorporation with several polymerases
(68).

Transcription of polymers containing modified nucleotides has
been used as an in vitro model of in vivo alterations which would
be expected to cause point mutations. Direct mispairing resulting
from a specific modification in a template is the simplest kind of
mutational mechanism. It should be noted that changes in "base
pairing" may be considered misincorporations only in a biological
sense, in that they do not yield a faithful copy of the template.
However, physicochemically the interaction may be correct. For
example, the substitutions found when C is in the imino form, or a
purine glycosyl bond is syn, are considered part of the explanation
for the background level of base changes observed which are both
transitions and transversions (60,61,69,70). Modifications which
change the proportion of a tautomer would presumably act this way,
e.g., hydroxylamine. Figure 6, taken from Fresco, et al. (69)
shows the simplest tautomeric conversions which do occur at a low
frequency as a result of the normal tautomeric equilibrium of
nucleotides.

When the presence of modified nucleotides in templates
increases the level of misincorporation above the background, it
has been assumed that it is based on at least a transient hydrogen
bonding which satisfies the polymerase (71,72). It is possible to
conclude that the polymerase will insert any base which can form
one or more hydrogen bonds into the transcript. However, when
specific base pairing sites are modified, the resulting ambiguity
may not require any hydrogen bond formation. In living organisms
the chemistry of base pairs, including tautomeric equilibria and
the equivalence of electron distributions, is tempered by the
biochemistry of a variety of enzymes with specific proofreadings,
alignment, and error levels.

Some types of modification have no measurable effect on
fidelity in transcription (Table III, left side). Alkylation of
ribose or substitutions at the C-5 of pyrimidines would not be
considered mutagenic on this basis. The replacement of oxygen by
sulfur also does not change the observed base pair, even though
sulfur is larger.

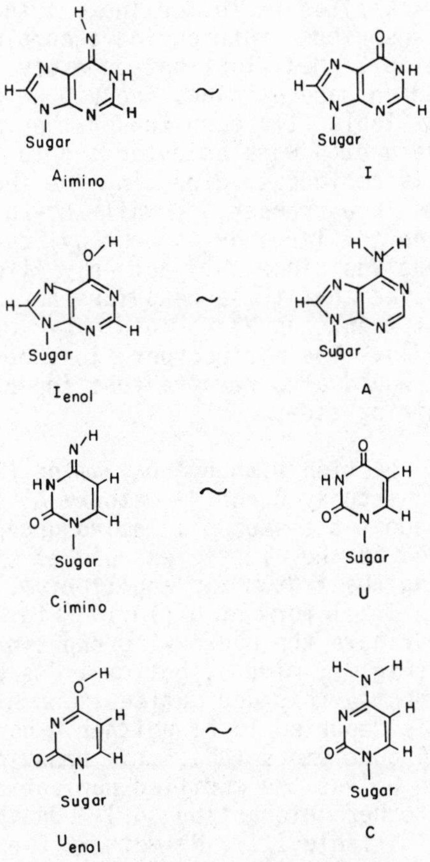

Figure 6. From Fresco et al. (69). Equivalence of the unfavored
           tautomeric forms (left side) to another base (right
           side) in the favored form. These shifts would lead to
           transitions.

The termination observed in replication experiments is not
often seen in these in vitro transcription experiments although it
has been inferred when very large groups are the substituent.
However, polymerases may pause at modified nucleotides as observed
in transcription of 16S RNA containing N^2-methyl G (73). The
pause is probably the time spent by the enzyme in choosing a
"proper" triphosphate. In transcription of synthetic polymers
containing N^2-methyl G, a level of ambiguity is found supporting
the concept that a poor fit is better than none (74).

It might be expected that alkylated bases substituted at

essential Watson-Crick sites would terminate transcription but instead they are transcribed, introducing a complete ambiguity (Table III, right side). Modifications on exocyclic groups, not necessarily involved in base pairing, produce a variety of mispairing patterns (Table III) even though they could all be non-mutagenic if the proton were oriented syn to the Watson-Crick side. N^4-Methyl C is ambiguous, directing the incorporation of all four nucleotides in a transcript, while N^4-acetyl C, N^6-hydroxyethyl A and the large N^6-isopentenyl derivative are non-mutagenic derivatives since they act only like the unmodified base. We have proposed that these behaviors are a measure of the ability of the substituents on N^4 to rotate syn or anti to the Watson-Crick side (74). The misincorporations observed with O-alkyl derivatives would also require that the alkyl groups be anti to the base pairing side.

The products of reaction with hydroxylamine (N^4-hydroxy C) and methoxyamine (N^4-methoxy C and N^6-methoxy A) illustrate the effect of rotation about the exocyclic amino group, but specifically are used as the classic example of effect of a modification shifting the tautomeric equilibrium. N^4-Hydroxy C prefers the imino (U like) form of C (10:1) (Fig. 7, top) but while in the amino form may have the hydroxyl group syn or anti to the base pairing sites (Fig. 7, middle, bottom). As a result, in transciption experiments, this derivative is completely ambiguous. While N^6-methoxy A is reported to be neither A nor G in transcription with C copolymers (38), this behavior would be difficult to explain unless the modified nucleotide were excluded by some stacking or other interaction (69). On the basis of transcription results (Table III), N^4-methoxy C exists only in the imino form with its substituent anti to the Watson-Crick side (Fig. 7, top).

The cyclic etheno derivatives, 3, N^4-etheno C and 1, N^6-etheno A, are products of chloroacetaldehyde modification. Because of its size and fluorescence, etheno C has been used as an analogue for A with several enzymes. It is not surprising, therefore, that it acts like A in transcription (Table III). It also acts like U and to a lesser extent, like G. However, it does not act like the parent nucleotide, C. Likewise, etheno A is not transcribed as C, but as U predominantly, even though this is a bulky adduct in a fixed position (75). Figures 8 and 9 illustrate why neither εA nor εC can direct G into a complimentary polymer. Even by rotating the sugar from the anti form (bottom of Fig. 8,9) it is not possible to bring the two bases together. However, a single hydrogen bond can be drawn between εA and A if the A is syn. Other single hydrogen bonds are theoretically possible to account for the observed misincorporations (top of Fig. 8,9).

Table III. Summary of Effect of Nucleoside Modification on
Transcription of Ribopolynucleotide Templates[a]

Base-Pairing Unchanged	Changed Pairing or Misincorporation
N^4-Acetyl C	N^4-Hydroxy C · <u>A</u>, G, U, <u>C</u>
	N^4-Methoxy C · <u>A</u>
	N^4-Methyl C · A, <u>G</u>, U, C
N^6-Methyl A	
N^6-Hydroxyethyl A	
N^6-Isopentenyl A	
	N^2-Methyl G · <u>A</u>, G, U, C
	Xanthosine · <u>U</u>
5-Halo, Methyl, Hydroxyl U, C	
2'-O-Methyl A, C, U	None known
7-Methyl G	
2-Thio U, C	O^2-Alkyl U · <u>A</u>, G, U, C
4-Thio U,	O^4-Alkyl G · <u>A</u>, <u>G</u>, U, C
	O^6-Alkyl U · <u>A</u>, <u>G</u>, U, C
	1-Methyl A · <u>A</u>, G, <u>U</u>, <u>C</u>
	3-Methyl C · <u>A</u>, G, <u>U</u>, <u>C</u>
	3-Methyl U · A, G, <u>U</u>, <u>C</u>
	1,N^6-Etheno A · <u>A</u>, U, C
	3,N^4-Etheno C · <u>A</u>, U, C
iso A	iso C · A
(3-ribosyladenine)	(2-amino-2-deoxyuridine)

[a]Most of these data were published in a different form by Singer
and Kröger (72). Additional data are in references 74-76.
Underlining indicates the preferred incorporation.

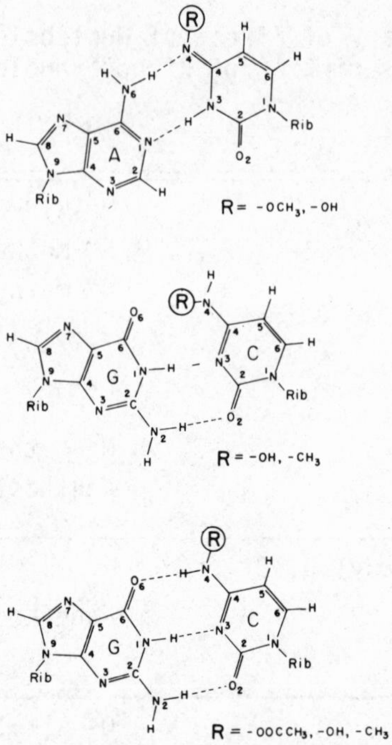

Figure 7. Effect on transcription of hydroxy and methoxy substituents on the N^4 of C. (Top) Base-pairing with A
when either derivative is in the tautomeric form and the
substituent is anti; (middle) Amino form of ho^4C with
substituent syn, causing ambiguity, and (bottom) Amino
form of ho^4C with substituent anti, allowing normal
G·C pairing.

It should be recalled that in these transcription experiments
the modification is in the polymer. The ability of the polymerase
to use a modified triphosphate can also be measured (38,77). This
is an interesting approach because triphosphates are more reactive
than nucleic acids in cells. The failure of many polymerases to
use these triphosphates may illustrate a partition
of labor with polymerases only checking the incoming substrate
while accepting the modified template.

The mutagenic effects of modified nucleotides in nucleic acids
studied in in vitro transcription systems can be summarized as
follows:

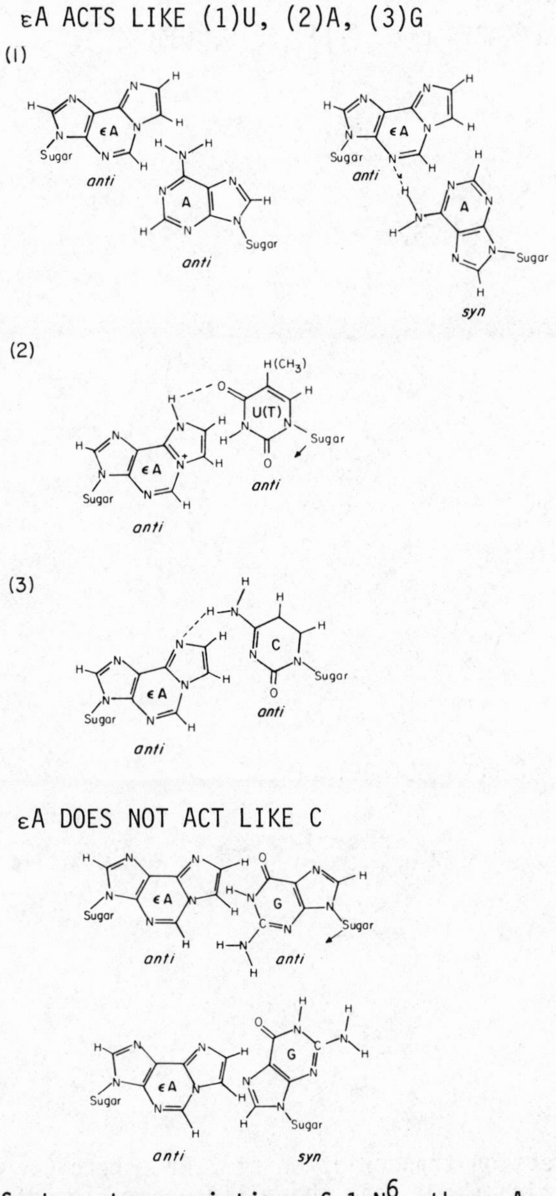

εA ACTS LIKE (1)U, (2)A, (3)G

(1)

(2)

(3)

εA DOES NOT ACT LIKE C

Figure 8. Effect on transcription of 1,N^6 etheno A. The top section shows the types of misincorporations observed. εA directing misincorporations of A (1) is likely to occur only when an adenosine is in the syn conformation. The bottom section indicates the two possible orientations of εA and G (top). When the G is in the normal anti conformation and (bottom) when G is rotated. An εA·G interaction is not likely to occur due to size, lack of any hydrogen-bonding potential, and possible charge repulsion.

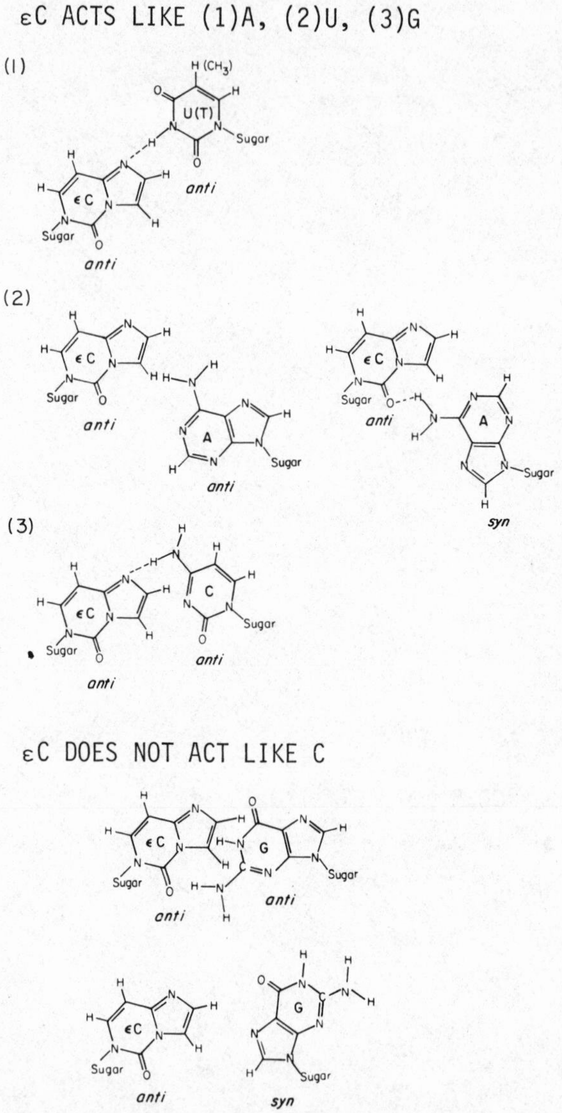

Figure 9.   Effect on transcription of 3,N^4 etheno C. The top
section shows that the misincorporation of A probably
involves the adenosine rotating into the syn
conformation while the other misincorporations may
occur according to (1) and (3). The bottom section is
similar to that in Fig. 8.

Consideration of in vitro transcription studies has led us to propose that small substituents blocking Watson-Crick sites will lead to ambiguity whenever hydrogen bonds of the appropriate number and length cannot be formed. The distortion may be due to steric hindrance or electron shielding of sites as well as loss of a receptor or donor by direct substitution. Stacking interactions and other neighborhood effects may determine the specific ambiguity observed when weak or no hydrogen bonds are formed. Thus, the mutagenic effects of unrepaired modifications of DNA can be elucidated in transcription experiments in vitro, assuming that the process of base interaction seen there is the same as that in replication, exclusive of repair or proofreading.

ACKNOWLEDGEMENT

This work was supported by Grant CA 12316 from the National Cancer Institute, National Institutes of Health.

REFERENCES

1. Montesano, R. and Bartsch, H.: Mutagenic and carcinogenic N-nitroso compounds: Possible environmental hazards. Mutat. Res., 32: 179-228 (1976).

2. Maher, V.M. and McCormick, J.J.: Mammalian cell mutagenesis by polycyclic aromatic hydrocarbons and their derivatives. In Polycyclic Hydrocarbons and Cancer, Ed. by H.V. Gelboin. Academic Press, New York, Vol. II, pp. 137-159 (1978).

3. Bartsch, H., Malaveille, C., Camus, A.-M., Martel-Planche, G., Brun, G., Hautefeuille, A., Sabadie, N., Barbin, A., Kuroki, T., Drevon, C., Piccoli, C. and Montesano, R.: Bacterial and mammalian tests: Validation and comparative studies in 180 chemicals. In Molecular and Cellular Aspects of Carcinogen Screening Tests, Ed. by R. Montesano, H. Bartsch and L. Tomatis, IARC Scientific Publication No. 27, pp. 179-241 (1980).

4. Pienta, R.J.: Transformation of Syrian hamster embryo by diverse chemicals and correlation with their reported carcinogenic and mutagenic activities. In Chemical Mutagens, Ed. by F. de Serres and A. Hollaender. Plenum Press, New York, Vol. 6, pp. 175-202 (1980).

5. Singer, B.: All oxygens in nucleic acids react with carcinogenic ethylating agents. Nature, 264: 333-339 (1976).

6. Singer, B.: The chemical effects of nucleic acid alkylation
    and their relationship to mutagenesis and carcinogensis.
    Prog. in Nucleic Acid Res. and Mol. Biol., 15: 219-284
    (1975).

7. Singer, B. and Fraenkel-Conrat, H.: The specificity of
    different classes of ethylating agents toward various sites
    in RNA. Biochemistry, 14: 772-782 (1975).

8. Goth, R. and Rajewsky, M.F. Molecular and cellular
    mechanisms associated with pulse-carcinogenesis in the rat
    nervous system by ethylnitrosourea: Ethylation of nucleic
    acids and elimination rates of ethylated bases from the DNA
    of different tissues. Z. Krebsforsch., 82: 37-64 (1974).

9. Lawley, P.D., Orr, D.J. and Jarman, M.: Isolation and
    identification of products from alkylation of nucleic
    acids: Ethyl- and isopropyl-purines. Biochem. J., 145:
    73-84 (1975).

10. Sun, L. and Singer, B.: The specificity of different
    classes of ethylating agents toward various sites of HeLa
    cell DNA in vitro and in vivo. Biochemistry, 14: 1795-1802
    (1975).

11. Singer, B.: Guest Editorial - N-nitroso alkylating agents:
    Formation and persistence of alkyl derivatives as
    contributing factors in carcinogenesis. J. Natl. Cancer
    Inst., 61: 1329-1339 (1979).

12. Shackleton, J., Warren, W. and Roberts, J.J.: The excision
    of N-methyl-N-nitrosourea-induced lesions from the DNA of
    Chinese hamster cells as measured by the loss of sites
    sensitive to an enzyme extract that excises
    3-methyl-purines but not O^6-methylguanine. Eur. J.
    Biochem., 97: 425-433 (1979).

13. Newbold, R.F., Warren, W., Medcalf, A.S.C. and Amos, J.:
    Mutagenicity of carcinogenic methylating agents is
    associated with a specific DNA modification. Nature, 283:
    596-598 (1980).

14. Beranek, D.T., Weis, C.C. and Swenson, D.H.: A
    comprehensive quantitative analysis of methylated and
    ethylated DNA using high pressure liquid chromatography.
    Carcinogenesis, 1: 595-606 (1980).

15. Bodell, W.J. and Singer, B.: Influence of hydrogen bonding
    in DNA and polynucleotides on reaction of nitrogens and
    oxygens toward ethylnitrosourea. Biochemistry, 18:
    2860-2863 (1979).

16. Lukashin, A.V., Vologodskii, A.V., Frank-Kamenetskii, M.D. and Lyubchenko, Y.L.: Fluctuational opening of the double helix as revealed by theoretical and experimental study of DNA interaction with formaldehyde. J. Mol. Biol., 108: 665-682 (1976).

17. McGhee, J.D. and von Hippel, P.H.: Formaldehyde as a probe of DNA structure. 4. Mechanism of the initial reaction of formaldehyde with DNA. Biochemistry, 16: 3279-3293 (1977).

18. Mandal, C., Kallenbach, N.R. and Englander, S.W.: Base-pair opening and closing reactions in the double helix. J. Mol. Biol., 135: 391-411 (1979).

19. Moschel, R.C., Hudgins, W.R. and Dipple, A.: Aralkylation of guanosine by the carcinogen N-nitroso-N-benzylurea. J. Org. Chem., 45: 533-535 (1980).

20. Brookes, P. and Lawley, P.D.: The reaction of mustard gas with nucleic acids in vitro and in vivo. Biochem. J., 77: 478-484 (1960).

21. Brookes, P. and Lawley, P.D.: The reaction of mono- and di-functional alkylating agents with nucleic acids. Biochem. J., 80: 496-503 (1961).

22. Brookes, P. and Lawley, P.D.: Effects of alkylating agents on T2 and T4 bacteriophages. biochem. J., 89: 138-144 (1963).

23. Chun, E.H.L., Gonzales, L., Lewis, F.S., Jones, J. and Rutman, R.J.: Differences in the in vivo alkylation and cross-linking of nitrogen mustard-sensitive and resistant lines of Lettre-Ehrlich ascites tumors. Cancer Res., 29: 1184-1194 (1969).

24. Roberts, J.J. and Warwick, G.P.: The reaction of β-propiolactone with guanosine, deoxyguanosine and RNA. Biochem. Pharmacol., 12: 1441-1442 (1963).

25. Maté, U., Solomon, J.J. and Segal, A.: In vitro binding of β-propiolactone to calf thymus DNA and mouse liver DNA to form 1-(2-carboxyethyl)adenine. Chem. Biol. Interact., 18: 327-336 (1972).

26. Lawley, P.D. and Jarman, M.: Alkylation by propylene oxide of deoxyribonucleic acid, adenine, guanosine and deoxyguanylic acid. Biochem. J., 126: 893-900 (1972).

27. Barbin, A., Bresil, H., Croisy, A., Jacquignon, P., Malaveille, C., Montesano, R. and Bartsch, H.: Liver microsome-mediated formation of alkylating agents from vinyl bromide and vinyl chloride. Biochem. Biophys. Res. Commun., 67: 596-603 (1972).

28. Green, T. and Hathway, D.E.: Interactions of vinyl chloride with rat liver DNA in vivo. Chem. Biol. Interact., 22: 211-224 (1978).

29. Sattsangi, P.D., Leonard, N.J. and Frihart, C.R.: 1,N^2-Ethenoguanine and N^2,3-Ethenoguanine. Synthesis and comparison of the electronic spectral properties of these linear and angular triheterocycles related to the Y bases. J. Org. Chem., 42: 3292-3296 (1977).

30. Kayasuga-Mikado, K., Hashimoto, T., Negishi, T., Negishi, K. and hayatsu, H.: Modification of adenine and cytosine derivatives with bromoacetaldehyde. Chem. Pharm. Bull., 28: 932-938 (1980).

31. Goldschmidt, B.M., Blazej, T.P. and Van Duuren, B.L.: The reaction of guanosine and deoxyguanosine with glycidaldehyde. Tetrahedron. Lett. No. 13, 1583-1586 (1968).

32. Van Duuren, B.L. and Loewengart, G.: Reaction of DNA with glycidaldehyde. Isolation and identification of a deoxyguanosine reaction product. J. Biol. Chem., 252: 5370-5371 (1977).

33. Ludlum, D.B., Krame, B.S., Wang, J. and Fenselau, C.: Reaction of 1,3-Bis(2-chloroethyl)-1-nitrosourea with synthetic polynucleotides. Biochemistry, 14: 5480-5485 (1975).

34. Bradley, M.O., Sharkey, N.A., Kohn, K.W., Layard, M.W.: Mutagenicity and cytotoxicity of various nitrosoureas in V-79 Chinese hamster cells. Cancer Res., 40: 2719-2725 (1980).

35. Tong, W.P. and Ludlum, D.B.: Mechanism of action of the nitrosourea I. Role of fluoroethylcytidine in the reaction of bis-fluoroethyl nitrosourea with nucleic acids. Biochem. Pharmac., 27: 77-81 (1978).

36. Brown, D.M.: Chemical reactions of polynucleotides and nucleic acids. In Basic Principles in Nucleic Acid Chemistry, Ed. by P.O.P. Ts'o. Academic Press, New York, Vol. II, pp. 1-90 (1974).

37. Brown, D.M. and Osborne, H.R.: The reaction of adenosine
     with hydroxylamine. Biochem. Biophys. Acta, 247: 514-518
     (1971).

38. Budowsky, E.I., Sverdlov, E.D., Spasokukotskaya, T.N. and
     Koudelka, J.A.: Mechanism of mutagenic action of
     hydroxylamine VIII. Functional properties of the modified
     adenosine residues. Biochim. Biophys. Acta, 300: 1-13
     (1975).

39. Shapiro, R., Dubelman, S., Feinberg, A., Crain, P. and
     McCloskey, J.: Isolation and identification of cross-links
     of nucleoside from nitrous acid treated deoxyribonucleic
     acid. J. Am. Chem. Soc., 99: 302-303 (1977).

40. Singer, B. and Fraenkel-Conrat, H.: The role of
     conformation in chemical and mutagenesis. Prog. Nucleic
     Acid. Res. and Mol. Biol., 9: 1-30 (1977).

41. Fraenkel-Conrat, H.: Reaction of nucleic acid with
     formaldehyde. Biochim. Biophys. Acta, 15: 308-309 (1954).

42. Shapiro, R., Law, D.C.F. and Weisgras, J.M.: A chemical
     probe for single-stranded RNA. Biochem. Biophys. Res.
     Commun., 49: 358-363 (1972).

43. Kimura, K., Nakanishi, M., Yamamoto, T. and Tsuboi, M.: A
     correlation between the secondary structure of DNA and the
     reactivity of adenine residues with chloroacetaldehyde. J.
     Biochem., 81: 1699-1703 (1977).

44. Strauss, B., Tatsumi, K., Karran, P., Higgens, N.P.,
     Ben-Asher, E., Altamirano-Dimas, M., Rosenblatt, L. and
     Bose, K.: Mechanisms of DNA excision repair in human
     cells. In Polycyclic Hydrocarbons and Cancer, Ed. by V.
     Gelboin. Academic Press, New York, Vol. 2, pp. 177-201
     (1978).

45. Roberts, J.J.: The repair of DNA mofidifed by cytotoxic,
     mutagenic, and carcinogenic chemicals. Advances in
     Radiation Biology, 7: 211-436 (1978).

46. Hanawalt, P.C., Cooper, P.K., Ganesan, A.K. and Smith,
     C.A.: DNA repair in bacteria and mammalian cells. Annu.
     Rev. Biochem., 48: 783-836 (1979).

47. Verly, W.G.: Prereplicative error-free DNA repair.
     Biochem. Pharmacol., 29: 977-982 (1980).

48. Lindahl, T.: DNA glycosylases, endonucleases for
     apurinic/apyrimidinic sites, and base-excision repair.
     Prog. in Nucleic Acids and Mol. Biol., 22: 135-192 (1979).

49. Frei, J.V., Swenson, D.H., Warren, W. and Lawley, P.D.:
     Alkylating of deoxyribonucleic acid in vivo in various
     organs of C57BL mice by the carcinogens N-methyl-N-
     nitrosourea, N-ethyl-N-nitrosourea and ethyl
     methanesulfonate in relation to induction of thymic
     lymphoma. Biochem. J., 174: 1031-1044 (1978).

50. Bodell, W.J., Singer, B., Thomas, G.H. and Cleaver, J.E.:
     Evidence for removal at different rates of O-ethyl
     pyrimidines and ethyl-phosphotriesters in two human
     fibroblast cell lines. Nucleic Acids Res., 6: 2819-2829
     (1979).

51. Singer, B., Spengler, S. and Bodell, W.J.: Tissue-dependent
     enzyme-mediated repair or removal of O-ethyl pyrimidines
     and ethyl purines in carcinogen-treated rats. Nature, in
     press (1981).

52. Singer, B. and Brent, T.P.: Human lymphoblasts contain DNA
     glycosylase activity excising N-3 and N-7 methyl and ethyl
     purines but not O^6-alkylguanine or 1-alkylademine. Proc.
     Natl. Acad. Sci. U.S.A., 78: 856-860 (1981).

53. Margison, G.P. and Pegg, A.E.: Enzymatic release of
     7-methylguanine from methylated DNA by rodent liver
     extracts. Proc. Natl. Acad. Sci. U.S.A., 78: 861-865
     (1981).

54. Pegg, A.E.: Enzymatic removal of O^6-methylguanine from
     DNA by mammalian cell extracts. Biochem. Biophys. Res.
     Commun., 84: 163-178 (1978).

55. Renard, A. and Verly, W.G.: A chromatin factor in rat liver
     which destroys O^6-ethylguanine in DNA. FEBS Letters,
     114: 98-102 (1980).

56. Karran, P., Lindahl, T., Griffin, B.: Adaptive response to
     alkylating agents involves alteration in situ of
     O^6-methylguanine residues in DNA. Nature, 280: 76-77
     (1979).

57. Olsson, M. and Lindahl, T.: Repair of alkylated DNA in
     Escherichia coli. Methyl group transfer from
     O^6-methylguanine to a protein cysteine residue. J. Biol.
     Chem., 255: 10569-10571 (1980).

58. Foote, R.S., Mitra, S. and Pal, B.C.: Demethylation of
     O^6-methylguanine in a synthetic DNA polymer by an
     inducible activity in Escherichia coli. Biochem. Biophys.
     Res. Commun., 97: 654-659 (1980).

59. Shooter, K.V. and Slade, T.A.: The stability of methyl and
     ethyl phosphotriesters in DNA in vivo. Chem. Biol.
     Interact., 19: 353-361 (1977).

60. Topal, M.D. and Fresco, J.R.: Complementary base pairing
     and the origin of substitution mutations. Nature, 263:
     285-289 (1976a).

61. Topal, M.D. and Fresco, J.R.: Base pairing and fidelity in
     codon-anticodon interaction. Nature, 263: 289-293 (1976b).

62. Streisinger, G., Okada, Y., Emrich, J., Newton, J., Tsugita,
     A., Terzaghi, E. and Inouye, M.: Frameshift mutations and
     the genetic code. Cold Spring Harbor Symp. Quant. Biol.
     XXXI, 77-84 (1966).

63. Witkin, E.M.: Ultraviolet mutagenesis and inducible DNA
     repair in Escherichia coli. Bacteriol. Rev., 40: 869-907
     (1976).

64. Hsu, W.-T., Lin, E.J.S., Harvey, R.G. and Weiss, S.B.:
     Mechanism of phage øX174 DNA inactivation by
     benzo[a]pyrene-7,8-dihydro-9,10-epoxide. Proc. Natl. Acad.
     Sci. U.S.A., 74: 3335-3339 (1977).

65. Grunberger, D. and Weinstein, I.B.: Biochemical effects of
     the modification of nucleic acids by certain polycyclic
     aromatic carcinogens. Prog. in Nucleic Acid Res. and Mol.
     Biol., 23: 106-149 (1979).

66. Moore, P.D., Rabkin, S.D. and Strauss, B.S.: Termination of
     in vitro DNA synthesis at AAF adducts in DNA. Nucleic
     Acids Res., 8: 4473-4484 (1980).

67. Chan, J.Y.H. and Becker, F.F.: Decreased fidelity of DNA
     polymerase activity during N-2-fluorenylacetamide
     hepatocarcinogenesis. Proc. Natl. Acad. Sci. U.S.A., 76:
     814-818 (1979).

68. Sirover, M.A. and Loeb, L.A.: Restriction of carcinogen-
     induced error incorporation during in vitro DNA synthesis.
     Cancer Res., 36: 516-523 (1976).

69. Fresco, J.R., Broitman, S. and Lane, A.-E.: Base mispairing
    and nearest neighbor effects in transition mutations. In
    Mechanistic Studies of DNA Replication and Genetic
    Recombination, Ed. by B. Alberts and C.F. Vox. Academic
    Press, New York, pp. 753-768 (1980).

70. Drake, J.W. and Baltz, R.H.: The biochemistry of
    mutagenesis. Annu. Rev. Biochem., 45: 11-37 (1976).

71. Singer, B.: Effect of base modification on fidelity in
    transcription. In Carcinogenesis; Fundamental Mechanisms
    and Environmental Effects, Ed. by B. Pullman, P.O.P. Ts'o
    and H. Gelboin. D. Reidel, Dordrecht, Holland, pp. 91-102
    (1980).

72. Singer, B. and Kroger, B.: Participation of modified
    nucleosides in translation and transcription. Prog. in
    Nucleic Acid Res. and Mol. Biol., 23: 151-194 (1979).

73. Youvan, D.C. and Hearst, J.E.: Reverse transcriptase pause
    at N^2-Methylguanine during in vitro transcription of
    Escherichia coli 16S ribosomal RNA. Proc. Natl. Acad. Sci.
    U.S.A., 76: 3751-3754 (1979).

74. Singer, B. and Spengler, S.: Ambiguity and transcriptional
    errors as a result of modification of exocyclic amino
    groups of cytidine, guanosine and adenosine. Biochemistry,
    20: 1127-1132 (1981).

75. Spengler, S. and Singer, B.: Transcriptional errors and
    ambiguity resulting from the presence of
    $1,N^6$-Ethenoadenosine or $3,N^4$-Ethenocytidine in
    polyribonucleotides. Nucleic Acids Res., 9: 365-373 (1981).

76. Singer, B. and Spengler, S.: Mutagenesis from a chemical
    perspective: Nucleic acid reactions, repair, translation
    and transcription. In Molecular and Cellular Mechanisms of
    Mutagenesis, Ed. by J.F. Lemontt. Plenum Press, in press
    (1981).

77. Engle, J.D. and von Hippel, P.H.: $d(m^6ATP)$ as a probe of
    the fidelity of base incorporation into polynucleotides by
    Escherichia coli DNA polymerase I. J. Biol. Chem., 253:
    935-939 (1978).

DISCUSSION

DR. GLICKMAN: I'm concerned about the extrapolation from
experiments using the RNA polymerase to DNA polymerase, and I
think the reasons for concern are obvious. In Table III I noticed
a few things which also bothered me. One of them is that the
5-halopyrimidines you have tested do not miscode, even though we
know, or at least believe, that 5-bromouracil is a miscoding base
analog. Do you have a comment on the mechanism by which 5-BU
might be working in E. coli and other organisms?

DR. SINGER: That I don't know. In one system that I am familiar
with, we did an experiment using 5-FU with a messenger RNA.
There, of course, you look for phenotypic changes, or amino acid
exchanges in the progeny, and none were observed. In fact, we
could get complete substitution, if we wanted to, by growing the
virus only with 5-FU, which is taken up and gives a perfectly
viable virus, without measurable mutation. I am aware of the fact
that 5-Bromo U has been used to mutate E. coli. But I cannot,
certainly from my experience, reconcile the two. One possible
mechanism is that 5-Bromo U has an increased tendency to
tautomerize. Based on the pKa, Fresco estimates 5-Bromo U to be
the enol form 6-10 times more frequently than T.

DR. GLICKMAN: I believe there are some obvious difficulties with
the base analogues as base analogues. Also, in the case of the
one hydrogen bond question, 2-aminopurine is a good base analogue;
I think there's no question that there is good evidence, with
purified polymerase and in vivo, that 2-aminopurine is a base
analogue. But still it pairs with a single hydrogen bond.

DR. SINGER: That's right. Let me go back to the first question
you asked about the polymerases. So far, in Ludlum's laboratory
two experiments have been done transcribing a DNA template
containing a modified nucleotide with a DNA-dependent RNA
polymerase, which I think is probably the best system to use. In
both those cases, some of the same ambiguities were reported as
with the system we use. We also intend to use other polymerases
to find whether the mispairing we observe is a function of a
specific polymerase. We presently have the first eukaryotic
RNA-dependent RNA polymerase to be isolated and described. The
use of eukaryotic RNA-dependent RNA polymerase to copy a
ribopolynucleotide should tell us something about whether or not
these interactions occur under what could be called normal
conditions. This enzyme, however, can also copy
deoxypolynucleotides which should make it more useful for our
purposes of exploring the role of polymerases in mutation.

DR. GLICKMAN: In talking about the 2-ethyl derivative of uracil you mentioned that it would be a transitory effect because of the frequency. I don't think the frequency gives any indication of a transitory pairing. More correctly you can say that it mispairs rarely. I think transitory pairing as a term is really only applicable to the present models of kinetic proofreading with DNA polymerase. I think in this case it's fair to say that this derivative pairs rarely, rather than transitorily.

DR. SINGER: I think nothing can mispair frequently and maintain the species. I'm not saying that in vivo mispairs are going to occur at anything like the rate we're observing. It's only that if it does occur in our system, it can possibly occur in replication.

DR. SHAPIRO: I just want to make some comments on the same theme. First, let me say as a chemist I also fell in love with the field of mutagenesis a long way back when certain elegant mechanisms could be drawn involving hydrogen bonds that clearly explained how mutations worked. But I think as we go along this picture has become more and more flawed. When you have results in which you have to try and rationalize some misreplication or mistranscription on the base of just one hydrogen bond, and when you start moving bases far away from Watson & Crick distances, you know it's about time to stop: I think the horse is limping so badly in this case it's just as well to shoot it. I can't believe the cases where the existence of just one hydrogen bond is assumed to represent what's really going on inside the system.

DR. SINGER: What I wanted to say is that if you can't even draw a hydrogen bond with the right geometry, it is not likely to occur. When we find by model building it's not possible to form even a single base pair, then you might as well forget it.

DR. SHAPIRO: I suspect that unless you have a molecule which has no NH's or OH's at all, you can always find a way to draw one hydrogen bond, if you wish. But what I'm suggesting is that in the cases where you can show only one hydrogen bond, where you don't show a good Watson-Crick geometry with at least two solid hydrogen bonds forming, then probably the mechanism involves something totally different. The enzyme specificity is involved and base to base hydrogen bonding is irrelevant in such a case.

DR. SINGER: The enzyme certainly plays a role. We are not simply looking at two bases seeing each other and saying one bond is enough. It's quite different and I'm sure that different enzymes have different abilities. In the case, for example, of

N^6-methyl A, Peter Von Hipple believes that is exists in the ratio of 20:1 in the syn conformation. That means that the methyl group is toward the Watson-Crick side, and yet N^6 methyl ADP substitutes perfectly for ADP in transcription. His belief is that the polymerase turns the methyl group around. I think that's perfectly valid, polymerase does do such things.

DR. SHAPIRO: I'm suggesting that in those cases where you are only able to draw one hydrogen bond, and you also distort the hydrogen-bonding distance, that probably what's happening in such cases is that the polymerase is doing everything and the one hydrogen bond is irrelevant.

DR. SINGER: When you get a very large substituent in a position of rotation, then you will find that the absolute extent of transcription is very low, and that means that if a substituent actually is oriented blocking, not just the Watson-Crick site but keeping the bases apart, no transcription occurs.

DR. THILLY: John Essigman, Gerry Wogan and I have tried to look at molar mutational efficiency as a black box; how much DNA alkylation goes in and how much mutation comes out. One shouldn't overlook the fact that on the order of 1% of the naturally-occurring DNA already has a methyl group on it somewhere. If we just multiply through the actual level of alkylation we examine in mutation experiments, we would suggest that it is possible that twice-alkylated bases are of sufficient frequency to be potentially important in mutation. It might, for instance, be interesting to look for dimethyl bases in which the first methyl arises naturally and the second by reaction with an exogenous agent.

DR. SINGER: Chemically, it's very hard to put a second methyl on the N^6 position of A.

DR. THILLY: Calculations using the free solution reaction rates suggest that the actual frequency of appearance of such things would be of sufficient magnitude to participate, if they were mutagenically efficient.

DR. SINGER: If you had an N^6, N^6-dimethyl adenine, and you looked to see how it was transcribed in vitro, what would you think it would do?

DR. THILLY: Probably prevent transcription.

DR. SINGER: Do you know what N^2-methyl G does?

DR. THILLY: I have no idea.

DR. SINGER: Well, there's an experiment by Youvan and Hearst using a naturally-occurring N^2-methyl G in a segment of 16S RNA which they transcribed with reverse transcriptase. They did kinetics, and found that transcription paused at the N^2-methyl G, but then it continued. The pause was approximately 3 seconds, which is a long pause, but transcription did continue.

Now, what was put in opposite N^2-methyl G? This they could not determine by sequencing, and the reason is likely to be exactly what we observed -- total ambiguity. That means that anything could be put in. It just took the polymerase a while to decide what to do with N^2-methyl G but finally a base was randomly inserted. Biologically, if this occurred I suspect that that molecule of nucleic acid would no longer be part of the replication scheme. I think the N^6,N^6-dimethyl A would also effectively slow transcription.

DR. THILLY: You misunderstand me. I just used the dimethyl as an example. Any second alkylation reaction with an already naturally methylated base is what we're trying to consider in a general sense.

DR. GLICKMAN: 6-methyladenine is restricted to prokaryotes. I don't believe there are any 6-methyl A modified bases in eukaryotes, only 5 methyl cytosine. In E. coli where mutational specificity is done with alkylating agents, both methylating and ethylating agents, you have three positions in the lacI gene which have 5-methyl C. Should those sites represent greater targets for mutagenesis, they would come up as hot spots with an alkyating agent, and the case of MMS, MMNG, EMS, and ENU they don't. They're not hotter for GC to AT transitions than other GC sites. Now, that could mean that such a dimethylated site had a different specificity and wasn't causing a transition, but if they would cause transitions they would be seen as hot spots. They're not seen. So there's no evidence biologically for the kind of mechanism operating which you've suggested.

DR. THILLY: That's absence of evidence, however. Your example may not be relevant to addition, deletion and insertion mutation which maybe frequent, as opposed to base-pair substitution in those systems.

DR. SINGER: By the way, let's be clear. I'm speaking of point mutations, only. This is the kind of systems with which I work in order to simplify the problem. Point mutations, however, are not necessarily predominant in vivo.

DR. MAHER: I wanted to ask you to explain a little more about your last statement, that is about a large, bulky derivative where the enzyme would stop, just pause, and then something would drop out.

DR. SINGER: If you have a large substituent, for example N^6-isopentenyl A you never get any misincorporation. If transcription occurs, it is always as A, but much of the time the substituent is rotated into the Watson-Crick site, in which case the 4 carbons sterically prevent incorporation of any base, which terminates transcription. What occurs as I indicated is true only with small substituents, unless the substituent is oriented far away from the base-pairing site. It's not applicable to modifications such as diolepoxide of benzo(a) pyrene on the N^2 of G.

PARTICIPANT: What do you know about the efficiency of methylating or alkylating agents between core DNA and linker DNA, for example.

DR. SINGER: It does not seem to be any different. You must also remember that the extent of alkylation that we are working with in vivo is about one alkyl group in 10^5 or 10^6 DNA bases, which means that it is necessary to use a very high specific radioactivity. There is a limit to the amount of radioactivity that can be used for both economic and biological reasons. This means we cannot detect, let's say, a difference of 10%, just to take a likely figure. Given the present data, the availability of sites in DNA to react does not seem to be affected by the milieu.

DR. ROSENKRANZ: Can you elaborate on the stability of alkylphosphotriesters?

DR. SINGER: The ribophosphotriester is unstable because of the attack of the free 2'-hydroxyl of the ribose on the P=O bond.

DR. ROSENKRANZ: I'm familiar with Todd's work, but one of the things he said is that triesterfied phosphate is unstable.

DR. SINGER: In RNA or in a ribopolynucleotide that is true but it's extremely stable in a deoxynucleotide. It takes 70% perchloric acid at 100° to break a triester in a deoxypolynucleotide. No hydrolytic enzyme that we use can break the bond.

DR. GLICKMAN: When you were talking about multiphasic removal of alkylation products you mentioned that you had some evidence for the induction of repair systems. Could you comment?

DR. SINGER: What we found was that in liver of rats given ethylnitrosourea, O^4-ethylthymidine and O^2-ethylcytidine were removed with kinetics that might be interpreted as indicating induction, whereas other derivatives were removed with what would be considered normal kinetics. We don't have such an extraordinary amount of data that I would want to say that this was absolute evidence of induction, but we have studied removal of

seven derivatives in several tissues and only O^4-EtT and
O^2-EtC in liver show such kinetics. In contrast, in brain DNA
from this animal, very little removal of these derivatives occurs.

DR. GLICKMAN: Your results could be explained just as easily by a
competitive situation, where another lesion is preferentially
removed by those enzymes.

DR. SINGER: We believe that the removal of the derivatives that
we're studying is due to several independent enzyme mechanisms.
We also came to that conclusion when studying persistence in
mammalian cells.

DR. SAMSON: The kinetics of O^6-alkyl G removal, an inducible
process, don't look like that.

DR. SINGER: No, it does not because it goes so fast. If you used
a system where it wasn't removed within the first hour, which is
the first time period in this study, then you might be able to see
an induction.

DR. LOEB: If you take a deoxynucleotide triphosphate and make it
a triester, will it be polymerized by a polymerase, that is, with
the alpha phosphoryl group as a triester, will the enzyme accept
it? Can you make a homopolymer, polynucleotide or DNA?

DR. SINGER: I haven't the foggiest idea. The enzymes will accept
almost anything but that I'm not sure.

DR. LOEB: Do you know that triesters don't miscode?

DR. SINGER: I was sure somebody was going to ask me something
like that. The problem is we can't synthesize a polymer that
contains only triesters and I don't believe in chemically
modifying polymers because multiple products result. It is
preferable to put in a specific derivative so you know that what
you're observing is due only to this. But we have not been able
to make the polymer you're speaking of. And so, unfortunately,
our evidence is again indirect. We've studied protein synthesis,
using a monocistronic messenger predominantly containing
phosphotriesters and we find the right protein is made, but the
amount is greatly diminished. The tentative conclusions are that
triesters do not miscode but probably affect ribosome binding.

GENETIC EFFECTS OF BISULFITE:

IMPLICATIONS FOR ENVIRONMENTAL PROTECTION

Robert Shapiro

Department of Chemistry
New York University
New York, N.Y. 10003

INTRODUCTION

Sulfur dioxide is unique among environmental substances because of the many routes of human exposure to it. Combustion of coal and oil releases SO_2 into urban atmospheres. It is added to foods, beverages, and pharmaceuticals and we produce it within our bodies as a product of the catabolism of sulfur containing amino acids. It is short-lived within us, as it is rapidly oxidized to sulfate by sulfite oxidase.

The many ways in which this substance reaches us are matched by the many names used to describe it. It exists as sulfur dioxide in the gaseous state. In water it hydrates to form H_2SO_3, sulfurous acid. This substance is too acidic to persist as such within living cells. At neutral pH, it dissociates to a mixture of bisulfite (HSO_3^-) and sulfite ($SO_3^=$). In more concentrated solutions, it dimerizes to some extent to metabisulfite ($S_2O_5^=$), also called pyrosulfite. It can exist in the solid state entirely as metabisulfite, and is sometimes sold commercially in this form. The mixture of forms present in a solution will depend on the pH, ionic strength, and other conditions of the solution, but not on the route used to prepare it. In this paper, I will refer to the gaseous form as sulfur dioxide, and the forms in solution as bisulfite, whatever the actual composition present.

The regulation of our exposure to sulfur dioxide and bisulfite (in the United States) is also divided into many parts. Exposure to atmospheric sulfur dioxide is controlled by the Environmental Protection Agency, which sets permissable levels under the terms

of the Clean Air Act.   Ingestion of bisulfite in foods and pharmaceuticals is regulated by Food and Drug Administration, and consumption in wines and beer by the Bureau of Alcohol, Tobacco Products, and Firearms, Department of the Treasury.   Each of these
forms of exposure to sulfur dioxide is a matter of economic importance, and its regulation has been a subject of controversy.   The
various agencies have responded to different sources of information
on the health effects of sulfur dioxide, and taken contrasting
regulatory positions.

In this review, the sources of our environmental exposure to
sulfur dioxide will be surveyed briefly, and the information on
its health effects provided by epidemiology and toxicology studies.
Genetic and biochemical studies will be brought up to date.   In a
final section, the implication of the genetic studies for future
regulatory policy will be discussed.   Only selected references of
importance will be cited.   A fuller, more completely documented
account may be obtained by consulting my earlier review (1), and
the recent one by Gunnison (2).

SULFUR DIOXIDE IN THE AIR (3)

Coal and oil contain organic sulfur compounds and inorganic
sulfides.   When they are burned as fuel, sulfur oxides are released into the atmosphere. Electrical power plants are the most
important source of sulfur dioxide emissions.   The amount of SO_2
released can be limited by using coal and oil of lower sulfur
content, desulfurization of fuel prior to combustion, or scrubbing
SO_2 from the emissions.   Each of these alternatives is expensive,
and the selection of a method of choice involves a balancing of
costs with the degree of desulfurization that is desired (4).
Heated controversy has accompanied the enactment, enforcement, and
revision of the Clean Air Act, with its attendant costs amounting
to billions of dollars (5).

A number of considerations have provided the motivation for
removing sulfur oxides from the air: the adverse ecological impact
of acid rain, damage to vegetation, and corrosive effects on materials.   However, the threat to human health has always been put
forward as the matter of greatest importance (5,6).

HEALTH EFFECTS OF INHALED SULFUR DIOXIDE

Sulfur dioxide is only one substance in the complex mixture
of pollutants that exists in urban atmospheres. Epidemiological
studies have indicated that air pollution can lead to increased
deaths and illnesses among those affected. Excess deaths in the
thousands, and hundreds, respectively were associated with short-

term high pollution incidents in London in 1952 and 1962 (8). Many
longer term epidemiological studies have shown a less dramatic, but
generally consistent, association of illness and death with enhanced levels of sulfur dioxide and particulates. These two classes
of pollutant inevitably occur in combination with one another, so
that it has been difficult to separate their effects. In certain
cases, however, the evidence has pointed to particulate matter as
the more important of the two. Thus, sulfur dioxide levels were
about the same in the pollution episodes in London in 1952 and 1962,
but particulates ("smoke") were less, and deaths lower, in the later
episode (8). A detailed analysis compared death rates in New York
City in the 1960's with those in the 1970's, after sulfur dioxide
levels had been reduced substantially (9). Deaths were found to be
related to weather changes, and, to a lesser extent, to particulate
levels, but not to sulfur dioxide concentrations.

Human volunteers have been exposed to SO_2 atmospheres several
times higher than those permitted by current air quality standards
(0.14 ppm over 24 hours). The principal effects observed, both
with normal subjects and those with lung disease, have been a re-
flex bronchoconstriction, and sometimes, mild irritation. Such
changes are readily reversible; the question has been raised
whether they should be considered adverse, and a reason for legislative action (7).

Guinea pigs and monkeys have been maintained for months in
atmospheres containing 5 ppm of SO_2. No adverse effects were detected. Similarly, industrial workers have shown little or no
health impairment, after years of employment involving exposure
to several ppm of sulfur dioxide.

While such studies suggest that the levels of sulfur dioxide
present in urban air pollution are harmless, the criticism has
been raised that the animal colonies and industrial groups do not
reflect the general population, since they exclude the very young,
the aged, and the infirm (10,11). It has further been suggested
that the health effects of sulfur dioxide are not readily observed
when subjects are exposed to it alone, but that it rather has a
synergistic effect, with other pollutants (12). No firm body of
evidence exists in support of this view at the present time, and
the matter is unresolved.

THE USE OF SULFUR DIOXIDE IN FOODS AND BEVERAGES

Sulfur dioxide has been used in the preparation of wines since
ancient times. It is currently used as a preservative in a variety
of substances, including wines, beers, preserved fruits and vege-
tables, fruit juices and syrups (13). It offers a combination of
useful properties, including the inhibition of microbial growth

and of discoloration due to enzymatic and chemical oxidation. In wine-making, sulfur dioxide selectively inhibits the growth of certain bacteria, allowing desirable yeast strains to dominate the fermentation. Until recently, there has been little information available about the mechanism of inhibition of microbial growth. The assumption has been made that covalent reactions were involved, which inactivated enzymes, mRNA, or other target molecules (1,13). We have gained more understanding of the process in recent studies with E. coli K-12 in our laboratory (14).

In our cultures, inhibition of growth is accompanied by a simultaneous cessation of protein and RNA synthesis. Inhibition of RNA synthesis is not an effect of the stringent control system, as it still occurs   in the presence of chloramphenicol, and in relaxed E. coli strains. Growth in the presence of bisulfite is stimulated by addition of certain combinations of amino acids. In this respect, the inhibition of growth by bisulfite resembles that produced by certain amino acids such as valine (15) or cysteine (16). In those cases, growth inhibition by one amino acid could be reversed by addition of another amino acid.

After a period of growth inhibition by bisulfite, E. coli K-12 adapts to its presence. It resumes growth at bisulfite concentrations that previously were inhibitory. A number of the inhibitory effects of bisulfite are also shown by cysteine, at lower concentrations. It appears that the inhibitory properties of bisulfite are due to its conversion to its conversion to cysteine, (and perhaps other substances). These metabolites inhibit growth through their effects on cellular control processes.

EXPOSURE TO INGESTED BISULFITE: HEALTH EFFECTS

In a number of chronic feeding studies with animals, no effects were observed at levels of about 96 mg(as SO_2)/kg day. At higher levels various types of injury were seen, including hyperplasia and inflamation of the stomach, blood in the faeces, an increase in kidney weight, and slight growth retardation (2). Bisulfite reacts with thiamine, and can induce deficiency in this vitamin by destroying it in foods. In the most careful animal studies, supplementary thiamine was provided, so that the symptoms due to bisulfite toxicity could be separated from those due to thiamine deficiency (7).

The data from these studies was adjusted by a hundred-fold safety factor by the World Health Organization, and used as the basis for a recommended daily intake of not more than 0.7 mg (as SO_2)/kg, for humans. This would amount to about 50 mg for a 70 kg person. It has been estimated that the average daily per capita consumption of bisulfite in this country (calculated as SO_2)

in foods is 2 mg, while an additional 5 mg is taken in through
wines and beer (mostly wine) (17). These averages were considered
almost meaningless, however, because of the wide variety that exists
in individual consumption patterns. A pint of wine may contribute
100 mg or more of bisulfite (17). An individual consuming such
wine could readily take in 2 mg/kg day of bisulfite, and thus have
only a fifteen-fold safety margin over the no-effect level in animals (18).

THE ROLE OF SULFITE OXIDASE (2)

Bisulfite is produced endogenously in mammals, as an intermediate in the catabolism of sulfur-containing amino acids. The
amount of bisulfite produced in this manner undoubtedly exceeds the
quantity taken from the environment. The terminal step in the process of amino acid catabolism is the oxidation of bisulfite to
sulfate, catalyzed by sulfite oxidase. This enzyme, a molybdenum-
containing protein (19) is widely distributed among a variety of organs; liver is the principal location, followed by kidney, heart,
and lungs. The enzyme exists in the intermembranous space of the
mitochondria.

Sulfite oxidase is not inducible, and there seems to be no
need for this capacity in normal animals. It has been estimated
that the quantity of enzyme present in rat tissues is capable of
oxidizing bisulfite supplied at a rate (calculated as SO_2) of
48g/kg body weight each day. The half life of bisulfite, administered intravenously to rats, has been calculated to be approximately one minute. In this short time, however, there is opportunity
for bisulfite to interact with tissue components. One reaction
that has been used as a measure of such events, is the formation of S-
sulfonates, by sulfitolysis of protein disulfide bonds:

$$RSSR' + HSO_3^{(-)} \longrightarrow RSSO_3^{(-)} + HSR$$

The S-sulfonate species in proteins have a half-life of perhaps
a day, which allows sufficient time for chemical analysis. After
ingestion, or intraperitoneal injection of bisulfite, such S-sulfo-
nates can be detected in the plasma. In a study involving inhala-
tion of SO_2 (3-10 ppm) by rabbits (20), considerable amounts of S-
sulfonates were found in the trachea, lesser amounts in the plasma,
and small quantities in the distal lobe of the lung. The quantities found in the lung were believed to result from reactions of
bisulfite transported in the plasma, rather than from sulfur dioxide
which had made its way through the air passages of the lung. No
enhancement of S-sulfonate levels was found in the walls of the
aorta, which indicated that the bisulfite in the plasma had been
consumed rapidly.

Because of the speed and capacity of sulfite oxidase, acute
toxicity studies, with large amounts of bisulfite, have more re-
flected the conditions needed to overwhelm the enzyme, rather than
the innate toxicity of bisulfite. One direct measure of bisulfite
toxicity has been obtained, however (2). When steady state concentrations of 0.7 - 1.0 mM were maintained in the plasma of rabbits
by continuous i.v. infusion, the animals died within 2-3 hours.

One difficulty in attempting to extrapolate animal bisulfite
toxicity studies to man has been the existence of interspecies
variations in in vivo sulfite oxidase levels. The rat, which is
commonly used in animal studies, has a level which considerably
exceeds that of the rabbit and rhesus monkey, and is perhaps ten-
fold greater than that of man (21). It has been concluded, from
this data, that normal rats are not a good model for the prediction
of the toxicity of bisulfite to humans. Sulfite oxidase-deficient
rats can be obtained, however, by treating the animals with a diet
low in molybdenum and high in tungsten (2,22). Such rats retain
only 1-2% of their normal quantities of sulfite oxidase and are
considered models for humans with one-tenth of their normal human
sulfite oxidase capacity. Free sulfite, from endogenous sources,
could be detected in the plasma of the treated rats, whereas none
could be detected in normal rats (23).

Several human cases have been known with a deficiency in the
ability to synthesize active sulfite oxidase (24). This condition
is associated with severe neurological abnormalities, and can lead
to early death. In one case, plasma levels of sulfite were measured
and found to vary from 2 to 130 μM according to the sulfur amino acid
content in the diet of the individual (a child).

MUTAGENESIS AT HIGH BISULFITE CONCENTRATIONS

Bisulfite undergoes a number of reactions with nucleic acids,
the best known of which is the deamination of cytosine to uracil
(25,26). This conversion is specific for single-stranded nucleic
acids (27) and proceeds most rapidly at pH 5 to 6, in bisulfite
solutions of 1 M, or higher concentration. As adenine and guanine
do not react under these conditions, the deamination is specific
for cytosine (28). The reaction has been widely used for chemical
modification of nucleic acids (29). In one particularly elegant
application (30), a pre-selected section of the genome of SV40
virus was rendered single-stranded by the use of appropriate enzymatic techniques, and bisulfite was used to convert cytosine to
uracil in the exposed region. When the treated virus was allowed
to replicate in vivo, it was found that point mutations had been
induced at the intended locations.

The very nature of this cytosine to uracil transformation
had made it apparent that it should be mutagenic, and it was predicted at an early stage that bisulfite would function as a specific
mutagen (25). This proved to be the case, when it was tested in
E. coli (31). The situation has turned out to be more complex
than anticipated, however. For further understanding, we must
consider the mechanism of this reaction, and that of a slower
background reaction, catalyzed by other buffers, or water alone,
that also converts cytosine to uracil (32).

THE MECHANISM OF DEAMINATION

The overall mechanism of the reaction is given in Fig. 1, in
the path I→II→III→IV, where R = H.

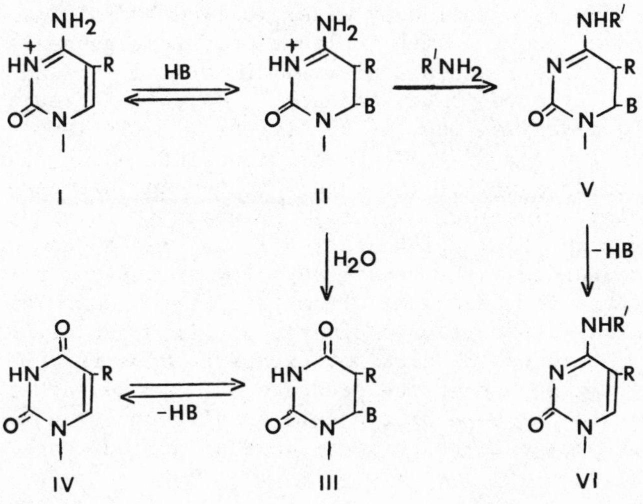

Fig 1. Deamination and transamination of cytosine derivatives.

Protonated cytosine, I, reacts by reversible addition across its
5,6-double bond to afford intermediate II. This equilibrium is
greater for bisulfite (B=-SO$_3^-$) than for other substances yet
tested, with an equilibrium constant of 11.2 1. mol^{-1} for the case

of cytidine, 25° (33). For water, on the other hand (B= -OH), the
equilibrium constant for addition is about 10^{-4} (34).

The dihydrocytosine intermediate, II, deaminates to dihydro-
uracil derivative III, in a step that is slower than the initial
addition. This step is subject to general base catalysis, with
bisulfite, sulfite, and pyrosulfite all acting as active catalysts
(35). Thus bisulfite plays an effective role in two separate steps
in the reaction scheme, and deamination rates rise sharply with
increasing bisulfite concentration.

The last step in the reaction is the loss of the added buffer
(or of water) to give uracil (III→IV). In the case of bisulfite,
the equilibrium constant for formation of the adduct III from
uracil is high, at neutral pH. Removal of the bisulfite is therefore needed to complete the conversion of III to IV (36,37). The
decomposition of the uracil-bisulfite adduct may require hours at
neutral pH, but it is speeded by alkali, or by increased concentrations of buffer.

Deamination of cytosine by water (called hydrolytic deamination) is also a reaction with implication for mutagenesis (32).
The pH-rate profile reaches a minimum at neutral pH, and the rate
is accelerated by acid, base, and heat (38). This reaction, like
the bisulfite catalyzed one, is specific (33). It is considered
to be responsible for GC → AT transitions induced by heat and by
acid in bacteriophage T4. The base involved in this case is 5-
hydroxymethylcytosine (40). Hydrolytic deamination of cytosine
to uracil is a reaction which apparently carries an unacceptably
heavy genetic burden, as a repair capacity exists for uracil in DNA
in many, perhaps all organisms, from bacteria to mammals. The initial step involves the release of free uracil by uracil-DNA-glycosylase (41). This enzyme does not excise thymine from DNA; in fact,
it has been suggested that the presence of thymine rather than uracil
in DNA is due to the need of the cell to distinguish a natural coding
base from the mutagenic reaction product of another base, for repair
purposes (42).

In cells with effective repair, then, hydrolytic deamination
of cytosine in DNA is not mutagenic. Mutations are produced,
however, by spontaneous deamination of the minor base, 5-methyl-
cytosine, to thymine. Thymine is not recognized by uracil-DNA-
glycosylase and escapes repair. As a result, in E. coli, 5-methyl-
cytosine residues are hotspots for spontaneous transition mutations
(43). Bacterial mutants have been isolated, however, which are
deficient in uracil-DNA-glycosylase, and cannot excise uracil
from DNA (ung⁻). In ung⁻ mutants, the rate of transition mutation
at cytosine residues is raised to the level observed at 5-methyl-
cytosine hotspots in the wild type (44).

Specific GC to AT transitions are induced in E. coli upon exposure of E. coli to 1 M bisulfite, at pH 5.2 (31). The strict
specificity is maintained both in forward and back-mutation studies
(1). The pH profile for mutagenesis is quite similar to that for
cytosine deaminations with a maximum at pH 5.2-5.3 (45). At pH 7,
the bisufite-induced mutation rate is just above the background.
Thus, in M HSO_3^-, 37°, the number of revertants obtained per 10^7
cells with an arginine-requiring mutant (A38, courtesy of Dr. F.
Mukai, NYU Medical School) were: pH 5.2, HSO_3^- ; 750, NaCl control
18; pH 6.2. HSO_3^- 240, NaCl control 18; pH 7.2, HSO_3^- 4.5, NaCl
control 1.5. Mid-log phase cultures showed a tenfold higher rate
of mutation than stationary phase cells. This result agrees with
the known chemical specificity of bisulfite for single-stranded
DNA. Such DNA would be expected to occur in greater amounts in
rapidly dividing cells.

A reasonable expectation is that the mechanism for bisulfite-
induced mutagenesis should parallel that for spontaneous hydrolytic deamination. Thus, deamination at cytosine residues should
be repaired, while reaction at 5-methylcytosine should escape repair and lead to mutations. This expectation was reinforced by
recent studies in which ung⁻ mutants of E. coli were treated with
bisulfite. The uracil-excision deficient mutants were much more
sensitive to killing that the wild type strain and log phase cultures were more sensitive than stationary phase (46,47).

Difficulties with this interpretation arise from a consideration of the chemistry of reaction of bisulfite with 5-methylcyto-
since. While hydrolytic deamination of 5-methylcytosine and cytosine proceed at roughly comparable rates (39), bisulfite-induced
deamination proceeds much more slowly with 5-methylcytosine than
with cytosine (48,49). The equilibrium for the formation of the
bisulfite adduct (II, Fig. 1 R=CH_3,B= SO_3^-) is less favorable for
5-methylcytosine. In addition, deamination leads to the formation
of two stereoisomers of the uracil adduct (III, Fig. 1, R=CH_3,
B= -SO_3^-), in which the methyl and sulfonyl groups are cis- and
trans to one another (50). One stereoisomer is resistant to elimination of bisulfite, and remains as the adduct, even under alkaline
conditions (48). For these reasons, bisulfite-catalyzed formation
of thymine from 5-methylcytosine in DNA may be about two orders
of magnitude less rapid than the corresponding conversion of
cytosine to uracil.

However, the inefficiency of this reaciton need not exclude
the possibility that bisulfite mutagenesis is due to deamination
of 5-methylcytosine. On the basis of detailed kinetic studies (51),
we can estimate that the rate of deamination of cytidine at pH 7.4,
in 1 M HSO_3^- is 10^3 faster than the background. Deamination of
5-methylcytosine under these conditions can then be estimated at

perhaps ten times faster than the background rate, a result consistent with the E. coli mutagenesis data discussed above. We must remember, of course, that the concentration of bisulfite within the cell was likely to be much less than 1 M, which makes the comparison of mutagenic data with chemical rates less exact.

An alternative mechanism for bisulfite mutagenesis has been provided (52). It was observed that the number of TrpA446 mutants induced by bisulfite was the same in ung⁻ and ung⁺ strains of E. coli. This was ascribed to the inactivation, by bisulfite and acid of uracil-DNA-glycosylase. Thus, ung⁺ cells were rendered ung⁻ by the treatment, and no difference was seen when a comparison was made with cells that were genetically ung⁻. In support of this idea, it was demonstrated that uracil-containing λ phage was plated with great efficiency on ung⁺ cells, after a bisulfite-acid treatment. Bisulfite mutagenesis in E. coli may therefore be due to a combination of cytosine deamination and repair. These results, however, appear to conflict with the others (40,41), described above, which state that ung⁻ cells are much more sensitive than ung⁺ to killing by bisulfite. A comparison of mutagenesis rates and inactivation rates in the same strains of ung⁺ and ung⁻ bacteria will be needed to resolve the issue.

Finally, a third possibility exists for the mechanism of mutagenesis by bisulfite. It may be due to miscoding by the intermediate adduct II, Fig. 1. Mutagenesis should then be the same in ung⁺ and ung⁻ cells, as such adducts are not repaired by uracil-DNA-glycosylase. The greater senstivity of ung⁻ to killing by bisulfite could be ascribed to failure to repair uracil residues in DNA. One point of difficulty would remain. It is unclear why miscoding by adducts should be mutagenic, while miscoding by uracil would be lethal.

GENETIC EFFECTS OF LOW CONCENTRATIONS OF BISULFITE

While the deamination of cytosine by high concentrations of bisulfite, at pH 5.6, has been useful as a biochemical and genetic tool, it is unlikely to be important with respect to human health. The above conditions are far from physiological ones. The existence of a uracil repair mechanism, and the relative unreactivity of 5-methylcytosine to bisulfite all serve to diminish the importance of bisulfite deamination. In E. coli, as we have seen, GC→AT mutagenesis is barely above background in 1M bisulfite, pH 7 (45). At a concentration of 0.1 M, no mutations were observed in E. coli (53).

However, various genetic effects, including mutations, have been observed at even lower bisulfite concentrations, at neutral pH, in other systems. These effects are presumably caused by reactions other than the deamination of cytosine. The results are described in two recent reviews (1,2) and will be summarized here.

Effects on Microorganisms and Plants

Mutations were induced by 10 mM bisulfite in Micrococcus
aureus and by 5 mM bisulfite in Saccharomyces cerevisiae (at pH 3.6,
but not at pH 5.5). Concentrations of 2 mM or greater impaired
DNA synthesis in cultures of Chlorella pyrenoidosa. Bisulfite
serves as a strong mutagen in barley, when applied to seeds at
7-10 mM (54) Depression of the mitotic index was produced by bisulfite concentrations above 2 mM in the root meristem of the puff
bean, Vicia faba, and by low concentrations of gaseous SO_2 in ger-
minating pollen tube cultures of Tradescantia.

Effects on Mammalian Cells and Mammals

Studies on cells in culture have been run using mouse liver
cells, chick embryo and mouse fibroblasts, Hela cells, human embry-
onic lung cells, and human lymphocytes. A variety of mammalian oocytes have also been examined. A number of significant effects were
observed in these studies at bisulfite concentrations that varied
from 0.1 to 10 mM. The effects included inhibition of DNA synthesis
and growth reduction of the mitotic index, and cytotoxicity. In human lymphocytes (55) and mammalian oocytes (56), chromosomal abnormalities were observed such as clumping, fuzziness, and breaks. In
addition, it has been shown that bisulfite induces dose-related sister chromatid exchange in Chines hamster ovary cells at levels down
to 0.03 mM.

Attempts to detect effects of the above types in laboratory
animals have yielded negative results thus far. The experiments
include a search for chromosomal abberations in the oocytes of
female mice (56), the dominant lethal test (57), the specific locus test (58), and a carcinogenesis study (59), all in mice. No
studies have been reported to date using sulfite-oxidase deficient
animals.

COMUTAGENESIS AND COCARCINOGENESIS

It has been suggested that sulfur dioxide may exert its health
effects in connection with other pollutants, in a synergistic
manner (12). For this reason, studies demonstrating comutagenic
or carcinogenic effects of bisulfite are of particular interest.

Comutagenesis has been demonstrated in Chinese hamster V79
cells (at 10 mM bisulfite) and E. coli (at 75 mm bisulfite) (53).
Ultraviolet mutagenecity was doubled in the hamster cells, and

increased tenfold the bacterial system, by bisulfite treatment
during or immediately after, irradiation. Experiments with repair-
deficient E. coli strains indicated that bisulfite acted by inhibiting repair. Recent studies have shown that DNA polymerase I is
inactivated by bisulfite. (60).

In an earlier study, a significant co-carcinogenic effect was
demonstrated, in rats exposed to atmospheres containing sulfur
dioxide and benz(a)pyrene (61). No cancers were produced among
15 rats exposed to SO_2 alone, and only one among 30 rats exposed
to benz(a)pyrene alone. Nine cancers were produced in 46 rats
exposed to both substances.

CHEMICAL REACTIONS OF POTENTIAL RELEVANCE

No chemical basis has been extablished for the above effects,
except for the case of comutagenesis in bacteria. If they are the
result of direct damage to DNA by bisulfite, then two known reactions
may be contributing causes.

The deamination sequence for cytosine described in Fig. 1,
can be diverted in the presence of an appropriate amine, to yield
an N^4-substituted cytosine (II→ V →VI, Fig. 1) (62). This process
is called transamination. It proceeds effectively at neutral pH,
with lower concentrations of bisulfite than those needed for deamination. When the ε-amino group of lysine, in a polypeptide or
protein, is involved, then nucleic acid–protein cross-linking
takes place (63). In some situations, the reaction apparently can
take place with cytosine in double-stranded DNA. Thus, such DNA
has been cross-linked to polylysine (64), and to the coat protein
of S_d phage (65). These reactions have been run at bisulfite concentrations of 0.15 M or greater, for experimental convenience,
but it is possible that they would take place at lower concentrations.

Bisulfite, at concentrations of 10 mM or less, participates in
free radical chain reactions, particularly in the presence of oxygen or certain metals (29). Species such as $\cdot SO_3$ and $\cdot OH$ are involved in these processes. Higher concentrations of bisulfite are less
effective as the substance then quenches its own chains. Damage
of an undeterminate type was produced in DNA upon treatment with
10 mM bisulfite, and Mn^{++}, at pH 7. Subsequent alkaline treatment
led to chain breakage (29). Exposure of Bacillus subtilis DNA to
10 mM bisulfite, and oxygen, led to the inactivation of its trans -
forming activity (24). In a study of the action of bisulfite upon
repair-deficient E. coli mutants, it was postulated that DNA
strand-breaking events, caused by free radicals, were taking place
(47).

Of course, some of the effects discussed above may be the result of reactions of bisulfite with proteins, rather than nucleic acids, such as the already mentioned sulfitolysis of disulfide bonds. Such a reaction could destabilize proteins responsible for the integrity of chromosomal structure, or inactivate enzymes by combining with a cofactor, such as NAD, or a flavin derivative, in an enzyme-cofactor complex (2). Finally, bisulfite may also damage enzymes through free radical reactions, which have been demonstrated for methionine and tryptophan (1,2).

THE RELATION BETWEEN THE GENETIC AND BIOCHEMICAL EFFECTS OF BISULFITE, AND ITS TOXICITY: REGULATORY IMPLICATIONS

As we have seen, the regulatory attitutes of various agencies towards sulfur dioxide vary considerably. One agency is responsible for enforcing, at a cost of billions of dollars, regulations designed to protect us from the adverse effects of SO_2 as a component of air pollution. Another agency is much more permissive toward the ingestion of much larger amounts of the same pollutant in wine. As a result, it is possible to take in more sulfur dioxide in one glass of wine containing the maximum legal amount (350 ppm) than in breathing air for 10 days, at the maximum level permitted by air quality standards (0.14 ppm, 24 hr. avg) (1,2).

This situation would be rational if the effects of sulfur sioxide were limited to the upper respiratory system, and specific to the tissues there: for example, a transient physiological response such as bronchoconstriction. This situation is possible, but it might prove difficult to justify expensive desulfurization measures if this were the only adverse health effect that would be determined by such measures.

A different picture emerges, however, when we consider the biochemical and genetic properties of sulfur dioxide, and its toxic effects on various systems. It is a pollutant with a great capacity for inflicting biological injury. It reacts readily with most of the molecules of importance in biochemistry: nucleic acids, proteins, including enzymes, carbohydrates, cofactors, and many others (1,66). Vegetation is damaged by exposure to small amounts of sulfur dioxide (7). Rabbits die when concentrations below 1 mM are maintained in their bloodstream. Low concentrations of bisulfite inflict genetic and biochemical damage upon cells in culture. We are normally protected from the toxicity of bisulfite by the sulfite oxidase in our tissues. Two separate situations can exist, however, in which this defense system may be less than perfect:

(a) Sulfite oxidase, while rapid, is not instantaneous in its action. Nor does it exist in large amounts in every tissue. Temporary, high concentrations of sulfur dioxide may produce biochemical damage at the sites of entry to our bodies. Inhaled sulfur dioxide is largely absorbed in the nose, with additional amounts reaching the trachea and bronchi (7). It is absorbed into the blood stream from the upper respiratory tract, and oxidized. Ingested bisulfite passes through the stomach and small intestine, before reaching the bloodstream. While in transit, bisulfite could react, by rapid and irreversible reactions, with the cellular constituents of the above organs. Possible effects would be somatic mutations, and interference with macromolecule synthesis, and enzyme functions. This would lead to inhibition of repair, as well as and intensification of damage caused by other environmental pollutants. Such reactions could act as the basis of the synergistic effects implied by epidemiological data, and demonstrated in comutagenesis and cocarcinogenesis studies.

(b) In a subset of the human population, there may also be adverse health effects due to chronic bisulfite exposure. As we have discussed, a few cases of severe human sulfite oxidase deficiency have been recorded. In one such case, an examination of the plasma showed that measurable levels of free bisulfite were present. Sulfite oxidase levels have not been measured systematically in the general population. It is reasonable to expect, however, that they vary over a certain range, and that a fraction of the population will suffer moderate sulfite oxidase deficiency. Low levels of free bisulfite might exist in the plasma and tissues of such individuals, and the levels would vary with their exposure to environmental bisulfite, and with the sulfur amino acid content of their diets. If either source of exposure were enhanced, sulfite oxidase deficient individuals would suffer a greater risk of genetic damage, and other toxic effects of bisulfite.

The above scenarios are possible, and must be taken seriously, though they are not yet proven. In order to evaluate their importance, more information is needed about the nature of the reactions responsible for the genetic and biochemical effects of low levels of bisulfite, and the variation of sulfite oxide capacities and of plasma bisulfite concentrations among humans.

In the interim, we must still make regulatory decisions. The two situations described above are speculative, but no more so than the idea that sulfur dioxide is harmful when inhaled, but harmless when ingested in much larger quantities. It would be rational for the three agencies involved with the regulation of our exposure to sulfur dioxide to adopt a unified approach, while the scientific uncertainties were being settled. A moderate reduction in the amounts present in wine would be a prudent initial step, as so much of our consumption comes from this single source.

ACKNOWLEDGEMENTS

The author's own research in this area, and the preparation of this manuscript, have been supported by NIH research grant ES-01033, National Institutes of Environmental Health Sciences.

REFERENCES

1. Shapiro, R.: Genetic effects of bisulfite. Mutation Res., 39: 149(1977)

2. Gunnison, A.F.: Sulfite toxicity: A critical review of in vitro and in vivo data. Food Cosmet. Toxicol., in press (1981)

3. American Chemical Society: Cleaning Our Environment, A Chemical Perspective, Washington, D.C., (1978)

4. Newell, V.A., Wyzga, R.E., and McCarroll, J.R.: Costs versus benefits of sulfur oxides and related particulate matter control. Bull. N.Y. Acad. Med., 54:1211(1978)

5. Carter, L.J.: Uncontrolled SO_2 emissions bring acid rain. Science, 204:1179(1979)

6. Lave, L.B.: Health benefits of abating air pollution. Bull. N.Y. Acad. Med., 54:1235(1978)

7. Committee on Sulfur Oxides, National Research Council: Sulfur Oxides. National Academy of Sciences, Washington, D.C.(1978)

8. Ellison, J.M. and Waller, R.E.: A review of sulphur oxides and particulate matter as air pollutants with particular reference to effects on health in the United Kingdom. Environ. Res., 16: 302(1978)

9. Schimmel, H.: Evidence for possible acute health effects of ambient air pollution from time series analysis: methodological questions and some new results based on New York City daily mortality, 1963-1976. Bull. N.Y. Acad. Med., 54:1052(1978)

10. Hickey, R.J., Clelland, R.C., Bowers, E.J. and Boyce, D.E.: Health effects of atmospheric sulfur dioxide and dietary sulfates. Arch. Environ. Health 31:108(1976)

11. Ferris, B.G.: Health effects of exposure to low levels of regulated air pollutants. J. Air Poll. Control Assoc., 28:482(1978)

12. Kassell, E.J.: Towards an optimum environment. In Environmental
    Factors in Respiratory Disease, Ed. by D.H.K. Lee. Academic
    Press, New York, p 237(1972)

13. Chichester, D.F., and Tanner, Jr. F.W.: Antimicrobial food
    additives. In Handbook of Food Additives 2nd ed., Ed. by T.E.
    Furia. CRC Press, Cleveland, p 115(1972)

14. Shapiro, R., Robakis, N.K. and Rossman, T.:Inhibition of growth
    and macromolecular synthesis in E. coli by bisulfite. Fed.
    Proc., 40: 1621(1981)

15. De Felice, M.,Levinthal, M., Iaccarino, M., and Guardiola, J.:
    Growth inhibition as a consequence of antagonism between re-
    lated amino acids: Effect of valine in Escherichia coli K12.
    Microbiol Rev., 43:42(1979)

16. Harris, C.L.: Cysteine and growth inhibition of Escherichia
    coli: Threonine deaminase as the target enzyme. J. Bacteriol.,
    145:1031(1981)

17. Institute of Food Technologists Expert Panel on Food Safety and
    Nutrition and the Committee on Public Information: Sulfites as
    food additives. Food Technol., 29:117(1975)

18. Life Science Research Office, Federation of American Societies
    for Experimental Biology: Evaluation of the Health Effects of
    of Sulfiting Agents as Food Ingredients. Bethesda, Maryland
    (1976)

19. Southerland, W.M., Winge, D.R. and Rajagopalan, K.V.: The do-
    mains of rat liver sulfite oxidase. J. Biol. Chem., 253:8747
    (1978)

20. Gunnison, A.F., Zaccardi, J., Dulak, L. and Chiang, G.: Tissue
    distribution of S-sulfonate metabolites following exposure to
    sulfur dioxide. Environ. Res., in press (1981)

21. Gunnison, A.F., Bresnahan, C.A. and Palmes, E.D.:Comparitive
    sulfite metabolism in the rat, rabbit, and rhesus monkey.
    Toxicol. Appl. Pharmacol., 42:99(1977)

22. Johnson, J.L., Rajagopalan, K.V. and Cohen, H.J.: Molecular
    basis of the biological function of molybdenum. Effect of
    tungsten on xanthine oxidase and sulfite oxidase in the rat.
    J. Biol Chem., 249:859(1974)

23. Gunnison, A.F., Farruggella, T.J., Chiang, G., Dulak, L. and
    Birkner, J.: A sulphite oxidase-deficient rat model: metabolic
    characterization. Fd. Cosmet. Toxicol., in press(1981)

24. Johnson, J.L., Wand, W.R., Rajagopolan, K.V., Duran, M. Beemer, F.A. and Wadman, S.K.: Inborn errors of molybdenum metabolism: Combined deficiencies of sulfite oxidase and xanthine dehydrogenase in a patient lacking the molybdenum cofactor. Proc. Natl. Acad. Sci. USA, 77:3715(1980)

25. Shapiro, R., Servis, R.E., and Welcher, M.: Reaction of uracil and cytosine derivatives with sodium bisulfite. A specific deamination method. J. Amer. Chem. Soc., 92:422(1970)

26. Hayatsu, H., Wataya, Y. Kai, K. and Iida, S.: Reaction of sodium bisulfite with uracil, cytosine, and their derivatives. Biochemistry, 9:2858(1970)

27. Shapiro, R., Braverman, B., Louis, J.B. and Servis, R.E.: Nucleic acid reactivity and conformation. II. Reaction of cytosine and uracil with sodium bisulfite. J. Biol Chem., 248: 4060 (1973)

28. Shapiro, R., Cohen, B.I. and Servis, R.I.: Specific deamination of RNA by sodium bisulfite. Nature, 227:1047(1970)

29. Hayatsu, H.: Bis;ulfite modification of nucleic acids and their constituents. Progr. Nucleic Acid. Res. Mol. Biol., 16:75(1976)

30. Shortle, D. and Nathans, D.:Local mutagenesis: A method for generating viral mutants with base substitutions in pre-selected regions of the viral genome. Proc. Natl. Acad. Sci. USA, 75:2170(1978)

31. Mukai, F., Hawryluk, I., and Shapiro, R.:The mutagenic specificity of sodium bisulfite. Biochem. Biophys. Res. Commun., 39: 983(1970)

32. Shapiro, R. and Klein, R.S.: The deamination of cytidine and cytosine by acidic buffer solutions. Mutagenic implications. Biochemistry 5:2358(1966)

33. Shapiro, R., DiFate, V. and Welcher, M.: Deamination of cytosine derivatives by bisulfite. Mechanism of the reaction. J. Amer. Chem. Soc. 96:906(1974)

34. Shapiro, R.: Damage to DNA caused by hydrolysis. In Chromosome Damage and Repair, Ed. by E. Seeberg and K. Kleppe. Plenum Press, New York, in press(1981)

35. Slae, S. and Shapiro, R.: Kinetics and mechanism of the deamination of 1-methyl-5,6-dihydrocytosine. J. Org. Chem. 43:1721 (1978)

36. Pitman, I.H. and Jain, N.B.: The covalent addition of bisulfite ion to N-alkylated uracils and thiouracils. Aust. J. Chem., 32: 545(1979)

37. Shapiro, R., Welcher, M., Nelson, V. and DiFate, V.: Reaction of uracil and thymine derivatives with sodium bisulfite. Studies on the mechanism and reduction of the adduct. Biochim. Biophys. Acta, 425:115(1976)

38. Garrett, E.R. and Tsau, J.: Solvolyses of cytosine and cytidine. J. Pharm. Sci., 61:1052(1972)

39. Lindahl, T. and Nyberg, B.: Heat-induced deamination of cytosine residues in deoxyribonucleic acid. Biochemistry, 13:3405(1974)

40. Baltz, R.H., Bingham, P.M. and Drake, J.W.: Heat mutagenesis in bacteriophage T4: The transition pathway. Proc. Nat. Acad. Sci. USA, 73:1269(1976)

41. Lindahl, T.: DNA glycosylases, endonucleases for apurinic/ apyrimidinic sites, and base excision-repair. Prog. Nucleic Acid Res. Mol. Biol., 22:135(1979)

42. Lindahl, T., Ljungquist, S., Siegert, W., Nyberg, B. and Sperens, B.: DNA-N-glycosidases; properties of uracil-DNA glycosidase from Escherichia coli. J. Biol Chem., 252:3286(1977)

43. Coulondre, C., Miller, J.H., Farabaugh, P.J. and Gilbert, W.: Molecular basis of base substitution hotspots in Escherichia coli. Nature, 274:775(1978)

44. Duncan, B.K. and Miller, J.H.: Mutagenic deamination of cytosine residues in DNA. Nature, 287:560(1980)

45. Moller, M. and Shapiro, R.: unpublished results

46. Hayakawa, H. Kumara, K. and Sekiguchi, M.: Role of uracil-DNA glycosylase in the repair of deaminated cytosine residues of DNA in Escherichia coli. J. Biochem.(Tokyo) 84:1155(1978)

47. Simmons, R.R. and Friedberg, E.C.: Enzymatic degradation of uracil-containing deoxyribonucleic acid. V. Survival of Escherichia coli and coliphages treated with sodium bisulfite. J. Bacteriol. 137:1243(1980)

48. Wang, R.Y-H, Gehrke, C.W and Ehrlich, M.: Comparison of bisulfite modification of 5-methyldeoxycytidine and deoxycytidine residues. Nucleic Acids Res., 8:4777(1980)

49. Shapiro, R., Slae, and Crane, L.E.: unpublished results

50. Shirigami, M., Kudo, I, Iida, S. and Hayatsu, H.: Formation of
    diastereomers of 5,6-dihydrothymine-6-sulfonate by deamination
    of 5-methylcytosine with bisulfite. Chem. Pharm. Bull. 23:302
    (1975)

51. Slae, S. and Shapiro, R.: Deamination of bytidine by bisulfite:
    mechanism at neutral pH. J. Org. Chem., 43:4197(1978)

52. Duncan, B.K.: personal communication

53. Mallon, R.G. and Rossman, T.G.: Bisulfite (sulfur dioxide) is a
    comutagen in E. coli and in Chinese hamster cells. Mutation
    Res., 88:125(1981)

54. Kak, S.N. and Kaul, B.L.: Mutagenic activity of sodium bisulphite
    in barley. Experentia, 35:739(1979)

55. Schneider, L.K. and Calkins, C.A.: Sulfur dioxide-induced lymph-
    ocyte defects in human peripheral blood cultures, Environ. Res.,
    3:473(1971)

56. Jagiello, G.M., Lin, J.S. and Ducayen, M.B.: SO_2 and its meta-
    bolite: effects on mammalian egg chromosomes. Environ. Res.,
    9:84(1975)

57. Generoso, W.M., Huff, S.W. and Cain, K.T.: Tests on induction
    of chromosome aberrations in mouse germ cells with sodium
    bisulfite. Mutation Res., 56:363(1978)

58. Russell, W.L. and Kelly, E.M.: Results from a specific-locus
    test of the mutagenecity of sulfur dioxide in mice. In Annual
    Progress Report for period ending June 30, 1975. Biology Divi-
    sion, Oak Ridge National Laboratory, Oak Ridge, Tennessee,
    p 119(1975)

59. Tanaka, T., Fujii, M. Mori, H. and Hirono, I.: Carcinogenicity
    test of potassium metalisulfite in mice. Ecotoxicol and Envrnm.
    Safety, 3:451(1979)

60. Rossman, T.:personal communication

61. Laskin, S., Kuschaer, M., Kellakumar, A. and Katz, G.V.: Com-
    bined carcinogen-irritant animal inhalation studies. In Air
    Pollution and the Lung, Ed. by E.F. Aharonson, A. Ben-David
    and M.A. Klingberg, John Wiley and Sons, New York, p 190(1976)

62. Shapiro, R. and Weisgras, J.M.: Bisulfite-catalyzed transamina-
    tion of cytosine and cytidine. Biochem. Biophys. Res. Commun.,
    40:839(1970)

63. Shapiro, R., and Gazit, A: Cross-linking of nucleic acids and
    proteins by bisulfite. In Protein Cross-linking: Biochemical
    and Molecular Aspects, Ed. by M. Friedman, Plenum Press, New
    York, p 633(1977)

64. Sklyadneva, V.B., Shie, M. and Tikchonenko, T.I.: Alteration
    of the DNA-secondary structure in the DNA polylysine complex
    evidenced by sodium bisulfite modification. FEBS Lett., 107:
    129(1979)

65. Sklyadneva, V.B., Chekanovskaya, L.A., Nikoleva, I.A., Tikcho-
    nenko, T.I.: The secondary structure of bacteriophage DNA in
    situ VII. The reaction of sodium bisulfite with intraphage
    cytosine as a probe for studying the DNA-protein interaction.
    Biochim. Biophys. Acta, 565:51(1979)

66. Petering, D.H. and Shih, N.T.: Biochemistry of bisulfite-sulfur
    dioxide. Environ. Res., 9:55(1975)

DISCUSSION

DR. SINGER: Bob touched on something which I think ought to be
emphasized: single-stranded nucleic acids react to a very much
higher extent that double-stranded ones. This is true for
depurination, deamination, reactions with formaldehyde, bisulfite,
or chloroacetaldehyde. The latter reactions are considered to
occur only in single stranded nucleic acid. With many reactions
there is sometimes a 100-fold difference depending on
strandedness, so that reactions which may be rather minor in
double-stranded nucleic acids are really major in
single-stranded. Originally, I believe, when the bisulfite
reaction was first published the deamination was to the sulfite.
When you treat it with alkali, you of course lose this, but
perhaps in vivo you do not lose the sulfite from the uracil, and
that is the modified base that is not repaired.

DR. SHAPIRO: First about the question of single versus
double-strandedness. I think, as far as anybody knows from
reactivity studies, that the single-strand specificity of
bisulfite is total. You can always expect some double-stranded
reactivity because of the thermodynamic helix-opening effect, but
none has been picked up yet, so bisulfite has been used as a
really marvelous reagent for the modification of nucleic acids.
One elegant example is the work of Shortle and Nathan, in which
they take SV-40, use a restriction enzyme to open one chain at a
particular site, digest a few bases away, exposing cytosines, and
then expose the SV-40 to bisulfite. After growth and repair in
vivo you get mutations exactly as expected, exactly of cytosine to
uracil; there's no doubt that C is going to U, ultimately T, and
that it's mutagenic. So, bisulfite is a very good single strand
reagent.

   With respect to the second question, this is another puzzle in
the mechanism. You do get the cytosine adduct with bisulfite and
that then forms the uracil adduct. This will take many hours to
decompose at pH-7 in chemical studies, yet, you get mutagensis in
many minutes. As far as we know, uracil DNA glycosylase does not
recognize cytosine or uracil adducts, so those adducts are
presumably in DNA for a time and they're another possible cause of
mutagenesis. But the uracil adduct, which is the one that's
really stable, seems to me biologically inert in various systems
we've looked at; it 's not recognized, it doesn't mispair, it just
presumably exists, must exist. Because a uracil bisulfite is so
bulky, it must exist in the syn conformation. So mispairing would
be out of the question and one would expect that the uracil
adduct, if it persisted in DNA would just be responsible for some
killing or for temporary inactivation.

DR. SINGER: Why syn?

DR. SHAPIRO: Because it's at the 6 position of pyrimidine; the
same effects also occur at the 8 position of purines. If you
build a model, there's enormous steric interference between the
6-sulfonate group and various parts of the sugar. A methyl group,
which is smaller than bisulfite, in this position will cause
rotation into the syn conformation; there are X-ray studies and
many others to show that. SO3 being enormous would surely be
expected to cause the same rotation. In addition to the bulk
there would be some negative repulsion between the sulfonate and
the phosphodiester backbone. One would therefore really expect
these adducts not to miscode and mispair and not be a cause of
mutagenesis, even though it would help to resolve some of the
contradiction within the genetic data if they were a cause of
mutagensis. There are always surprises available for us: we
wouldn't expect that a 3-methyl C should cause mispairing either,
because putting a methyl right where the Watson-Crick hydrogen
bonding should be might be expected to just obliterate it, yet you
seem to get mispairing. So there may be surprises here but there
would be surprises in the genetics. The chemistry is very clear
about what it predicts for these adducts.

DR. BRIDGES: Is there any information about the effects of the
TexA or uvrA genes on bisulfite mutagenesis?

DR. SHAPIRO: No, not that I know of.

DR. BRIDGES: You have interpreted everything in terms of
chemistry, but studies with those mutants could give you a very
fair picture of whether in fact the rec-lex pathways are involved,
which you would expect if the adduct remains. This is one way in
which it could be mutagenic.

DR. GLICKMAN: After Bob Shapiro visited our lab a month ago, we
have started to look at specificity in lacI and also look at umuC
strains with this in mind. But there is, as far as we've
discussed and as far as the computer searches we've done, no
information in the literature about rec-lex and sodium bisulfite
mutagenesis.

DR. SHAPIRO: Let me just throw in one nugget of data, which we
started studying in our own lab because no one else was working
with the chemistry. I much prefer other people to do the
complicated genetics. Let me say in all humility I no longer
assume that for any reagent the chemistry can necessarily explain
what's happening biologically. But if you treat organisms at pH 5
with bisulfite and then instead of plating them immediately at pH
7 you just shake them for awhile in buffer, the number of mutants

that you get then when you plate them goes down, down to about 1/3 the original amount. I understand that from the UV field this is called liquid-holding recovery. So apparently some repair may be taking place during this holding period at pH 7. This does indicate that whatsoever is causing the mutations in bacteria is subject to some repair system. I can't tell you specifically what it might be, except of course the uracil glycosylase may be involved.

DR. WALKER: Does the uracil glycosylase work as well with GU pairs as it does with AU pairs?

DR. SHAPIRO: Apparently it just goes for uracil, regardless of the identity of the base that its opposite to.

DR. WALKER: It should be possible to find it in AU pairs because you can now get uridine triphosphate into DNA, if you knock out the uracil glycosylase and use a mutant that makes very high concentrations of the uridine triphosphate.

DR. SHAPIRO: Hayakawa gave the bisulfite treatment to DNA and then showed that uracil glycosylase was causing changes in it which led to fragmentation in alkali, so presumably uracil glycosylase was removing the uracil.

DR. GLICKMAN: Your question though is, if you have a GC pair going to a GU pair, as compared to an AT going to AU, is the U removed? In one case the U is across from a G and in another case across from an A. The answer from the specificity experiments in the lacI ung system would be yes, because if you take an ung- strain the mutational frequency goes up five-fold and they're all accounted for by GC to AT's at GC sites. In the ung- situation that would suggest that mutation at those sites is due to the glycosylase not removing the U which is across from a G. So normally it should do that.

DR. RIPLEY: However, in an experiment we did with phage T4, in which the phage lacked the normal T4 modification of cytosine but had ordinary cytosine in its DNA, we expected to create deaminated C by heat treatments. We were unable to show any effect of ung- on heat-induced GC to AT transition mutations.

DR. WALKER: There is a difference between working on the mono nucleotide pool versus working on the whole DNA; you come up with different pairs.

DR. GLICKMAN: I think Bob's point is that the contradiction with Duncan's results is not really a contradiction, it's confusion. Unless you know more about the genetic system that he was using and the site that he was looking at, a reversion of something I would presume, it's really very difficult to compare the survival results from one lab and the mutagenesis lab in another.

DR. WALKER: There is no contradiction if the lethal event is different that the pre-mutagenic event which is quite a common kind of phenomenon for a lot of these chemicals.

DR. GLICKMAN: Do you know the genetic system?

DR. SHAPIRO: Not in detail.

DR. GLICKMAN: Did he say anything to you about killing?

DR. SHAPIRO: I didn't see any extensive data on killing. I mean, if we ignore that one set of results, everything else becomes perfectly consistent.

DR. GLICKMAN: It might be helpful to live besides an electricity generating station burning fuel on one side of you and a nuclear reactor on the other, because if you're getting SO_2 and that's raising your cystine content in the cell, that's going to protect you from the ionizing radiation of the nuclear reactor on the other side of you.

DR. SHAPIRO: If water is a mutagen, my solution is to drink more wine.

DR. SARASIN: Like you, I am very concerned about bisulfite. Since you pronounced the magic word of adaptation, do you know if it's possible for example to measure the sulfite oxidase activity from people who drink wine every day and from people who never drink wine to see if there is a difference in activity?

DR. SHAPIRO: This would be worthwhile doing. I talked to two people who study sulfite oxidase. Al Gunnison at NYU and Dr. K.V. Rajagopalan at Duke. They both assure me unfortunately that it is rather difficult to measure sulfite oxidase levels. You need a rather large tissue sample from an individual. Gunnison thinks it would have to be done from skin; Rajagopalan thinks you can use fatty tissue. Presumably one could do autopsies on recently deceased wine drinkers and non-wine drinks and measure it. It's not so easy to do. Much more importantly, I think it would be very interesting to see how much other variation there is in sulfite oxidase deficiency genetically among the human population so we know if there's any subpopulation that is particularly at risk to genetic damage by bisulfite.

DR. SAMSON: Have you tried to look at the basis of this induced resistance to bisulfite in the bacteria? In other words, are you inducing some mechanism like the oxidase that will detoxify the bisulfite or are you inducing DNA repair? One way one could go about doing this would be to try and transform these induced cells with DNA that has already been damaged by sulfite and see if that DNA would survive in these cells.

DR. SHAPIRO: I don't think the resistance is a genetic thing because they lose it shortly thereafter.

DR. SAMSON: What if you've induced some kind of repair?

DR. SHAPIRO: My own guess would be that growth inhibition in bacteria is the consequence of some protein synthesis, a trickle of protein synthesis, that slows down or represses this rapid conversion of bisulfite to cysteine or other substances. So it's not that bisulfite itself is blocking growth or doing other genetic things, but rather its conversion products. What you do is you modulate that conversion. When you add chloramphenicol, things get worse. Now, bisulfite stops protein synthesis but there may be a tiny trickle going on which allows some new factors that regulate metabolism to be synthesized. You put in chloramphenicol and now you've completely shut down protein synthesis of the cells and they cannot adapt readily at all. I think this is likely to be what's going on, but of course there is an enormous amount of work to be done in that area.

DR. WALKER: Do you think they adapt to cysteine as well as growth on bisulfite?

DR. SHAPIRO: They recover. If you give them cysteine, there is a period of growth inhibition whose length depends on the amount of cysteine you give, and then they go back to growing in the cysteine.

DR. GLICKMAN: Is that a mass effect where they all go back and grow or is that a subpopulation effect?

DR. SHAPIRO: It's a good question. You're asking if this is perhaps selection of mutational events. Well, the lag can be as little as 10 minutes with very little bisulfite or it can be almost 24 hours with high doses. You can adjust the lag to any amount you want by the adjustment of the concentration. You can get it down to 5 minutes, if you wish.

DR. SINGER: Can I ask a completely different kind of question, because I think it should be answered. That is, who set them, and why were limits set on the exposure to sulphur dioxide when, if I understand you correctly, it does not cause any damage in animals.

DR. SHAPIRO: There are two agencies involved in regulating it. One is the FDA and the other is the Treasury. I can find nothing about where the Treasury got its limits from, even though it controls most of our exposure to sulphur dioxide. My guess is they just copied it from the FDA. And the FDA worries about the destruction of thiamine in foods. It doesn't want bisulfite to be present in foods at levels which would give you a vitamin deficiency. So this has been the primary reason. Too much sulphur dioxide in the wines makes them reek of it, which sets a limit. In wine-tasting terminology, the wine is sulphurous, which means if it's getting pretty close to that limit it's a bad wine. In fact, if you don't have too much bacterial contamination on your grapes you can, with care, make wine using very little or no sulphur dioxide. But if you have grapes that are really loaded with bacteria which is common in very hot countries, you need to literally pour the sulphur dioxide in.

DR. WALKER: Since you brought this discussion up, you should probably say something about acid rain, at least for the record, because that may be one of the more serious problems with burning high sulphur fuels. Maybe the pH changes are more profound in the ecological system than anything else that we're talking about.

DR. SHAPIRO: Of course, sulphur dioxide goes to sulphur trioxide and that causes acid rain. You get a pH drop and this has adverse ecological effects. But in most cost benefit analyses, the value attributed to damage due to acid rain together with the value due to material corrosion by sulphur dioxide has been given a very minor role compared to the human health effect. Most of the costs attributed to sulphur dioxide and most of the justification for its regulation have been built on the human health effects. In fact, some of the acid rain problem arises because, in order to avoid getting sulphur dioxide concentrations in urban places, industry has taken to building enormous smokestacks, a thousand feet and higher, which sends it way out into the upper atmosphere. As a result, a thousand miles away you get acid rain.

If, in fact, there are no bad health effects on human beings due to sulphur dioxide, the prescribed environmental thing to do would be to take down those tall smokestacks and let the SO_2 be emitted locally--say in New York City--well, I say this seriously. Acid rain is a problem where you have unbuffered soils which is in the Adirondacks. It is not a problem near New York City and much of the United States where you have well-buffered soils. So when you build a very tall smokestack, you protect the people in New York City from supposed health damage; on the other hand, you kill fish in the Adirondacks. If the people from New York City are not actually suffering health damage from SO_2, the prescribed thing to do would be to lower the smokestack size. So things get very involved, as you may gather.

SESSION II

INFIDELITY AND SPECIFICITY

MODERATOR: FRED SHERMAN

INFIDELITY OF DNA REPLICATION AS A BASIS OF MUTAGENESIS[1]

Thomas A. Kunkel, Roeland M. Schaaper
Elizabeth James and Lawrence A. Loeb
Gottstein Memorial Cancer Research Laboratory
Department of Pathology SM-30
University of Washington
Seattle, WA 98195

INTRODUCTION

Errors in DNA replication must be very infrequent, so as not to alter essential genetic information; yet they must occur rarely so as to permit species divergence through mutations. On the basis of rates of appearance of spontaneous mutations, 10^{-7} to 10^{-11}, DNA replication in vivo is assumed to be highly accurate (1). If one considers the multiplicity of genes demonstrated to affect mutation rates, it seems reasonable that this phenomenal accuracy is achieved through a multi-step process. We have focused on the fidelity by which DNA polymerases copy synthetic polynucleotides and natural DNA and the effects of mutagenic compounds on this accuracy. The underlying hypothesis is that infidelity by DNA polymerase, either on normal or damaged DNA, is a major determinant in mutagenesis.

RESULTS

Until recently, assays of the fidelity of DNA synthesis have depended either on measuring incorporation of non-complementary nucleotides using synthetic polynucleotide templates containing only one or two bases (2,3), or on measuring the incorporation of nucleotide analogs (4). It has been assumed that the results obtained with these model systems are similar to those that would be exhibited during copying natural

[1]This work was supported by grants from the National Institutes of Health (CA-24845 and CA-24498) and the National Science Foundation (PCM 76-80439).

63

DNA templates containing all four bases. However, it is known
that during polymerization with artificial template-primers,
there is slippage of the primer relative to that of the template.
Changes in fidelity may result from this slippage, an event which
presumably does not occur during copying of natural DNA templates.
Also, homopolymers and repeating heteropolymers containing only
one or two bases may be atypical in that a single non-complementary
nucleotide can occupy a looped-out structure. With synthetic
templates, the purity of the template, substrates, and DNA
polymerases themselves must be carefully monitored, particularly
with respect to high fidelity reactions. Any DNA contaminant in
the reaction mixture could act as a template-primer for the
incorporation of the "non-complementary" nucleotide substrate.
Without careful product analysis, this would mimic mis-
incorporation. To circumvent these limitations, we have
designed an assay using a natural DNA template (5,6).

φX174 Fidelity Assay

Our assay for measuring the fidelity of DNA synthesis on
single-stranded φX174 DNA is diagrammed in Figure 1. When φX174
am3 DNA primed with a Z-5 restriction fragment is copied in vitro
by E. coli DNA polymerase I using Mg^{2+} as a metal activator, the
reversion frequency is slightly greater than that obtained with
uncopied DNA. In order to increase the number of revertants, we
can selectively bias the concentration of each of the nucleotides
in the polymerase reaction mixture. A compilation of the error
rates of different DNA polymerases after extrapolation to equal
nucleotide concentration is shown in Table I.

Table I. Accuracy of Purified DNA Polymerases

DNA Polymerase	Error Rate	3'→5'-Exonuclease
Avian Myeloblastosis Virus	1/1000 - 1/5000	No
Rat Hepatoma-β	1/6000	No
HeLa-γ	1/6000	No
Calf Thymus-α	1/30,500	No
E. coli Pol I	1/680,000	Yes
Bacteriophage T_4	< $1/10^7$	Yes
E. coli Pol III Holoenzyme	< $1/10^7$	Yes

In general, eucaryotic DNA polymerases are not very accurate -- the error rates of a C for a T substitution at position
587 being 1/3000 to 1/30,000 (7). Presumably, there are other

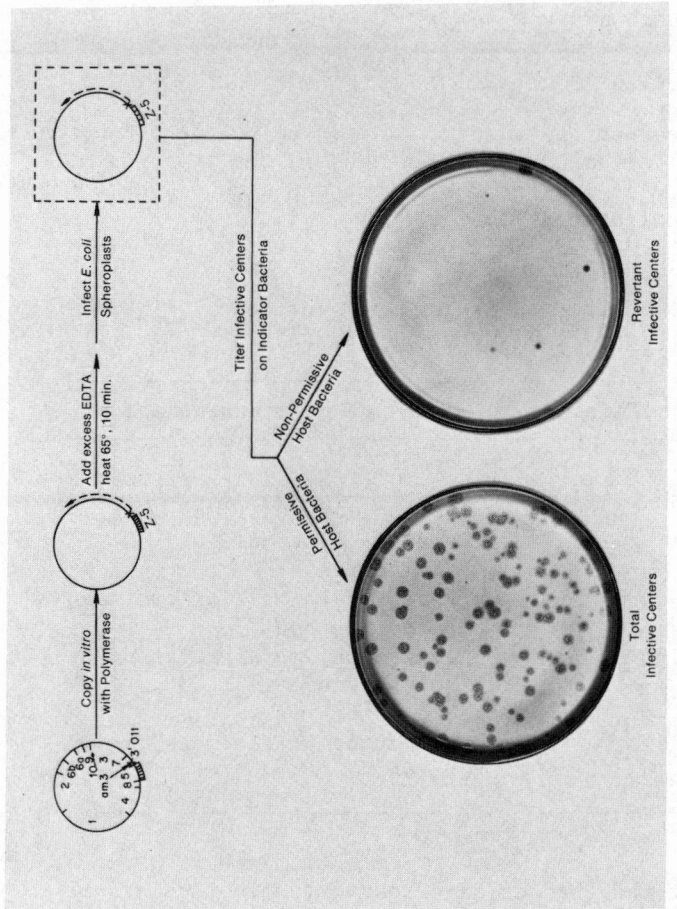

Figure 1. φX174 Fidelity Assay. In this assay, the template is a mutant single-strand circular DNA molecule from bacteriophage φX174, containing a single nucleotide substitution. Synthesis is initiated at a fixed point on the template, using as a primer the DNA restriction fragment whose 3'-terminus is located 83 nucleotides away from the mutation, an A for a G at position 587. After copying past the amber mutation *in vitro*, the frequency of erroneous nucleotide substitutions in the daughter strand is quantitated from the reversion frequency of the amber mutation to wild type phage after infecting spheroplasts and plating on permissive and non-permissive E. coli strains. Measurements can be carried out by either the infective centers or progeny phage methods (6).

factors that work in concert with DNA polymerases to increase
accuracy in eucaryotic cells. These could include DNA binding
proteins, separate proofreading exonucleases and mis-match correcting enzymes, as well as other mechanisms that have not been conceived. Procaryotic DNA polymerases are more accurate (error rates
10^{-5} to 10^{-7}), presumably due to proofreading by the associated
$3' \rightarrow 5'$ exonucleolytic activity. E. coli single-stranded binding
protein has been demonstrated to increase fidelity by 20-fold (8).
Thus, the DNA polymerase and binding protein are adequate to carry
out synthesis in the test tube with an accuracy approaching that
achieved by procaryotes in vivo.

Infidelity by Mutagenic Agents

In addition to variations in error rates due to the polymerase
itself (Table I), a limited number of changes in the reaction
increase the frequency of mis-incorporation, and each of these could
be of biological significance (Table II).

Table II. Effect of Exogenous Agents on
the Fidelity of DNA Synthesis

Little or No Effect on Fidelity	Enhanced Mis-incorporation
Template (length)	Ratio of incorrect to correct nucleotide substrates
Type of primer	Alkylation of template (small groups)
pH (6-8)	Different metal activators for DNA polymerase
Urea (< 4M)	Nonactivating metal mutagens or carcinogens
Anions	Intercalators
Caffeine	Depurination of Template
Ethanol	Steroids

Firstly, most of the changes have been determined using polynucleotide templates (3) and only a few [metals (9), nucleotide
ratios (6), intercalation and depurination (10)] have been veri-
fied using the ϕX fidelity assay. Secondly, most changes in the
reaction condition have no discernible effect on accuracy, including
pH (6 to 8) and anions. Lastly, agents that reduce fidelity with
one DNA polymerase invariably do so for all. In general, those
agents that increase infidelity are mutagenic. We would like to
consider two aspects of this relationship: altered pool size and
depurination.

Alterations in Nucleotide Pools

The ratio of incorrect to correct substrates is a particularly important consideration. Increasing the concentration of the incorrect nucleotide results in an increased frequency of mis-incorporation; and conversely, increasing the concentration of the correct nucleotide reduces the frequency of mis-incorporation. This competition is in agreement with evidence of Englund, et al. (11), for a single NTP substrate site on E. coli DNA polymerase I. Studies by Bernstein et al. (12) initially suggest that alteration of relative NTP pools in T_4-infected cells can result in enhanced mutagenesis. Infidelity at the level of DNA polymerization due to biased nucleotide pools could be central to toxicity and mutagenesis by a number of agents. These included toxicity by thymidine (13), BudR mutagenesis (14), potentiation of mutagenesis by alkylating agents (15), as well as manifestation of disease with deficits in nucleotide metabolism (16).

Depurination and Mutagenesis

Depurination could be a common intermediate in mutagenicity by diverse pathways. Depurination is the release of purine bases from DNA through breakage of the N-glycosylic bond which connects the purine base to the sugar-moiety of the sugar-phosphate backbone. Evidence suggests that this process occurs spontaneously at significant rates. From measurements on the rates of depurination in acid and at elevated temperatures, an in vivo rate constant has been estimated by Lindahl and Nyberg (17) to be $3 \times 10^{-11} \times s^{-1}$. This corresponds to 20,000 apurinic sites per mammalian cell per 24 hour period. The resulting apurinic site is relatively stable, with an estimated half-life of several hundred hours (18). Thus, if unrepaired, depurination could represent a major and clearly intolerable loss of genetic information from the cell. A search for enzymes which might recognize and repair apurinic sites quickly led to the discovery of the ubiquitous and abundant apurinic endonucleases which initiate excision repair of apurinic sites. In addition, activities have been described which insert purines back into the DNA (19) and have been designated as insertases. However, we would argue that a certain sub-fraction of the apurinic sites remain unrepaired and are present at the time of DNA replication. Such lesions could result in mis-incorporations and contribute to spontaneous mutation frequencies. Secondly, rates of depurination are enhanced several orders of magnitude by modification of purines at N-3 and N-7 positions (20). Since a large number of chemical carcinogens modify these positions, substantial depurination may be expected after exposure to these agents. Damage-specific N-glycosylases of which many are being discovered (21) may also contribute in this process.

Early in vivo studies on heat, acid and alkylation mutagenesis did focus on apurinic sites as premutagenic lesions, since apurinic sites seemed to be a common and relatively frequent product in all three treatments (22,23). However, later, more detailed studies mainly with bacteriophage T₄ (24,25) argued against the apurinic site and in favor of other simultaneously induced lesions as the major cause of mutations.

In order to assess the mutagenic consequences of depurination, we first made use of the homopolymer fidelity assay. In these studies, poly [d(A-T)], poly [dA] or poly [d(G-C)] templates were replicated with E. coli DNA polymerase I, avian myeloblastosis virus (AMV) DNA polymerase, or DNA polymerase-β from human placenta (26, 27). The fidelity assay measures the frequency by which these enzymes incorporate non-complementary (incorrect) nucleotides. Depurination of these templates and also of natural DNA (vide infra) can be done in a controlled manner by means of heat, a low pH or a combination of both. Mis-incorporation of dGTP on a poly [d(A-T)] template was found to increase with the extent of depurination, indicating the increased error proneness of DNA synthesis as a function of depurination. The response was linear with the number of depurinated sites introduced, and is identical for E. coli Pol I or AMV DNA polymerase--enzymes with marked differences in their intrinsic accuracy. Extensive control experiments showed that this increased mis-incorporation occurs as single-base substitutions throughout the newly made DNA. From these data, the model emerged that errors arise as single-base substitutions opposite apurinic sites in the template strand. This model was in part confirmed by a nearest-neighbor analysis of the product DNA, which occurred opposite the position of a template adenine in the case of poly [d(A-T)] and the position of a guanine in the case of poly [d(G-C)]. Secondly, treatment of the depurinated DNA with alkali before replication-- thus breaking the DNA at positions of apurinic sites (18)--abolished the mutagenic effect by 83-98%.

The mutagenic consequences of depurination were also studied with the φX174 system (28). Here one measures replication errors at the middle position of the am3 TAG amber codon, which will revert the amber phage to wild type or pseudo-wild type. Table 3 shows the change in error rate for Pol I as a result of depurination. This experiment was performed with a 10-fold dCTP pool bias to increase its sensitivity. Am3 is expected to be a good tool to study depurination mutagenesis, since reversion occurs exclusively at the adenine of the amber codon which should be susceptible to removal by depurination. Table 3 demonstrates the mutagenicity of depurination in this system and shows a proportionality between the number of apurinic sites per φX template and the frequency of mis-incorporation by the polymerase. Again, treatment of the depurinated template with alkali before replication abolishes the enhancement in mutagenicity resulting from depurination.

Table 3.   Effect of Depurination on the Fidelity of Copying
ΦX174 am3 DNA by E. coli DNA Polymerase I

Apurinic Sites per Molecule	Nucleotides Added per Template	Permissive Titer (X 10^7)	Non-Permissive Titer (X 10^2)	Reversion Frequency (X 10^{-6})	Error Rate
0	0	25.7	4.80	1.87	--
0	520	25.9	10.4	4.02	1/90,700
2.2	425	5.15	14.0	27.2	1/7,700
2.2 (Alkali)	416	9.98	1.62	2.71	1/232,000
5.5	338	0.25	1.68	67.2	1/2,980

ΦX174 DNA was depurinated and then used as a template in the fidelity assay. The method for calculating error rate is given by Kunkel and Loeb (6). Pretreatment of the depurinated template with alkali is as described by Schaaper and Loeb (29).

If one transfects depurinated ΦX174 am3 DNA into spheroplasts without in vitro copying, replication of the viral circle and production of the replicative form (RF) will occur in vivo by the E. coli DNA polymerase III holoenzyme complex. Extrapolation from the Pol I data as presented above would suggest that the holoenzyme complex might copy over apurinic sites and thus a mutagenic response might be observed. This is not the case. The major effect of depurination observed with normal spheroplasts is inactivation, to the extent that approximately one apurinic site per circle constitutes a lethal event. No increase in reversion frequency above the background level of 10^{-6} is observed (29). This puts an upper limit to the frequency of the mutagenic process of less than 1 in 1,400 per apurinic site. This lack of mutagenicity can be interpreted in two ways: either depurinated molecules are destroyed in vivo by hydrolysis of the apurinic sites by one of the three apurinic endonucleases found to date in E. coli; or alternatively, the Pol III holoenzyme complex is not able to copy over apurinic sites. The following arguments suggest the latter possibility. Transfection of depurinated single-strand ΦX DNA in the xth⁻ mutants, which lack the exonuclease III/endo VI enzyme, which constitutes 90% of the apurinic endonuclease activity in a crude extract, does not change the survival (unpublished results). This is in agreement with findings that most known E. coli apurinic endonucleases are specific for double-stranded DNA (30). Secondly, the E. coli Pol III holoenzyme is a much more faithful enzyme than Pol I (31) and therefore might not incorporate a base as easily at an apurinic site.

We succeeded, however, in demonstrating mutagenesis of depurinated DNA in vivo by introducing the error prone SOS response (32,33) in the spheroplasts. This was done by UV irradiation and allowing expression of the SOS phenotype in the bacteria before converting them to spheroplasts. Under these conditions, a strong response (i.e., up to a 10-20-fold increase in reversion frequency above background) is observed using templates containing 1 to 5 apurinic sites per circle (Figure 2). The increase is proportional to the number of apurinic sites introduced.

As demonstrated with synthetic polynucleotides (26) and natural DNA in vitro (4), exposure of depurinated DNA to alkali abolishes the mutagenic response. In Table 4, non-depurinated and depurinated DNA was incubated in alkali prior to transfection of spheroplasts. The mutagenic response observed in UV irradiated spheroplasts was abolished, presumably by selective alkaline hydrolysis at apurinic sites. Also, the same reversal of mutagenicity can be obtained upon treatment of the depurinated DNA with highly purified HeLa cell apurinic endonuclease (to be published in detail elsewhere). Thus, the conclusion seems justified that apurinic sites can be the direct cause of mutations

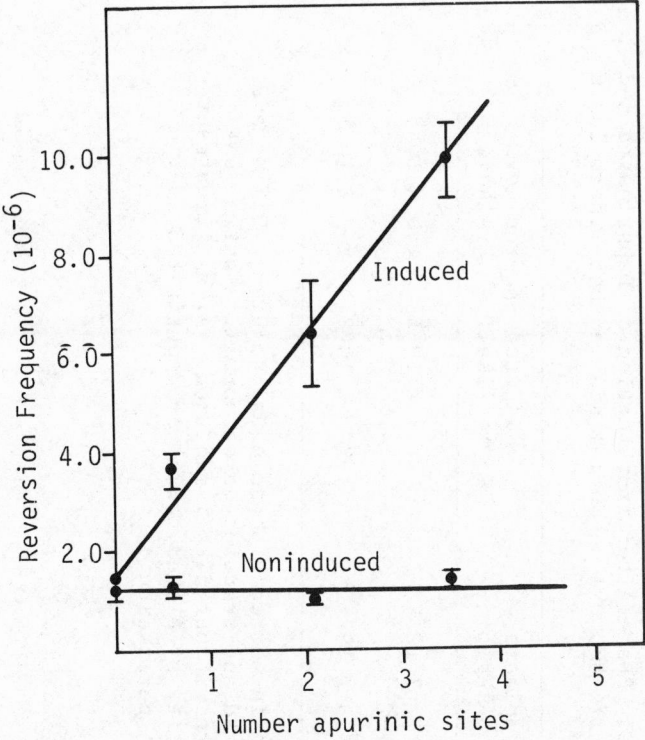

Figure 2. Reversion frequency among am3 progeny phaee as a function of the number of apurinic sites introduced. Single-stranded am3 DNA was depurinated and transfected into spheroplasts from normal or SOS-induced bacteria. The vertical bars represent the range of duplicate transfections. (Taken from Schaaper and Loeb (29) with permission.)

in vivo. In our case, we have shown this to be true on a single-stranded template. This may resemble rather closely the situation in a replication fork in which unrepaired apurinic sites from double-stranded DNA are faced by a DNA polymerase in a single-stranded region. The outcome would entirely depend on the ability of the enzyme to copy over such sites. If one assumes that the increased mutagenic response in SOS-induced cells is due to increased bypass of apurinic sites -- which is at the heart of the current SOS hypothesis (although no definite proof for this has been presented) -- then one can use the mutagenesis data to calculate bypass frequencies of the enzyme. In SOS-induced spheroplasts, it can be as high as 1/50 as compared to a

Table 4. Reversion Frequency Among Progeny Phage After Transfection of Depurinated and Alkali-Treated Single-Strand ΦX174 am3 DNA on Normal and Induced Spheroplasts

Apurinic Sites	Normal Spheroplasts		Induced Spheroplasts	
	No Alkali	Alkali	No Alkali	Alkali
0	1.2×10^{-6}	1.2×10^{-6}	1.2×10^{-6}	1.5×10^{-6}
4.5	1.5×10^{-6}	1.3×10^{-6}	20×10^{-6}	2.3×10^{-6}

After depurination, DNA samples were incubated with an equal volume of 0.2 N NaOH for 2.5 hr at room temperature, neutralized and transfected as normal. Control DNA samples were incubated at room temperature without alkali addition. Alkali treatment had no effect on the phage producing ability of either normal or depurinated DNA. (Adapted from Schaaper and Loeb (29) with permission.)

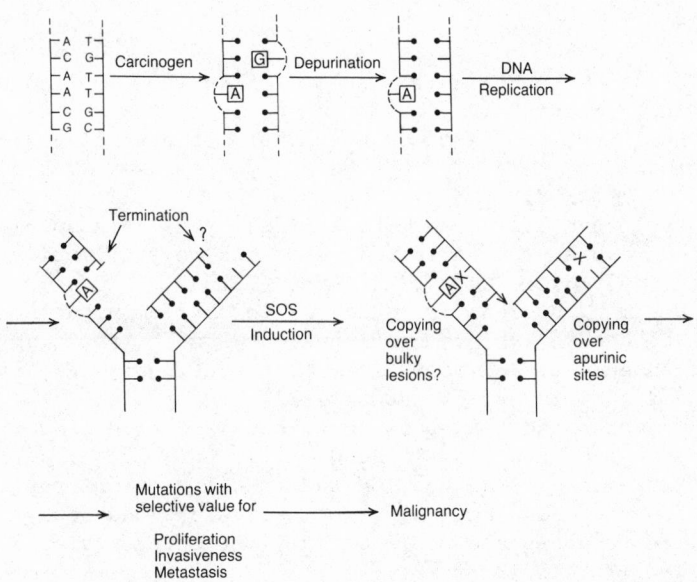

Figure 3. Mutagenesis by chemical carcinogens via depurination.

value of less than 1/1,400 in normal spheroplasts (29).

     If the above speculations pertain to mammalian cells, they
provide a mechanism by which mutagenesis can result from
modification of bases, particularly purines, by bulky chemical
carcinogens. This mechanism is diagrammed in Figure 3. Modification of purines by bulky carcinogens has been shown to terminate
DNA in vitro (34) and in vivo (35). In such a model, halting the
replication fork at the site of a modified base or perhaps even at
an apurinic site would induce an SOS response which permits the
replication to proceed past apurinic sites. Apurinic sites might
result from spontaneous hydrolysis of glycosylic bonds or from
enhanced hydrolysis of purines modified by chemical carcinogens,
particularly at the N-3 and N-7 positions of guanine and adenine.
Synthesis past apurinic sites would involve the non-template
directed insertion of nucleotides resulting in mutagenesis. This
mechanism is notably attractive for mutagenesis by aflatoxin B_1,
for which the principal adduct formed is on the N-7 position of
guanine (36). Conceivably, a bulky adduct could stall DNA
replication and induce an alteration in the DNA replication
complex. The altered replicating complex might proceed only
after removal of the adduct by depurination. This removal could
be spontaneous or catalyzed by a glycosylase that recognizes
the adduct.

REFERENCES

1. J.W. Drake, "The Molecular Basis of Mutation", Holden Dag,
    San Francisco (1970).
2. Z.W. Hall and I.R. Lehman, An in vitro transversion by a muta-
    tionally altered T₄ induced DNA polymerase, J. Biol. Chem.,
    36:321 (1968).
3. N. Battula and L.A. Loeb, The infidelity of avian myeloblastosis
    deoxyribonucleic acid polymerase in polynucleotide replica-
    tion, J. Biol. Chem., 249:4086 (1974).
4. M.F. Goodman, R. Hopkins, and W.C. Gore, 2-Aminopurine-induced
    mutagenesis in T₄ bacteriophage: A model relating mutation
    frequency to 2-aminopurine incorporation in DNA, Proc. Natl.
    Acad. Sci., USA, 74: 4806 (1977)
5. L.A. Weymouth and L.A. Loeb, Mutagenesis during in vitro DNA
    synthesis, Proc. Natl. Acad. Scil, USA, 75: 1924 (1978).
6. T.A. Kunkel and L.A. Loeb, On the fidelity of DNA replication:
    Effect of divalent metal ion activators and deoxyribonucleo-
    side triphosphate pools on in vitro mutagenesis, J. Biol.
    Chem., 254:5718 (1979).
7. T.A. Kunkel and L.A. Loeb, The fidelity of mammalian DNA
    polymerases, Science, in press (1981).
8. T.A. Kunkel, R. Meyer, and L.A. Loeb, Single-strand binding
    protein enhances the fidelity of DNA synthesis in vitro,
    Proc. Natl. Acad. Sci., USA, 76:6331 (1979).
9. L.K. Tkeshelashvili, C.W. Shearman, R.A. Zakour, R.M. Koplitz,
    and L.A. Loeb, Effects of arsenic, selenium and chromium on
    the fidelity of DNA synthesis, Cancer Res., 40:2455 (1980).
10. R.M. Schaaper, T.A. Kunkel, and L.A. Loeb, Depurination as a
    possible mutagenic pathway for cells, Gatlinburg Symposium
    on mutagenesis, in press (1981).
11. P.T. Englund, J.A. Huberman, T.M. Jovin, and A. Kornberg,
    Enzymatic synthesis of deoxyribonucleic acid. XXX.
    Binding of triphosphates to deoxyribonucleic acid poly-
    merase, J. Biol. Chem., 244:3038 (1969).
12. C. Bernstein, H. Bernstein, S. Mufti, and B. Strom, Stimulation
    of mutation in phage T₄ by lesions in gene 32 and by
    thymidine imbalance, Mutat. Res., 16:113 (1972).
13. G. Bjursall and P. Reichard, Effects of thymidine on deoxy-
    ribonucleoside triphosphate pools and deoxyribonucleic
    acid synthesis in Chinese hamster ovary cells, J. Biol.
    Chem., 248:3904 (1973).
14. R.L. Davidson and E.R. Kaufman, Bromodeoxyuridine mutagenesis
    in mammalian cells is stimulated by thymidine and suppressed
    by deoxycytidine, Nature (London), 276:722 (1978).
15. A.R. Peterson, J.R. Landolph, H. Peterson, and C. Heidelberger,
    Mutagenesis of Chinese hamster cells is facilitated by
    thymidine and deoxycytidine, Nature (London), 276:508 (1978).

16. E.R. Giblett, J.E. Anderson, F. Cohen, B. Pollara, and H.J.
     Meuwissen, Adenosine deaminase deficiency in two patients
     with severly impaired cellular immunity, Lancet, 2:1067
     (1972).
17. T. Lindahl and B. Nyberg, Rate of depurination of native
     deoxyribonucleic acid, Biochemistry, 11:3610 (1972).
18. T. Lindahl and A. Andersson, Rate of chain breakage at
     apurinic sites in double-stranded deoxyribonucleic acid,
     Biochemistry, 11:3618 (1972).
19. W.A. Deutsch and S. Linn, An apurinic DNA binding activity
     from cultured human fibroblasts that specifically inserts
     purines into depurinated DNA, Proc. Natl. Acad. Sci., USA,
     76:1089 (1979).
20. P.D. Lawley and P. Brookes, Further studies on the alkylation
     of nucleic acids and their constituent nucleotides,
     Biochem. J., 89:127 (1963).
21. B. Singer and T.P. Brent, Human lymphoblasts contain DNA
     glycosylase activity excising N-3 and N-7 methyl and
     ethyl purines but not 0^6-alkylguanines or 1-alkylguanines,
     Proc. Natl. Acad. Sci., USA, 78:856 (1981).

22. E. Bautz and E. Freese, On the mutagenic effect of alkylating
     agents, Proc. Natl. Acad. Sci., USA, 46:1585 (1960).
23. E.B. Freese, Transitions and transversions induced by
     depurinating agents, Proc. Natl. Acad. Sci., USA, 47:540
     (1961).
24. P.D. Lawley and C.N. Martin, Molecular mechanisms in alkyla-
     tion mutagenesis: Induced reversion of bacteriophage 4rII
     AP72 by ethylmethane sulphonate in relation to extent and
     mode of ethylation of purines in bacteriophage deoxy-
     ribonucleic acid, Biochem. J., 145:85 (1975).
25. J.W. Drake and R.H. Baltz, The biochemistry of mutagenesis,
     Annu. Rev. Biochem., 45:11 (1976).
26. C.W. Shearman and L.A. Loeb, Depurination decreases fidelity
     of DNA synthesis in vitro, Nature, 270:537 (1977).
27. C.W. Shearman and L.A. Loeb, Effects of depurination on the
     fidelity of DNA synthesis, J. Mol. Biol., 128:197 (1979).
28. T.A. Kunkel, C.W. Shearman, and L.A. Loeb, Mutagenesis in
     vitro by depurination of ΦX174 DNA, Nature, in press (1981).
29. R.M. Schaaper and L.A. Loeb, Depurination causes mutations
     in SOS-induced cells, Proc. Natl. Acad. Sci., USA, 78:1773
     (1981).
30. T. Lindahl, DNA glycosylases, endonucleases for apurinic/
     apyrimidinic sites and base excision-repair, Prog. Nucleic
     Acid Res. Mol. Biol., 22:135 (1979).
31. L.A. Loeb, T.A. Kunkel, and R.M. Schaaper, in: "Mechanistic
     Studies on DNA Replication and Genetic Recombination,"
     Vol. XIX, Academic Press, New York (1980), pp. 735-751.

32. M. Radman, Phenomenology of an inducible mutagenic DNA repair pathway in Escherichia coli: SOS repair hypothesis, in: "Molecular and Environmental Aspects of Mutagenesis," L. Prakash, F. Sherman, M.W. Miller, C.M. Lawrence, and H.W. Taber, eds., Thomas, Springfield, IL (1974), pp. 128-142.

33. E.M. Witkin, Ultraviolet mutagenesis and inducible DNA repair in Escherichia coli, Bacteriol. Rev., 40:869 (1976).

34. P. Moore and B.S. Strauss, Sites of inhibition of in vitro DNA synthesis in carcinogen- and UV-treated ΦX174 DNA, Nature, 278:664 (1979).

35. W.T. Hsu, E.J. Lin, R.G. Harvey, and S.B. Weiss, Mechanism of phage ΦX174 DNA inactivation by benzo[a]pyrene-7, 8-dihydrodiol-9, 10-epoxide, Proc. Natl. Acad. Sci., USA, 74:3335 (1977).

36. J.M. Essigmann, R.G. Croy, A.M. Nadzan, W.F. Busby, V.N. Reinhold, G. Büchi, and G.N. Wogan, Structural identification of the major DNA adduct formed by aflatoxin B_1 in vitro, Proc. Natl. Acad. Sci., USA, 74:1870 (1977).

DISCUSSION

DR. SINGER: Taking up your last point, aflatoxin, as you know, forms an adduct at the N7 of G which is extraordinarily labile and is lost by two mechanisms, one of which is depurination. There is unpublished evidence -- not from my laboratory but informed gossip -- that there are a number of other bulky carcinogens which form very fleeting adducts at the N7 of G, and that the net result is they depurinate very fast. How do you feel this would fit in with the idea of how bulky carcinogens cause errors?

DR. LOEB: We'd like to think, with very little evidence, that a number of bulky carcinogens favor increased rates of depurination. We have no evidence in eukaryotes or in bacteria that depurinated sites accumulate in cells. In the usual analysis of DNA with alkaline sucrose gradients, one would not be able to tell the differences between single-stranded breaks and depurinated sites. There's a large amount of evidence for accumulation of single-stranded breaks under physiological situations in eukaryotic cells. We now need to determine whether or not any of these represent apurinic sites.

The model we presented is advantageous because the polymerase is poised at a bulky carcinogen that can be rapidly depurinated and then be immediated copied by the polymerase before repair can occur. The key experiment is to prove that apurinic sites are present as an intermediate in eukaryotic cells.

DR. SAMSON: When you depurinate the DNA, are any other lesions produced?

DR. LOEB: Yes, there are probably several other lesions. The strongest evidence we have for at least the majority (we never go back to zero) being apurinic sites is that in all systems tested-- homopolymers, ØX in the test tube and ØX in cells-- the majority of mutagenesis can be abolished by preincubtion with alkali, and by preincubation with AP-endonuclease. One could argue circuitously and say that other lesions are responsible for mutagenesis; however, such lesions would have to be hydrolyzable by AP-endonucleases and also alkaline labile. Considering the high specificity of enzymes, it is unlikely that any lesion other than an apurinic site would be responsible for most of the observed mutagenesis.

DR. L. PRAKASH: How do you account for the specificity of depurination-induced mutagenesis?

DR. LOEB: For the moment we can't account for specificity. But you have to remember that experiments with specific mutations were carried out under situations that might not induce SOS repair. We don't know the status of SOS repair in the heat-mutagenesis experiments in the literature. The only time we've been able to demonstrate mutagenicity so far is in SOS-induced cells.

DR. GLICKMAN: It is easy to see how heat mutagenicity in a GC site gives you an AT, but in your system you're looking at TAG and there's no C target. You might wish to comment on the specificity as far as you know it.

DR. LOEB: The question is what's inserted opposite an apurinic site. We know many substitutions, but we don't yet have the key experiments. We're looking at a TAG sequence and we're asking what's inserted opposite and A because we can't look at the G. Changes opposite the G do not produce viable mutations. In 17 of 18 substitutions the insertion is an A opposite an A. The problem so far is that we've looked at only one codon. What we need is a batch of amber codons in which we can look at G and these experiments are in progress.

DR. SINGER: Larry, let me ask again for the record. We do not know, either you or I, that the mismatched base is inserted opposite the modified or wrong base. And I haven't thought of the experiment that will prove it; have you since we last talked?

DR. LOEB: Yes, we thought of a batch of experiments. They're being attempted. The idea is that one needs to produce a site-specific change on a template. With depurination there are easy ways of doing it. One essentially needs to introduce one depurinated site on a template at a specific location. And that can be done by incorporating a single uracyl and treating with uracyl glycosylase.

DR. SINGER: The question is pertinent only because we don't know where the perturbation is.

DR. SOBELL: What is the relative incidence of frameshifts versus base pair substitutions?

DR. LOEB: The only way I can answer the question at the moment is with homopolymer templates or heteropolymers, which may not be the same as natural DNA templates. For example, with a template such as alternating poly [d(A-T)], depurinate is so that there are depurinated sites in place of As, and ask what's inserted opposite the apurinic site. Normally, a T would be inserted. The least frequent thing that's inserted is a G, one for every 600 apurinic sites copied. You can do it with any polymerase; the results are the same. One C is incorporated for every 20 apurinic sites and one A for every 2.5 sites. You can't ask how frequently a T goes in because it is the normal base, but by subtraction it would be very frequent, about 1 per every 2 apurinic sites.

DR. SOBELL: The thing is, that polymer system may not be at all like normal DNA. It is full of bubbles and you're getting slippage of the polymerase.

DR. LOEB: That's right, we're fully aware of it. One needs to design a system with natural DNA in which all substitutions can be monitored. However, poly [d(G-C)] gives results similar to poly [d(A-T)]. Yet it is unlikely that a preference for correct nucleotide insertions would occur with DNA templates.

DR. MAHER: I have studies in which we were trying to find out at which point in the DNA replication the number of unexcised polycyclic hydrocarbon adducts formed by the diol epoxide of benzo(a)pyrene remaining in a normally-excising human cell would be equal to the number present in a repair-minus xeroderma pigmentosum (XP) cell, so that we could relate that to an equivalent biological effect observed in the two strains given very different doses of this agent. We monitored DNA replication by BrdUrd shift. We exposed the two kinds of cells to radioactive labeled diol epoxide, a compound which gives only a single adduct located on the N^2 atom of guanine, allowed the cells to replicate the DNA in synchrony, and then assayed the number of adducts remaining on the parental strand of non-replicated and newly-replicated DNA at various times during S phase. The cells replicated their DNA and the newly-replicated DNA which had shifted to a higher density in the CsCl gradient still contained bound hydrocarbon derivative residues. XP cells do not remove these adducts and yet replicate their DNA with the adducts present. How do these data fit in with the idea that the polymerases stop and wait for the bulky adduct to drop off before proceeding?

DR. LOEB: It doesn't.

DR. SHAPIRO: Because you ended rather quickly, I just wanted to go over your major carcinogenesis theme again. The idea is that the normal polymerase pauses at the adduct and then the adduct falls off, and then the polymerase quickly copies over before the gap can be repaired?

DR. LOEB: We would like to think that a normal polymerase pauses at bulky adducts. In bacteria, the replication complex is then somehow changed to be that of an SOS-induced type. We argue that the altered polymerase is poised at the adduct until the adduct is removed by depurination. So our thoughts are opposite to the position that a replication complex bypasses an adduct in SOS-induced cells. We argue for an apurinic intermediate.

DR. BAMBARA: I'd just like to ask what happens when you use the alpha-phosphothiol inhibitors on the range of polymerases that don't have 3' to 5' exonuclease?

DR. LOEB: There's no change.

DR. BAMBARA: In rate or fidelity?

DR. LOEB: There's a slight change in rate; there's no change in fidelity of the polymerases we examined. With AMV and DNA polymerase-β, the studies indicate no change in fidelity. There is only one experiment with no change using DNA polymerase -α and -γ. So we're not very comfortable with the alpha and gamma results yet.

DR. DOUBLEDAY: How do you square your apparently rather efficient replication in vitro of those AP sites with the fact that a single site appears to be lethal in vivo? Are they labile in the cell? Have you looked at that at all?

DR. LOEB: In the test tube, all polymerases slow up but they copy over apurinic sites at varying efficiencies. In the cell, we don't see it with the normal replicating mechanisms; we see it only with SOS-induced conditions. There are two likely interpretations. One is that during transfection, glycosylases in the cell destroy much of the apurinic DNA. The other is that the polymerase in normal cells, the holoenzyme, cannot go past apurinic sites and only a modified holoenzyme can. With respect to the first possibility, we have looked at strains with a mutation for one of the two apurinic endonucleases, but never at the double mutant. We can't detect enhanced survival of ØX in those mutants, which argues for the second possibility; that is, the normal replicating enzyme being much more stringent than purified polymerases in going past apurinic sites, and perhaps for modified complexes allowing bypass.

DR. DOUBLEDAY: When you induced SOS, am I right in believing that you didn't have any significant effect on survival?

DR. LOEB: That's correct. It's at most 10% and we don't make anything of it. One can calculate that we shouldn't have had any effect because we're only looking at one in 100 of the ØX producing progeny. One can calculate that the frequency of mutations per apurinic site introduced, even in SOS, is one in 50.

DR. SINGER: One of the things that occurs as the result of chemical modification in vivo is that the triphosphates or any other monomer react to a very much higher extent than any polymer. You would therefore have a pool which would have a considerable number of modified triphosphates. How would that affect the fidelity?

DR. LOEB: There are people who believe that modifying triphosphates in cells could lead to mutagenic event. I personally don't think that makes very much sense, at least if you modify triphosphates by adding bulky groups. Polymerases do not polymerize these, even under conditions where you force the enzyme. For example, if one takes an AAF modified nucleotide and measures a single addition in the presence of manganese using AMV polymerase, the most error-prone situation we know, we cannot demonstrate detectable incorporation. The polymerase fails to accept the bulky modified substrate.

DR. SINGER: But it does small ones.

DR. LOEB: It will accept small adducts, no question! It will be the same as the template situation. For small ones it's reasonable, for large ones it seems very, very unattractive.

DR. SOBELL: Especially when its forced the purine to be in the so-called syn configuration, which may in fact be unacceptable to the enzyme.

DR. LOEB: We've picked the world's most sloppy situation; that is the error rate at infinite pool biases with only the modified nucleotide being present.

DR. SINGER: You also pick the wrong triphosphate for the experiment.

DR. LOEB: We assume a small methylated group could be forced in. Many natural methylated nucleotides are polymerized.

DR. SHAPIRO: Those alpha-thiophosphate derivates: are they themselves mutagenic if administered alone?

DR. LOEB: There's only one system at the moment that's suitable. You have to have a monophosphate going into the cell, and at the moment there's very little evidence in any eukaryotic situation where you can measure mutation rates as a function of adding monophosphates to cells. There is one system that allows it, and that's being designed. But there's really very little good evidence that you can put a monophosphate into a eukaryotic cell and keep it intact; if you open up cells to do that (for example, create eukaryotic spheroplasts), it's not very simple to measure mutagenesis.

DR. SHAPIRO: What about prokaryotic cells?

DR. LOEB: We haven't tried. There's one eukaryote, yeast that's a possibility. There are mutants in yeast for uptake of monophosphates.

DR. THILLY: Shouldn't you be able to open eukaryotic cells, such as CHO cell, with a little bit of amphotericyn B? There should be a reversible opening without any effect on the survival.

DR. LOEB: And then try measuring mutagenesis, yes. One would like to determine if there's proof-reading in eukaryotic cells, and one has to measure proof-reading in eukaryotic situations where you can put the monophosphates in. So we're trying in whatever way we can.

DR. GLICKMAN: The value of such experiments is very difficult to assess. Unless you can show how this agent is working, it doesn't mean very much. Modified bases can work via a number of pathways and you can't show that its proof-reading by simply adding the base, and getting mutants.

DR. LOEB: It's the opposite that is important. If you get a negative with the base going in, you can prove proof reading doesn't occur. So if I substitute a base 100% with alpha-thiol, it should be viable and I see no mutagenesis or very little. I've eliminated proof-reading as a viable pathway in eukaryotes. I can't prove the other; I agree with you completely.

DR. WALKER: Is the diester bond with the thiol there intrinsically harder to cleave? Is it chemically more stable, or are you looking at some interaction of the enzyme where they're easily hydrolyzed. If it were the case that enzyme recognition wasn't so great or something like that, then you really couldn't tell anything in the eukaryotic system. If it's tougher to cleave then maybe you can eliminate --

DR. LOEB: What we believe, without evidence, is that it's not tougher to cleave. What we do know from the work of Fritz

Eckstein is that an alpha-thiol group has stereospecificity, and when a polymerase puts it in, it does it by inversion. We'd like to think that, when the exonuclease tries to reemove it, it can't because the sulfur at a particular position in place of an oxygen prevents hydrolysis by the exonuclease.

DR. WALKER: You could make the formal argument in the absence of information that a eukaryotic proof-reading system might be able to do that even though the prokaryotic can't.

DR. LOEB: Yes. A separate exonuclease might remove an inverted terminal nucleotide.

THE SPECIFICITY OF INFIDELITY OF DNA POLYMERASE

Lynn S. Ripley

Laboratory of Molecular Genetics
National Institute of Environmental Health Sciences
Research Triangle Park, North Carolina 27709

FIDELITY AND INFIDELITY

A complex network of metabolic processes is responsible for the accurate production of progeny DNA. The infidelity of these processes when measured as mutations per base pair in DNA varies dramatically among organisms (1) and seems likely to reflect unique combinations of fidelity determinants. Unfortunately, it is nearly impossible to measure fidelity directly, but instead it is the summation of accurate and inaccurate processes producing a mutation frequency which can be experimentally determined.

The experimental perturbation of DNA metabolism frequently increases infidelity. By studying the detailed properties of the perturbation and the resulting mutations, it is sometimes possible to identify and characterize the components important to fidelity. Experimental perturbations increasing infidelity frequently include altered proteins, nucleotide substrates, or reaction conditions. Even when infidelity is enhanced in a perturbed process, it is in fact very difficult to know whether the enhancement is mediated by mechanisms which decrease normal fidelity, or whether the increase reflects the introduction of a new component of infidelity which is largely irrelevant to normal fidelity. The distinction is subtle but bears upon the question of whether a particular mechanism is one by which fidelity is achieved, or one by which infidelity occurs. The analysis of either mechanism may be valuable, but when sufficiently understood, the distinction should be made.

83

In a few instances perturbations introduced into a system actually increase fidelity. As in the case of decreased fidelity, it is difficult to know if the increased fidelity is mediated by the same mechanism responsible for normal fidelity. However, when increased fidelity occurs in vivo, it is an indication that the studied process acts upon a substrate which significantly contributes to infidelity during normal DNA metabolism. Identification of the sources of avoided mutations identifies significant elements of DNA metabolic fidelity. An example of increased in vivo fidelity is that sometimes displayed by mutant DNA polymerases. These polymerase, commonly known as "antimutators", were first identified in bacteriophage T4 and have been the subject of numerous in vivo and in vitro studies. This paper describes our current knowledge of the properties of these polymerases and attempts, from their fidelity characteristics, to identify some intermediates of both base substitution and frameshift mutation which might be subject to the action of DNA polymerase fidelity components.

BACTERIOPHAGE T4 AND DNA METABOLISM

The complexity of DNA metabolism is expressed at three levels: One is the multitude of proteins involved. In bacteriophage T4, conditional mutations affecting DNA metabolism have been isolated in more than 20 genes (2). A second complexity is represented by physical interactions between DNA metabolic proteins. Some of these are strong and direct such as the tight coupling between the protein products of genes 44 and 62 (3). These products, along with that of gene 45, are believed to traverse the DNA in the company of the DNA polymerase and have been designated "polymerase accessory" proteins (4). Association of the DNA polymerase with the product of gene 32 (P32) has also been measured in vitro (5,6). A third complexity is represented by the observation that a number of genes act in more than one DNA metabolic process, and thus may interact through metabolic intermediates with each other in qualitatively different ways. Many of the genes whose products are required for in vitro DNA synthesis (4) are also implicated in DNA recombination and repair (7,8,9,10,11). Some of the interactions between genes of DNA metabolism have been deduced from the in vivo interactions of mutant alleles of these gene (11). Although the specific nature of the interactions is not always apparent in such studies, complexity is ever present.

It has been nearly impossible in vivo to sharply separate the influence of any gene on recombination processes from its influence on replication. For example, it appears that recombination intermediates may be intimately involved in the initiation of DNA synthesis (11) and recombination may be required for the formation of the concatemeric DNA thought to be an essential precursor for

T4 DNA packaging (12). The T4 DNA polymerase is directly involved
in both replication and recombination, but its impact in each
process with respect to fidelity is largely untested (13).

T4 DNA POLYMERASE AND THE FIDELITY OF DNA METABOLISM

The role of DNA polymerase is clearly central to DNA metabolism.
When mutants of T4 DNA polymerase (selected for their conditional
lethality) (14) were tested for their influence on mutation rates,
large effects have sometimes been found (15,16). Some polymerase
alleles increased frequencies of a wide spectrum of mutants, and
were thus called "mutators". Other polymerase alleles were called
"antimutators" because they decreased mutation frequencies.
Although these polymerases had antimutator effects on some mutation
frequencies, they increased other mutation frequencies, and had no
effect upon still other mutation frequencies. Thus, despite the
common designation of these polymerases as antimutators, it is more
accurate to reserve this adjective to their effects (i.e.,
antimutator effects) since in other instances the polymerases have a
very different fidelity effect.

The magnitude of DNA polymerase effects on error rates at
particular mutant sites may be very large. For example, an rII
ochre codon exhibits a 600-fold higher rate of reversion per
replication in the presence of the L56 polymerase allele (an
allele which increases spontaneous errors at this site compared to
the wild type) than it does in the presence of the CB120 polymerase
allele (an allele which decreases spontaneous errors at this site)
(17).

However, T4 DNA polymerase does not act alone in vivo but in
concert with other replication proteins. The roles of these
proteins in fidelity have also been examined, typically by measuring
the influence of conditional mutant alleles of these genes upon
mutation rates. Increased infidelity has consistently been
demonstrated (18,19,20,21,22,23), but the magnitude of these
effects upon infidelity are considerably smaller than those
exhibited by some DNA polymerase alleles. The absence of consistently large effects on mutation rates by other mutant replication
proteins suggests a central role for T4 DNA polymerase in the
fidelity of DNA replication. No studies have yet determined
whether the fidelity effects which are produced by the DNA
replication proteins other than the polymerase are mediated through
their interaction with DNA directly, or by indirect effects perhaps
mediated through the DNA polymerase itself.

For example, the T4 helix-destabilizing protein, P32, is
clearly implicated in the same DNA metabolic processes as the DNA

polymerase (11). There is evidence for P32 interactions with DNA
ligase, DNA polymerase, and recombination nucleases (11). Furthermore, in vitro studies demonstrate that P32 cooperatively binds to
single-stranded DNA (24), facilitates displacement synthesis by
T4 DNA polymerase (25), and is required for DNA synthesis by
purified proteins after the first round of replication on single-
stranded DNA templates (4,6). But despite the intimate involvement
of this protein in DNA metabolism, when three different mutant
alleles of gene 32 known to be defective in one or more of the
specific interactions mentioned above were tested for their influence
on mutation frequencies, only small effects were observed (21):
each gene 32 allele produced an average increase of less than 7-fold
on the reversion frequencies of 15 base pair substitutions. No
specificity based on type of base substitution was observed among
these mutator effects. When the same gene 32 alleles were tested
for their influence on the reversion frequencies of four frameshift
mutations, even smaller effects were seen (increases or decreases of
2-fold or less). Thus, the strong influence of T4 DNA polymerase
upon mutation rates may be unique.

THE INFLUENCE OF MUTANT DNA POLYMERASES ON BASE SUBSTITUTION
FREQUENCIES

Because DNA polymerase catalyzes the incorporation of
nucleotides into progeny DNA, many studies of base substitution
mutation have naturally focussed upon the fidelity of mutant DNA
polymerases. A survey by Speyer (26) identified several mutant
alleles of T4 DNA polymerase which increased forward r mutation
frequencies. An examination of the specificity of two of these
mutant alleles (L56 and L98) revealed that each strongly promoted
both transition mutations (15,27) and transversion mutations (27,28).
Another survey (16) demonstrated that some polymerase alleles
reduced the reversion frequencies of certain rII mutations, although
no significant change was observed in the magnitude of forward r
mutation frequencies. These latter polymerase alleles do not
uniformly reduce all base substitution mutation frequencies.
Instead the antimutator effects of these polymerases depend both
upon the particular substitution mutation measured and upon the
mutagen (and hence presumably the mechanism) responsible for the
mutation. Examples are given in Table 1 of mutation frequency
effects produced by the CB120 enzyme, but similar results have been
seen in the presence of three other frequently studied polymerase
alleles, L141, L42 and CB87. Among spontaneously arising base
substitutions, CB120 polymerase produces an antimutator effect upon
A·T → G·C transitions, but almost no effect upon G·C → A·T transition
frequencies, and in fact, produces weak mutator effects upon some
transversions. However, in contrast to spontaneous G·C → A·T
transition mutations, base analogue-induced transitions at G·C sites
are decreased in frequency by the presence of the CB120 alleles.

TABLE I. The influence of the <u>CB120</u> DNA polymerase on mutation
frequencies compared to the wild-type polymerase.

Mutational Pathway	Relative mutation frequency	Reference
A·T → G·C		
Spontaneous	7 - 170 x decrease	16
AP-induced	10 - 50 x decrease	16
BU-induced	10 - 140 x decrease	55
Thymineless	100 x decrease	38
G·C → A·T		
Spontaneous	1 - 2 x <u>increase</u>	16
BU-induced	60 - 70 x decrease	16
EMS-induced	3 - 4 x decrease	16,38
Hydroxylamine	1 - 2 x <u>increase</u>	16,38
Transversions		
Spontaneous	1 - 7 x <u>increase</u>	27

The multiple effects of a polymerase mutant upon different base
substitutions suggests a multifunctional basis for polymerase
effects <u>in vivo</u>. This paper will explore two different approaches
which may in part expose the basis of these multiple effects. The
first approach is the detailed <u>in vitro</u> characterization of the
enzymatic activities and substrate specificities of polymerases.
The second is a more detailed characterization of the intermediates
of these polymerase-mediated mutations and hence the potential
substrates for <u>in vivo</u> fidelity and infidelity.

<u>IN VITRO</u> CHARACTERIZATION OF DNA POLYMERASES

Mutant polymerases having antimutator properties exhibit
several novel properties <u>in vitro</u>. In a study which compared the

L141 and L42 enzymes (which can produce antimutator effects), the
L56 and L98 enzymes (which can produce mutator effects), and the
wild-type enzyme, a correlation was found between fidelity in vivo
and the ratio of 3' → 5' exonuclease activity of the enzyme compared
to its polymerase activity (29). The L141 and L42 polymerases are
relatively high in exonuclease activity compared to polymerase
activity, while the L56 and L98 polymerases are relatively low in
exonuclease activity compared to polymerase activity. This
correlation when coupled with earlier experiments of Brutlag and
Kornberg (30) suggesting an editing role for the 3' → 5' exonuclease
activity of the E. coli polymerase I enzyme, led to the model that
the mutator or antimutator phenotypes of T4 polymerases were due
to the decreased or increased editing potential of the polymerases.

Detailed studies of the CB120 (31,32) and L141 proteins (33)
do show that in vitro both polymerases have higher nucleotide
turnover activities relative to their polymerase activities than
does the wild-type enzyme. However, these mutant polymerases
differ from the wild type in a number of additional ways. The CB120
polymerase, when compared to the wild type, has a lower specific
activity for incorrect nucleotides and a reduced ratio of incorrect
nucleotide utilization to correct nucleotide utilization. This
difference suggests that the fidelity of replication by this enzyme
is enhanced by increased fidelity of nucleotide selection. Indeed,
the fidelity of nucleotide selection is also implicated in the
mutator effects produced by the L88 polymerase, which is unaltered
in relative levels of nucleotide turnover and nucleotide incorporation (32,34).

Another characteristic of the L141 and CB120 enzymes is
demonstrated by their altered abilities to utilize particular DNA
templates. In the case of CB120, the template specificity is
consistent with the interpretation that the enzyme cannot move
through hydrogen-bonded DNA: it fails to utilize poly d(A·T)
templates, which readily form hydrogen-bonded secondary structures.
The enzyme is strongly stimulated on these templates by P32, the
helix-destabilizing protein of T4, which removes much of this
secondary structure. However, even in the presence of P32, the
CB120 enzyme uses these templates less effectively than the wild-
type enzyme under the same conditions. These template-dependent
differences coupled with nucleotide turnover measurements, led to
the suggestion that the increased nucleotide turnover was due to a
reduced rate of DNA polymerization which thereby increased the time
available for editing incorporated bases. These observations are
consistent with increased fidelity predicted in a "kinetic proof-
reading" model proposed by Hopfield (35).

The following sections of the paper examine the in vivo
characteristics of four polymerase alleles CB120, CB87, L141
and L42. These polymerases share the property of altering mutation

frequencies for particular mutations in a characteristic way (i.e.,
the polymerase produces either a mutator, an antimutator, or no
effect upon a particular mutation). The antimutator effects provide
a particularly useful probe of the quantitative importance of
particular mutational intermediates of spontaneous and base analogue-
induced base substitution mutation. The effect of these polymerase
mutants upon frameshift mutation is also substantial: specific
fidelity effects are seen, and are likely to reflect both the
mechanism(s) by which these polymerases achieve fidelity, and
different mechanisms of frameshift mutation.

POTENTIAL INTERMEDIATES OF BASE SUBSTITUTION MUTATION

   The heterozygous intermediate of transition mutations at A·T
and G·C sites are superficially similar, being either A·C or G·T at
either site. When considered as intermediates of DNA synthesis,
however, the unique properties of each intermediate become apparent
(see Figure 1). The potential intermediates of transversion

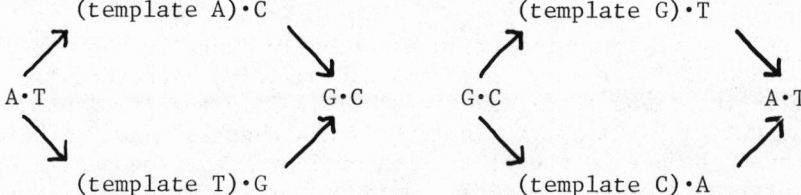

FIGURE 1.   Intermediates of polymerase-mediated base substitution
            mutations. A·T → G·C transition intermediates are shown
            on the left; G·C → A·T transition intermediates are on
            the right.

mutation are similarly unique. Mispaired intermediates are
potentially unique in an additional manner because of alternative
tautomeric forms available to the bases. For example, if the A·C
intermediate is hydrogen bonded, it is required that either the A
or the C adopt the minor tautomeric form. If * denotes the minor
tautomer, then the A·C pair alone has the potential for four
different intermediates of base substitution mutation: (template
A*)·C, (template A)·C*, (template C*)·A, and (template C)·A* (36).

The relative contributions of the four intermediates shown in
Figure 1 to polymerase-mediated transition mutation are unknown.
The problem is difficult to approach experimentally since the
heterozygous intermediates occur at very low absolute frequencies
in DNA, confounding most chemical approaches; and genetic experiments
do not easily distinguish A·C from G·T intermediates. It is furthermore difficult to assess the contribution of polymerase error to
those heterozygotes which might be detected in vivo. For example,
heterozygotes are also produced as intermediates of recombination
between base substitution mutants and the wild type. The
composition of such recombinational heterozygotes is of course
irrelevant to the question of the frequency of particular intermediates of base substitution.

One approach to the question of the relative importance of
different intermediates to base substitution mutation is to vary
the deoxynucleotide triphosphate pool sizes. The predicted result
of such an approach is that, when a high concentration of an
incorrect dNTP is provided relative to the correct dNTP, an
increased error frequency should result from the increased
probability with which the incorrect dNTP occupies the active site
of the enzyme. Both in vivo and in vitro experiments of this type
have now been reported, and in fact have demonstrated that increased
error frequencies can be produced.

In vivo thymine starvation drastically curtails the net
production of T4 progeny, but simultaneously and specifically
promotes mutations at A·T sites (27,37). This specificity is
consistent with the above predictions and suggests that mutations
may indeed be due to the production of increased frequencies of
A·C intermediates in transition mutation or A·A or A·G intermediates
of transversion mutation. Thymineless mutagenesis in T4 appears to
be quite different from that observed in E. coli; in T4 it is
largely unaffected by misrepair alleles while in E. coli it is
strongly dependent on repair functions (37).

Mutant polymerase alleles influencing spontaneous base
substitution frequencies also influence thymineless mutagenesis.
CB87 and CB120 produce antimutator effects (38) while L56 and L88
produce mutator effects (27) on both spontaneous and thymineless-induced transitions at A·T sites. If thymineless mutagenesis
in vivo does proceed through the formation of (template A)·C
intermediates, these polymerase mutants produce (or allow) very
different frequencies of these pairs. The antimutator effects of
CB87 and CB120 suggest a strongly decreased frequency of (template
A)·C pairs. It is unfortunately not possible from such in vivo
experiments to directly distinguish between polymerase effects due
to altered base-insertion fidelity and effects due to altered proof-
reading. The net effect on mutation frequencies may thus contain

elements of both. It may be especially interesting in this regard
that the quantitative characteristics of the mutator effects
expressed by the L56 and L88 alleles are different, since the L56
enzyme is thought to be primarily altered in its proof-reading
properties while L88 appears to be primarily altered in its fidelity
of base-insertion. The L88 enzyme produced a mutation frequency
during thymine starvation which was equivalent to the sum of the
increased spontaneous mutation frequency produced by the L88 allele
compared to the wild-type enzyme and the thymineless-induced
increase in mutation in the presence of the wild-type polymerase.
Thus, it is possible that the L88 enzyme is not significantly more
prone to producing mutations during thymine starvation than is the
wild-type enzyme. On the other hand, the L56 allele did produce a
thymineless-induced increase in mutation which was larger than the
sum of the two separately measured kinds of mutation. The basis
for the difference between the effects of the L56 and L88 polymerases upon thymineless mutagenesis awaits further illucidation.

     Disrupting cytosine metabolism in T4 also increases mutation
frequencies in vivo (19,22). These experiments used ts alleles of
gene 42 (the hydroxymethylase gene of T4) to lower dHMCTP concentrations. However, complications involving other potential roles
of the gene 42 product in T4 DNA synthesis make it quite possible
that factors other than altered availability of dHMCTP might play
a role in such mutagenesis. No measurements of mutant DNA
polymerase effects on this type of mutagenesis has been reported.

     The fidelity of in vitro DNA synthesis of ØX174 templates by
the wild-type T4 DNA polymerase in conjunction with other replication proteins is also sensitive to dNTP pool biases (39,40).
A·T → G·C mutations increased in a manner parallel to increased dGTP
concentration relative to dATP, suggesting that G·T intermediates
predominated A·T site transitions under these conditions. On the
other hand, A·T → G·C mutations were largely unaffected when the
dCTP concentration was higher than dTTP concentrations. (A very
small increase was seen at the most extreme bias which could be
experimentally produced.) Thus in the in vitro system the
contribution of the A·C intermediate to transition appears to be
small. The low contribution of A·C intermediates to A·T transitions
during in vitro pool-bias experiments is not easily rationalized in
view of in vivo thymineless mutagenesis, since it would seem that
the 100-fold or greater biases achieved in vitro would be larger
than those in vivo. However, it is now possible to conduct the in
vitro experiment under conditions where the background mutation
frequency is 10-fold lower, thus allowing a substantial increase in
the sensitivity of the assay to pool bias effects (N. Sinha, personal
communication); such experiments would be particularly useful in
evaluating the potential role of polymerases in mediating A·T site
transitions via A·C intermediates.

     Serious complications may exist in the interpretations of any
of these experiments. The in vivo experiments suffer from the
deficiency that effects of the perturbation other than dNTP pool
size alterations might contribute substantially to increased mutation
frequencies. The in vitro experiments suffer from the inherent
problems involved in extrapolating to the in vivo replication system.
For example, there is in vitro evidence for the physical association
of enzymes involved in T4 DNA precursor synthesis which can produce
concentration gradients of deoxyribonucleotides (41). It is thus
a possibility that dNTP's are not provided to the DNA polymerase
as "pools" but that such a complex maintains a high local
concentration of DNA precursors during DNA replication. Some in
vitro experiments have also suggested a second complicating factor,
the influence of the bias on fidelity due to an effect on proof-
reading via the speed with which the succeeding base is inserted
(42,43). This effect is a direct prediction of kinetic proof-reading
models (44). The T4 replication system described by Sinha and
Haimes (40) does not appear to exhibit strong "next-base-effects".

BASE ANALOGUE-INDUCED TRANSITION INTERMEDIATES AND THE INFLUENCE OF
T4 DNA POLYMERASE

     Although studies of transition intermediates arising during
DNA synthesis are frustrated by the difficulty of measuring A·C or
G·T intermediates, the intermediates important to base analogue-
induced transitions can be measured both chemically and genetically.
Thus, base analogue studies provide an approach to distinguishing
polymerase effects upon the formation of particular mispaired
intermediates. Two frequently studied base analogues are 2-amino-
purine (AP) and 5-bromouracil (BU); Figure 2 demonstrates the
pathways of AP- and BU-induced transitions. For examples, AP
intermediates involved in A·T → G·C mutations can be distinguished
by measurements which separate the frequency of AP incorporation
from the frequency of misreplication at that site after incorporation,
measures of (template T)·AP and (template AP)·C intermediates,
respectively. Alternatively, AP-induced mutation at G·C sites
depends primarily upon (template C)·AP pair formation.

     When taken together, three recent experiments (17,45,46)
probing the antimutator effects of some T4 DNA polymerase alleles
on transition mutations induced by 2-aminopurine or 5-bromouracil
provide some information about the quantitative contribution of
various base analogue-containing intermediates. The influence of
antimutator effects on particular intermediates along these mutational
pathways sometimes differ substantially. These differences are
reconciled by a model suggesting that the antimutator polymerases
exhibit differential fidelity for various intermediates even when
reciprocal template and primer-terminus orientations of the same
base pair are compared [e.g., (template AP)·C versus (template C)·
AP].

AP-INDUCED MUTATION

A·T → (TEMPLATE T)·AP   →   (TEMPLATE AP)·C → G·C

G·C → (TEMPLATE C)·AP   →   (TEMPLATE AP)·T → A·T

BU-INDUCED MUTATION

A·T → (TEMPLATE A)·BU   →   (TEMPLATE BU)·G → G·C

G·C → (TEMPLATE G)·BU   →   (TEMPLATE BU)·A → A·T

FIGURE 2.   Intermediates of base analogue-induced transition
            mutations. Underlined intermediates are important
            quantitatively to in vivo measured mutant frequencies.
            Those not underlined may be increased or decreased by
            the action of mutant polymerase, but the result is not
            expected to produce a measurable change in mutant
            frequency.

    Two of the experiments address the question of how antimutator
effects upon AP-induced transitions at A·T sites are partitioned
between effects upon the incorporation of AP into DNA [formation
of (template T)·AP pairs] and effects upon the frequency with
which AP templates direct the insertion of C into DNA [formation of
(template AP)·C pairs]. These partial reactions are described in
Figure 2. In the first experiment (45) phage coding the wild-type
or the L141 polymerase were grown in the presence of AP and the
quantity of AP in the packaged T4 DNA was measured. This average
AP incorporation was then used to partition the frequency of
mutation at a particular rII mutant site (at which A·T → G·C
mutations could be measured) between the rate of insertion of AP
into DNA [(template T)·AP] and the rate with which template AP
produced mutations [(template AP)·C]. The measurement suggested
that the in vivo reduction of AP-induced mutation at A·T sites in
the presence of the L141 polymerase is due primarily to decreased
frequencies of (template AP)·C pairs and only secondarily to
decreased frequencies of (template T)·AP pairs.

    In the second experiment, (17) genetic techniques allowed a
measure of the relative frequencies of (template T)·AP and (template
AP)·C pairs at the same DNA site, produced by the wild-type

polymerases and by two mutant polymerases (CB87 and CB120) which
reduce AP-induced mutagenesis. In this experiment, AP-induced
mutation was measured at the first base pair of an rII ochre (UAA)
codon. Unlike the first experiment, average frequencies of AP
incorporation into total DNA did not have to be extrapolated to a
particular site. The experiment measured the two pairing reactions
in the following manner. Phage carrying the rII UAA mutation, and
either the wild-type polymerase or a mutant polymerase allele, were
grown in the presence of AP to allow the incorporation of AP into
phage DNA. The packaged progeny phage (now containing AP in their
DNA) were then introduced into fresh cells in the absence of
exogenous AP. During this second cycle of growth, mutations
accumulated at the ochre site due to replication errors at the
occasional AP residues in the DNA. During the second cycle, the
error rates of various polymerases were compared by coinfecting the
cells with additional phage which provided either the same or a
different DNA polymerase. Although the coinfecting phage provided
polymerase to the infection, they carried deletions in the rII
region, and thus could not contribute to the revertants of the
UAA mutant which were the measure of (template AP)·C pairs.

The relative contribution to replication of the DNA containing
the UAA mutant by the competing polymerases could be calculated by
comparing the influence of the two polymerases on the spontaneous
reversion frequency of the ochre codon. Increased revertant
frequencies after second-cycle infections were corrected for background revertant frequencies, and in cases where different polymerases were present, were normalized to the error rate which would
have been expected if only the most error-prone polymerase had been
present. Comparisons of the frequency of errors by the different
polymerases on the same template measured the relative frequencies
of formation of (template AP)·C pairs. The relative frequency of
(template T)·AP pairs was measured by comparing the frequency of
errors produced by a single polymerase on templates having differing
frequencies of (template T)·AP pairs by virtue of having been replicated in the presence of AP by different polymerases.

Table II summarizes the results of both experiments and
demonstrates that mutant alleles which reduce AP mutagenesis compared
to the wild type polymerase do so primarily by reducing (template
AP)·C pair frequencies; smaller reductions in (template T)·AP pairs
were found. The difference between the frequencies of (template
AP)·C pair frequency estimates for the CB120 and the CB87 polymerases was caused by a difference in a maximum estimate of the
(template AP)·C frequencies in the presence of these polymerases
rather than by a difference in a measurement of those frequencies.
The estimate was required because no increase in mutation frequency
was detected over the background level when these polymerases were
responsible for replication in the second cycle. Thus the lower

relative frequency for (template AP)·C pairs for the CB87 polymerase
may simply reflect its higher background level of mutation when
compared to CB120. In view of the differences in the experimental
approaches the results for the CB120 polymerase are remarkably
similar to those for the L141 polymerase.

TABLE II.   The relative influence of wild-type and mutant
             polymerases on the frequency of (template T)·AP
             and (template AP)·C intermediates in AP-induced
             A·T → G·C mutation

Polymerases Compared	Ratio of the Frequencies of AP-Mutagenesis Intermediates	
	A·T → AP·T	AP·T → AP·C
Wild-Type / CB87	8	>11
Wild-Type / CB120	16	>50
Wild-Type / L141	4	100

The relative effects of the CB87 and CB120 mutant and the wild-
type polymerases were measured genetically at the rII UAA site
UV183 (17). The relative effects of the L141 mutant polymerase
and the wild-type were measured chemically for the A·T → AP·T
step. The AP·T → AP·C step was measured by applying the
chemically derived A·T → AP·T measurement to the total of these
two steps measured genetically at the rII mutant site UV199
(45).

The third experiment (46) compared the influence of polymerase
alleles upon both G·C-site and A·T-site mutations induced by base
analogues. The frequency of base analogue-induced mutation in the
presence of the wild-type polymerase and the CB87 allele could be
compared at six different rII sites. Each site was defined by a
set of nonsense codons at which both A·T- and G·C-site mutations
could be measured by interconversion of codons. The examples shown
in Table III are the conversions UAA → UGA and UGA → UAA; (other
A·T site transitions were measured, with similar results to those
shown for UAA → UGA.)

TABLE III.   A comparison of the effect of the CB87 mutant DNA
             polymerase to the wild-type polymerase on AP-induced
             and BU-induced transition mutations

RATIOS OF MUTANT FREQUENCIES

(Wild-type polymerase / CB87 polymerase)

| Mutant | AP-induced | | BU-induced | |
Site	A·T → G·C	G·C → A·T	A·T → G·C	G·C → A·T
rYH115	25	1.2	270	9.4
rYH119	45	6.9	76	0.3
rYH122	7	5.6	6000	0.4
rYH132	130	8.4	320	29
rYH320	120	1.4	160	3.3
rYH341	28	18	8500	16

Data of Ronen et al. (46) was used to calculate the ratio of the
frequencies of mutation produced in the presence of the CB87
and the wild-type polymerases for AP- and BU-induced mutation.
A·T → G·C mutations were measured by the conversion of UAA mutants
to UGA mutants, while G·C → A·T mutations were measured by the
conversion of UGA mutants to UAA mutants.   Both conversions were
examined at 6 different mutant sites in the rII region.   Spontaneous
background mutation frequencies were subtracted before the ratios
were calculated.

In the case of AP-induced mutation, the CB87 polymerase clearly
exerted larger effects upon A·T-site mutation than upon G·C-site
mutation when compared to the wild-type.   The results of the
Goodman et al (45) and Ripley (17) experiments predict that the
influence of the CB87 polymerase upon A·T-site mutation is largely
attributable to the influence of CB87 upon (template AP)·C pairs.
On the other hand, the mutations measured at the G·C sites are
primarily due to (template C)·AP pairs (Figure 2).   Thus, among
the three steps of mutagenesis measured, small polymerase effects
are seen upon the A·T → AP·T and G·C → AP·C steps, where incorpo-

ration of AP is directed by a normal template base. Only in the
AP·T → AP·C step is the very large antimutator effect seen. It is
in this step, in contrast to the others, that the incorporation of
normal dNTP by modified template (AP) is measured.

Because less is known about BU-induced mutation, conclusions
must be drawn more tentatively; however, the influence of the CB87
polymerase upon BU-induced transitions is consistent with the same
pattern of fidelity effects upon mutational intermediates as seen
in the case of AP-induced mutation. The net effect of CB87 on
A·T-site, BU-induced transitions is again substantially larger than
its effect upon G·C-site, BU-induced transitions. Effects on
mutation at G·C sites is attributable to effects on (template G)·BU
intermediates (Figure 2). Effects on mutation at A·T sites is the
sum of effects upon the (template A)·BU and (template BU)·G intermediates. No measurement of the effects upon these two individual
intermediates has yet been undertaken. If strong antimutator effects
are seen only when normal dNTP's are incorporated by modified
templates, analogous to the observation with AP, then a much larger
influence on the frequency of (template BU)·G intermediates is
predicted.

The model above is based on observations in which base analogue-
induced mutation is less subject to antimutator effects at G·C sites
than at A·T sites. The observation that spontaneous G·C-site
mutations are not subject to antimutator effects could be significant
in this regard. An alternative explanation for the lower G·C-site
mutation effects may be that antimutator effects are weaker when
intermediates contain either (template G) or (template C). However,
this explanation does not simultaneously explain the assymetric
division of antimutator effects between the (template T)·AP and
(template AP)·C intermediates, and might in fact be viewed as
weakly contradictory since (template AP)·C intermediates would seem
to be more like those intermediates containing (template G) than
the usual intermediates of A·T-site transitions.

Although the question has remained largely unaddressed, it
should be recalled that these same polymerases which produce antimutagenic effects upon transition mutations can also produce
mutagenic effects upon spontaneous transversions. It seems possible
that this mutagenic effect upon transversions reflects the perturbed
recognition of transversion intermediates by these polymerases, since
even among the different intermediates of base substitution
mutagenesis, these polymerases produce quantitatively distinct
fidelity effects. Since there is almost no information about
transversion mutation, other explanations are possible. Approaches
using in vitro systems, or the development of base analogues which
could be used to probe transversions in vivo, might well provide a
better understanding of the effects of polymerases upon transversion
mutation.

The assymetric patterns of antimutator effects upon intermediates of 2-aminopurine-induced mutation are inconsistent with
fidelity models (43) in which the determinant of the antimutagenic
effect senses only simple hydrogen bonding strength, since the
fidelity effects upon (template AP)·C intermediates are clearly
different from those upon (template C)·AP intermediates. A variety
of complex possibilities would still allow hydrogen-bonding
strengths to be major determinants of fidelity without requiring that
such reciprocal intermediates have identical fidelity. Furthermore,
there is an additional potential for complexity provided by the
separate consideration of the fidelity mechanisms acting during the
incorporation of a base into DNA versus those acting during a
subsequent proof-reading step.

For example, during the addition of a dNTP to the primer
terminus, the initial interaction of the dNTP with the polymerase
("on-rate") is thought on the basis of E. coli polymerase I
experiments to be largely independent of template specificity (47).
(In fact, it is this assumed characteristic which allows the dNTP
bias experiments described previously.) It is rather the "off-
rate" which is governed by template specificity, perhaps via
hydrogen-bonding strengths. Although it is generally assumed that
this template specificity is mediated by direct hydrogen bonding
between the incoming dNTP and the template, there is in fact no
evidence for this relationship. It is possible that hydrogen
bonding is probed indirectly via separate interactions between the
polymerase, the template and substrate bases, with the net result of
these separate interactions producing the characteristically
accurate addition of a dNTP to the newly synthesized DNA strand.
Even if hydrogen bonding between template and substrate bases is
direct and constitutes the major fidelity determinant, other
interactions with the polymerase could still lead to assymetric
fidelity properties for reciprocal intermediates. Such interactions
might include specific interactions of the polymerase with positions
of the base not directly involved in hydrogen-bond formation with
the template, but nonetheless influencing the strength of hydrogen
bonds to the template. It seems unlikely that the binding of the
dNTP is exactly identical to the binding of the template base within
the active site of the polymerase.

It is conceivable that stacking energies provided by the
interaction of a newly incorporated base with previously incorporated bases in the newly synthesized strand could provide a
mechanism by which infidelity might be promoted during DNA replication. A recent experiment (48) suggests that these stacking
interactions can be expressed assymetrically and promote infidelity
of base incorporation. In vitro replication fidelity (measured as
nucleotide turnover) of T4 DNA polymerase alone on homopolymer
templates was consistent with the suggestion that the addition of a

base to the primer was strongly influenced by nearest-neighbor base-
stacking interactions at the primer terminus. The quantitative
relevance of such a measurement to the in vivo replication system
(dependent upon numerous proteins in addition to the polymerase, and
using DNA rather than a homopolymer) remains to be determined.
Nearest-neighbor effects are certainly seen in vivo, although, it is
far from clear that they are attributable to base-stacking interactions (49).

The role of hydrogen bonding as the major determinant for
proof-reading fidelity is not strongly established. The view seems
to emanate originally from the Brutlag and Kornberg (30) experiments,
in which the bases subject to editing by the $3' \rightarrow 5'$ exonuclease
of the polymerase were in fact not hydrogen-bonded to the template.
However, the experimental conditions of these measurements are
highly artificial, and the characteristics of removal of a base
just misinserted by a continuously polymerizing enzyme could be
quite different. Our lack of knowledge in this area is vast.
Experimentally, proof-reading by the exonuclease is measured by the
production of dNMP from dNTP (turnover). While the exonuclease is
clearly capable of this reaction, it remains undetermined whether
the dNMP produced in such reactions is formed primarily from this
process acting in a proof-reading mode. Even the relative position
of a proof-reading step within the sequence of events surrounding
incorporation of bases is unknown. Although mathematical models of
proof-reading processes frequently assume that translocation of the
polymerase to the succeeding template base occurs before proof-
reading, there is no direct evidence which bears upon this question.

THE INFLUENCE OF MUTANT DNA POLYMERASES ON FRAMESHIFT MUTAGENESIS

Although a great deal is known about the influence of T4 DNA
polymerase upon base substitution mutation, far less is known
about its role in DNA infidelity exhibited as frameshift mutation.
Recent studies in our laboratory demonstrate that mutant DNA
polymerase alleles profoundly influence the frequency of addition
and deletion mutations (50). Our observation of antimutator effects
by some polymerase alleles for certain frameshift mutations suggests
that we may be able to identify quantitatively important intermediates of frameshift mutation through these studies. Determining
the role of the polymerase in base addition or deletion infidelity
is difficult, however. The relative importance of different DNA
metabolic processes, the nature of the DNA intermediates upon which
the polymerase acts, and the mechanisms by which mutagens produce
frameshifts are not understood. We have begun a systematic
examination of the influence of a number of DNA polymerase alleles
upon frameshift mutation in the rII genes of bacteriophage T4 as
an approach to determining the role of polymerase in frameshift
mutation.

We have examined a number of different polymerase alleles
hoping to discover correlations between frameshift mutation patterns
and properties of the enzyme sufficient to suggest links between
frameshift mutation and particular fidelity mechanisms. More
specifically, we are examining the nature of the frameshift mutations
produced in the presence of different polymerases, believing that
certain polymerases might produce particular types of mutational
outcomes (e.g., additions versus deletions, mutations of preferred
sizes, or mutations at particular DNA locations). The mutational
outcomes might further suggest important intermediates in the
production of frameshift mutations.

There have been scattered reports in the literature documenting
increased frameshift mutation frequencies in the presence of some
polymerase alleles (16,18). However, other attempts to measure
increased frameshift mutation frequencies were unsuccessful (15).
The CB87 and CB120 polymerase alleles produce antimutator effects
upon frameshift mutation at some rII sites, but the effect is much
smaller than upon spontaneous A·T site transitions (16).

In our initial experiments we surveyed a large number of
mutant polymerase alleles to determine whether frameshift fidelity
effects were common. Among 19 alleles, located at recombinationally
distinct sites of gene 43, more than half produced significant
increases or decreases in frameshift frequency measured as the
reversion of the rIIA frameshift mutation r131 (50). Table IV shows
the results for two alleles (L141 and L42) which reduced frameshift
frequency for r131 and two alleles (L56 and L88) which increased it.
We have identified five alleles which reduce reversion frequencies
of r131. Four of these, previously described in this paper (L141,
L42, CB87, and CB120), have well documented antimutator effects upon
some base pair substitutions. The fifth, A71, has not been tested
for its influence upon base pair substitution, but the A69 allele,
indistinguishable from A71 by recombination tests, has been shown
to produce antimutator effects upon A·T-site transition mutations
(27). Thus, there is a correlation between antimutator effects
measured by the reversion of the r131 frameshift mutation and
antimutator effects measured on some transition mutations.

Approximately a third of all spontaneous, forward rII mutations
(in the presence of the wild-type polymerase) are frameshift mutations
mapping at the r131 site (51). This frameshift hot spot, from
correlations between the wild-type DNA sequence and the genetic map,
appears to be located in a sequence of 6 A's (52). Similar
correlations place two other spontaneous frameshift hot spots,
r114 and r117 in the rIIB gene, in sequences of 6 A's. We therefore
tested the influence of these same polymerase alleles on the
reversion of frameshift mutations located at these sites (Table IV).
Although quantitative difference were found, antimutator effects
were clearly present at these frameshift sites as well. Although

we do not yet know whether it is the same property of these polymerases which is responsible for increased fidelity of spontaneous
frameshift and base substitution mutation at A·T sites, the polymerase alleles and mutational targets examined to data do not separate the two antimutator effects.

TABLE IV. The influence of DNA polymerase mutants on spontaneous
           revertant frequencies of rII frameshift mutantsresiding
           in runs of A's

POLYMERASE	REVERTANT FREQUENCY x 10^8			RELATIVE FREQUENCY (MUTANT POLYMERASE TO WILD-TYPE)
	r131	r114	r117	
WILD-TYPE	28	27	49	
L141	4	13	35	1 - 7x decrease
L42	5	9	26	2 - 6x decrease
L88	310	100	310	4 - 11x increase
L56	2300	330	1100	12 - 82x increase

The revertant frequencies for three frameshift mutations mapping at
the r131, r114, and r117 loci were measured in each polymerase
background after growing multiple stocks from a small number of
phage (approximately 10^3) to approximately 2×10^{10} phage at 30°C.
The reported frequencies are averaged from at least four stocks.

       A second intriguing similarity can be seen between antimutator
effects on frameshift and on base substitution mutations. In both
cases, although particular mutations (A·T-site transitions and
A-run frameshifts) are subject to antimutator effects other mutations
are not reduced in frequency and in fact are often promoted. This
latter effect was examined by studying frameshift mutations arising
in a DNA sequence within the rIIB gene in the L141 and wild-type
polymerase backgrounds.

       Frameshift mutagenesis were studied in this region by measuring
the production of frameshift mutations over a relatively long DNA
sequence (approximately 140 base pairs). This can be accomplished

in the rIIB gene because, when assayed by the restriction system
provided by the rex gene product of lambda lysogens, this region
is largely insensitive to amino acid changes. This permits the
phenotypic suppression of an initial frameshift mutation by a
second frameshift mutation if the combination of the two frameshifts
restores the normal reading frame and does not produce any in-frame
nonsense codons. Because the wild-type DNA sequence is known (52),
and thus also the location of out-of-frame termination codons
(potential barriers to mutual suppression by the two frameshift
mutations) (53), the DNA target can be estimated for frameshift
mutations whose positions relative to the barriers is known. Thus,
for example, the frameshift mutation FC47 can be suppressed by
mutations of -1 ± 3N bases occuring in a target sequence of
approximately 124 bases, the boundaries of the target being defined
by barriers (50).

TABLE V.   A comparison of suppressors of the FC47 frameshift
           mutation arising in the presence of the wild-type or the
           L141 polymerase.

SUPPRESSORS OF FC47 Mutation Frequency (x 10^8)			
POLYMERASE	Site 1	Site 2	Site 3	All Sites
Wild-type	8.3	0.7	7.6	33
L141	24	24	330	580
RATIO	3	35	43	18

Total suppressor frequencies were measured in a total of 48 stocks
grown from a low innoculum (approximately 1000 phage) to 2 x 10^{10}.
One suppressor frameshift was isolated from each stock and characterized. The frequency of mutation at each site was determined
from the ratio of isolated frameshift mutations mapping at sites 1,
2 and 3 and the total frameshift mutations characterized multiplied
by the frequency of FC47 suppressors. Site 1 is defined by failure
of the suppressor frameshift to recombine with the FC11 frameshift;
site 2 by failure of the suppressor to recombine with the FC88
frameshift; and site 3 by failure of the suppressor to recombine
with the FC36 frameshift.

Frameshift suppressors of FC47 isolated in the presence of either the wild-type or L141 polymerases have been mapped by recombination tests. These mapping data demonstrate a DNA region within the target for FC47 reversion wherein L141 produces frameshift frequencies approximately 40-fold larger than those produced by the wild-type enzyme. In contrast, other frameshifts within this target are not strongly affected by the mutant polymerase. The properties of three frameshift sites are described in Table V. Sites 2 and 3 show the mutator effect of the L141 enzyme, while the frequency of site 1 mutations is not strongly affected by this polymerase. These three sites are non-overlapping and are defined by the failure of a particular suppressor frameshift to recombine with a marker frameshift which defines that site. Not all mutations mapping within a site are identical, however, since they can show differences in recombination with outside markers and in spontaneous revertant frequencies.

We have noted two particularly interesting aspects of the wild-type DNA sequence within the target for reversion of the FC47 frameshift. One is the location of a tandem repeat of six bases (separated by two bases) close to site 3. Such a repeat might efficiently generate additions of eight base pairs between the repeats, due to misalignment of complementary DNA strands during DNA synthesis, recombination or repair; on the basis of frame, this addition would indeed be expected to suppress FC47. However, since more than one type of frameshift occurs frequently at site 3 in the presence of L141, not all of the L141 effect could be explained by this particular mutation. The second interesting property of the wild-type sequence is the potential to form a stable secondary structure involving approximately 40 bases including both sites 2 and 3 (50) (Figure 3). Whether, in fact such a structure actually forms in vivo is unknown, and the mechanism(s) by which it might promote frameshift mutation is also unknown. An intriguing property of the structure is its imperfections (i.e., unpaired side loops): such residues might easily constitute DNA substrates subject to unusual but predictable deletions or additions caused by nuclease digestion of loops followed by ligation, or by DNA synthesis templated from within the secondary structure. Alternatively, these looped structures may be particularly sensitive to aberrant forms of DNA recombination, providing a semi-stable structure which is particularly sensitive to invasion by single-stranded DNA. The repeated sequence is located within this potential structure, with both of its ends in unpaired regions. The role of recombination in spontaneous frameshift mutation is far from clear in T4. There is weak evidence for a correlation between spontaneous frameshift mutation and recombination between flanking markers (54) but proflavin-induced frameshift mutations do not appear to be correlated with recombination (56). While mutant T4 DNA polymerase alleles can influence recombination (57), the effects tend to be small, and no measurement has yet been made of specific effects upon recombination in this region by the L141 polymerase.

In the absence of more information about the nature of frameshift mutational mechanisms it is difficult to guess from the types

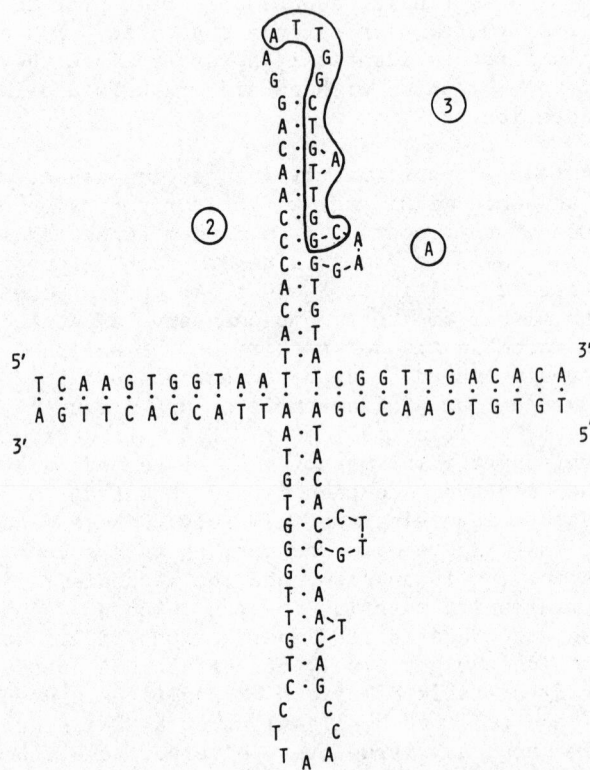

FIGURE 3. A potential secondary structure in the rIIB gene of T4 containing sites at which frameshift mutations are enhanced by the L141 polymerase compared to the wild-type. The circled bases in the structure, are the tandem repeat sequence. Mutations in site 2, lie in the stem structure to the left marked 2; mutations in site 3 lie near or in the tandem repeat. A marks an imperfection of the secondary structure, which might be a substrate of frameshift (see text).

of in vitro studies done to date what characteristics of a mutant enzyme might cause it to produce increased or decreased frameshift frequencies at different sites. As noted earlier, the CB120 polymerase is particularly defective during in vitro replication on templates having secondary structure, but whether this defect is characteristic of the other polymerase which display antimutator effects is unknown.

The determination of the DNA sequences of the frameshift mutations in the FC47 target is our current approach to deducing possible mechanisms of frameshift mutation. By precisely defining the character of the frameshift mutations with respect to both their composition and location, we hope to link the final mutational outcome to particular aspects of infidelity, especially those revealed by altered DNA polymerases.

REFERENCES

1. Drake, J. W.: Comparative rates of spontaneous mutation. Nature, 221:1132 (1969)

2. Wood, W. B. and Revel, H. R.: The genome of bacteriophage T4. Bacteriol. Rev., 40:847 (1976)

3. Barry, J. and Alberts, B.: In vitro complementation as an assay for new proteins required for bacteriophage T4 DNA replication: Purification of the complex specified by T4 genes 44 and 62. Proc. Natl. Acad. Sci. (U.S.A.), 69:39 (1972)

4. Liu, C. C., Burke, R. L., Hibner, U., Barry, J. and Alberts, B. M.: Probing DNA replication mechanisms with the T4 bacteriophage in vitro system. Cold Spring Harbor Symp. Quant. Biol., 43:469 (1979)

5. Huberman, J. A., Kornberg, A. and Alberts, B. M.: Stimulation of T4 bacteriophage DNA polymerase by the protein product of T4 gene 32. J. Mol. Biol., 62:39 (1971)

6. Burke, R. L., Alberts, B. M. and Hosoda, J.: Proteolytic removal of the COOH terminus of the T4 gene 32 helix-destabilizing protein alters the T4 in vitro replication complex. J. Biol. Chem., 255:11484 (1980)

7. Bernstein, H.: Repair and recombination in phage T4. I. Genes affecting recombination. Cold Spring Harbor Symp. Quant. Biol., 33:325 (1968)

8. Broker, T. R. and Lehman, I. R.: Branched DNA molecules:
    Intermediates in T4 recombination. J. Mol. Biol., 60:131
    (1971)

9. Harm, W.: On the control of UV-sensitivity of phage T4 by
    the gene x. Mutation Res., 1:344 (1964)

10. Maynard-Smith, S. and Symonds, N.: Involvement of bacterio-
    phage T4 genes in radiation repair. J. Mol. Biol., 74:33
    (1973)

11. Mosig, G., Luder, A., Garcia, G., Dannenberg, R. and Bock, S.:
    In vivo interactions of genes and proteins in DNA replication
    and recombination of phage T4. Cold Spring Harbor Symp.
    Quant. Biol., 43:501 (1979)

12. Streisinger, G., Emrich, J. and Stahl, M. M.: Chromosome
    structure in phage T4, III. Terminal redundancy and length
    determination. Proc. Natl. Acad. Sci. (U.S.A.), 57:292 (1967)

13. Speyer, J. F. and Rosenberg, D.: The function of T4 DNA
    polymerase. Cold Spring Harbor Symp. Quant. Biol., 33:345
    (1968)

14. Epstein, R. H., Bolle, A., Steinberg, C. M., Kellenberger, E.,
    Boy de la Tour, E., Chevalley, R., Edgar, R. S., Susman, M.,
    Denhardt, G. H., and Lielausis, A.: Physiological studies
    of conditonal lethal mutants of bacteriophage T4D. Cold
    Spring Harbor Symp. Quant. Biol., 28:375 (1963)

15. Speyer, J. F., Karam, J. D. and Lenny, A. B.: On the role of
    DNA polymerase in base selection. Cold Spring Harbor Symp.
    Quant. Biol., 31:693 (1966)

16. Drake, J. W., Allen, E. F., Forsberg, S. A., Preparata, R.
    and Greening, E. O.: Genetic control of mutation rates in
    bacteriophage T4. Nature, 221:1128 (1969)

17. Ripley, L. S.: Influence of diverse gene 43 DNA polymerases
    on the insertion and replication in vivo of 2-aminopurine
    at A·T base-pairs in bacteriophage T4. J. Mol. Biol. (in
    Press) (1981)

18. Bernstein, H.: Reversion of frameshift mutations stimutated
    by lesions in early function genes of bacteriophage T4. J.
    Virol., 7:460 (1971)

19. Chiu, C. and Greenberg, G. R.: Mutagenic effect of temperature-
    sensitive mutants of gene 42 (dCMP hydroxymethylase) of
    bacteriophage T4. J. Virol., 12:199 (1973)

20. Koch, R. E. and Drake, J. W.: Ligase-defective bacteriophage
    T4 I. Effects on mutation rates. J. Virol., 11:35 (1973)

21. Koch, R. E., McGaw, M. K. and Drake, J. W.: Mutator mutations
    in bacteriophage T4 gene 32 (DNA unwinding protein). J.
    Virol., 19:490 (1976)

22. Williams, W. E. and Drake, J. W.: Mutator mutations in
    bacteriophage T4 gene 42 (dHMC hydroxymethylase). Genetics,
    86:501 (1977)

23. Watanabe, S. M. and Goodman, M. F.: Mutator and antimutator
    phenotypes of suppressed amber mutants in genes 32, 41, 44, 45
    and 62 in bacteriophage T4. J. Virol., 25:73 (1978)

24. Alberts, B. M. and Frey, L.: T4 bacteriophage gene 32: A
    structural protein in the replication and recombination of
    DNA. Nature, 227:1313 (1970)

25. Nossal, N. G.: DNA synthesis on a double-stranded DNA template
    by the T4 bacteriophage DNA polymerase and the T4 gene 32 DNA
    unwinding protein. J. Biol. Chem., 249:5668 (1974)

26. Speyer, J. F.: Mutagenic DNA polymerase, Biochem. Biophys.
    Res. Commun., 21:6 (1965)

27. Ripley, L. S.: Transversion mutagenesis in bacteriophage T4.
    Molec. Gen. Genet., 141:23 (1975)

28. de Vries, F. A. J., Swart-Idenburg, Ch. J. H. and de Waard, A.:
    An analysis of replication errors made by a defective T4 DNA
    polymerase. Molec. Gen. Genet., 117:60 (1972)

29. Muzyczka, N., Poland, R. L. and Bessman, M. J.: Studies on
    the biochemical basis of spontaneous mutation I. A comparison
    of the deoxyribonucleic acid polymerases of mutator,
    antimutator, and wild type strains of bacteriophage T4. J.
    Biol. Chem., 247:7116 (1972)

30. Brutlag, D. and Kornberg, A.: Enzymatic synthesis of
    deoxyribonucleic acid. XXXVI A proofreading function for
    the 3' → 5' exonuclease activity in deoxyribonucleic acid
    polymerases. J. Biol. Chem., 247:241 (1972)

31. Gillin, F. D. and Nossal, N. G.: Control of mutation frequency
     by bacteriophage T4 DNA polymerase I. The CB120 antimutator
     DNA polymerase is defective in strand displacement. J. Biol.
     Chem., 251:5219 (1976)

32. Gillin, F. D. and Nossal, N. G.: Control of mutation frequency
     by bacteriophage T4 DNA polymerase II. Accuracy of nucleotide
     selection by the L88 mutator, CB120 antimutator, and wild type
     phage T4 DNA polymerases. J. Biol. Chem., 251:5225 (1976)

33. Lo, K. and Bessman, M. J.: An antimutator deoxyribonucleic
     acid polymerase I. Purification and properties of the enzyme.
     J. Biol. Chem., 251:2475 (1976)

34. Hershfield, M. S.: On the role of deoxyribonucleic acid
     polymerase in determining mutation rates. Characterization
     of the defect in the T4 deoxyribonucleic acid polymerase
     caused by the tsL88 mutation. J. Biol. Chem., 248:1417
     (1973)

35. Hopfield, J. J.: Kinetic proofreading: a new mechanism
     for reducing errors in biosynthetic processes requiring
     high specificity. Proc. Natl. Acad. Sci. (U.S.A.), 71:4135
     (1974)

36. Topal, M. D. and Fresco, J. R.: Complementary base pairing
     and the origin of substitution mutations. Nature, 263:285
     (1976)

37. Smith, M. D., Green, R. R., Ripley, L. S. and Drake, J. W.:
     Thymineless mutagenesis in bacteriophage T4. Genetics,
     74:393 (1973)

38. Drake, J. W. and Greening, E. O.: Suppression of chemical
     mutagenesis in bacteriophage T4 by genetically modified DNA
     polymerases. Proc. Natl. Acad. Sci. (U.S.A.), 66:823 (1970)

39. Hibner, U. and Alberts, B. M.: Fidelity of DNA replication
     catalysed in vitro on a natural DNA template by the T4
     bacteriophage multi-enzyme complex. Nature, 285:300 (1980)

40. Sinha, N. K. and Haimes, M. D.: Molecular mechanisms of
     substitution mutagenesis. An experimental test of the Watson-
     Crick and Topal-Fresco models of base mispairings. J. Biol.
     Chem., (in Press) (1981)

41. Reddy, G. P. V. and Mathews, C. K.: Functional compartmentation
     of DNA precursors in T4 phage-infected bacteria. J. Biol.
     Chem., 253:3461 (1978)

42. Fersht, A. R.: Fidelity of replication of phage ØX174 DNA by
     DNA polymerase III holoenzyme: Spontaneous mutation by
     misincorporation. Proc. Natl. Acad. Sci. (U.S.A.), 76:4946
     (1979)

43. Clayton, L. K., Goodman, M. F., Branscomb, E. W. and Galas,
     D. J.: Error induction and correction by mutant and wild
     type T4 DNA polymerases. Kinetic error discrimination
     mechanisms. J. Biol. Chem., 254:1902 (1979)

44. Bernardi, F. and Ninio, J.: The accuracy of DNA replication.
     Biochimie, 60:1083 (1978)

45. Goodman, M. F., Hopkins, R. and Gore, W. C.: 2-Aminopurine-
     induced mutagenesis in T4 bacteriophage: A model relating
     mutation frequency to 2-aminopurine incorporation in DNA.
     Proc. Natl. Acad. Sci. (U.S.A.), 74:4806 (1977)

46. Ronen, A., Halevy, C. and Kass, N.: Site specificity and
     variability in the mutator and antimutator effects of phage
     T4 gene 43 mutants. Genetics., 90:647 (1978)

47. Englund, P. T., Huberman, J. A., Jovin, T. M. and Kornberg, A.:
     Enzymatic synthesis of deoxyribonucleic acid. XXX Binding
     of triphosphates to deoxyribonucleic acid polymerase. J.
     Biol. Chem., 244:3038 (1969)

48. Topal, M. D., DiGuiseppi, S. R. and Sinha, N. K.: Molecular
     basis for substitution mutations. Effect of primer terminal
     and template residues on nucleotide selection by phage T4 DNA
     polymerase in vitro. J. Biol. Chem., 255:11717 (1980)

49. Koch, R. E.: The influence of neighboring base pairs upon
     base-pair substitution mutation rates. Proc. Natl. Acad.
     Sci. (U.S.A.), 68:773 (1971)

50. Ripley, L. S. and Shoemaker, N. B.: Polymerase infidelity
     and frameshift mutation. In Molecular and Cellular Mechanisms
     of Mutagenesis, Ed. by J. F. Lemontt and W. M. Generoso.
     Plenum Press, New York (1981)

51. Benzer, S.: On the topography of the genetic fine structure.
     Proc. Natl. Acad. Sci. (U.S.A.), 47:403 (1961)

52. Pribnow, D., Sigurdson, D. C., Gold, L., Singer, B. S., Brosius,
     J., Dull, T. J., and Noller, H. F.: The rII cistrons of
     bacteriophage T4: DNA sequence around the intercistronic
     divide and positions of genetic landmarks. J. Mol. Biol.,
     (in Press) (1981)

53. Barnett, L., Brenner, S., Crick, F. H. C., Shulman, R. G. and
     Watts-Tobin, R. J.: Phase-shift and other mutants in the
     first part of the rIIB cistron of bacteriophage T4.
     Philosoph. Trans. Royal Soc. London, Ser. B, 252:487 (1967)

54. Strigini, P.: On the mechanism of spontaneous reversion and
     genetic recombination in bacteriophage T4. Genetics, 52:759
     (1965)

55. Drake, J. W. and Allen, E. F.: Antimutagenic DNA polymerases
     of bacteriophage T4. Cold Spring Harbor Symp. Quant. Biol.
     33:339 (1968)

56. Lindstrom, D. M. and Drake, J. W.: Mechanics of frameshift
     mutagenesis in bacteriophage T4: Role of chromosome tips.
     Proc. Natl. Acad. Sci. (U.S.A.), 65:617 (1970)

57. Allen, E. F., Albreght, I. and Drake, J. W.: Properties of
     bacteriophage T4 mutants defective in DNA polymerase.
     Genetics, 65:187 (1970)

DISCUSSION

DR. SHAPIRO: Is there any independent evidence for the
cross-structure you postulate (figure 3), or did you just use it
to try and rationalize the mutagenesis data.

DR. RIPLEY: There's no evidence that it in actual fact exists. I
think that there are some ways in which we can go about testing
whether it may or may not be important. We are looking at whether
we'll affect frameshift mutation frequency throughout this region
if we disrupt the stem structure. We have an ochre mutation which
we believe to be at this location which we have converted to an
opal mutation, therefore disrupting two base pairs. Now we're
looking at the effect of this disruption by isolating frame shift
mutations in this background to see whether there is an influence
on the frequency of mutation in this region.

DR. SHAPIRO: I was just thinking that such open loops and so on
would be very hot spots for single-strand reagents.

DR. RIPLEY: I certainly think that's a possibility. There's some
intriguing potential in this regard with respect to a particular C
which is a hot spot for base pair substitution mutation, induced
by hydroxylamine and nitrous acid.

DR. SHAPIRO: Nitrous acid isn't that single-strand specific, but
methoxylamine is specific.

DR. RIPLEY: Of course, these mutagenesis experiments are in
vitro--I mean, packaged T4 experiments, in which T4 DNA is
metabolically inactive. These same sites are also hot spots for
spontaneous mutation. So I don't know what the nature of that hot
spot is; all I'm saying is there is a hot spot in at least one of
these loops. Also two further bases tend to be hot as well, so
it's not clear exactly what may or may not be going on. One thing
to keep in mind about loop structures is that such a semistable
structure might provide a way to think about single-strand
invasion during recombination processes; this repeat then is in
both ends, and may well be a hot spot for recombination. Maybe
what we're assaying here are infidelity effects of byproducts of
recombination. But we haven't yet made any measurements on
recombination effects in this region.

DR. SHERMAN: Have you looked at chemically-induced patterns of
mutagenesis with the same system?

DR. RIPLEY: No, not yet.

DR. L. PRAKASH: How do you account for the anti-mutator activity
of the mutant DNA polymerase at some sites, but the opposite

effect at others; the same enzyme being an anti-mutator and mutator?

DR. RIPLEY: The mechanism of frameshift mutation at the different sites may possibly be different. Alternatively, the nature of the interaction between the polymerase and the particular local DNA sequence might be important. A possible unifying idea is that the reason anti-mutators are anti-mutators for this one particular set of frameshifts is that they share a common property of having an unusual kind of DNA, that is runs of A's that may be more like poly-AT than ordinary DNA. I think that very little is known about the specificity of the interaction of polymerases with their templates and its effect on fidelity.

Differences in mechanism might be related to differences in the metabolic process responsible. Frameshifts arising as aberrant recombinant templates might respond to a mutant polymerase differently from frameshifts arising directly out of DNA synthesis errors. For example, frameshifts could be viewed as errors of translocation of the polymerase. That is, a process involved in DNA replication directly, the actual physical process of inserting a base; it either skips over or adds an extra one. That's why I'd like to know, for example, whether they're plus or minus frameshifts. That kind of process might be more directly related to the base pair substitution process than say some mechanism involving recombinaton, where the polymerase may play a very different fidelity, or infidelity, role, as the case may be, in a recombination intermediate. I think it's an open question, otherwise.

DR. SHAPIRO: Has work been done to actually isolate and characterize some of these mutator or anti-mutator polymerases?

DR. RIPLEY: Yes, all of the four I mentioned have all been characterized to some extent in vitro. Antimutators all have the properties I mentioned of increased turnover, generally attributed to excess exonuclease activity. Their other altered properties are not well defined and one doesn't know what to make of them with regard to their quantitative importance. It's known, for example, that the excess turnover is very sensitive to the presence of other proteins. When you do T4 DNA replication in the presence of what are called by the Alberts-Group the "5-protein" or "7-protein" reactions, in which the extra accessory proteins plus gene-32 protein participate, the properties of DNA replication are quite different.

DR. L. PRAKASH: Can you give us some examples of other proteins which interact with the polymerase and change the fidelity of replication, especially the cleaving of incorrect bases?

DR. RIPLEY: Other than gene-32? Essentially all of them have been tested in vivo, at least a few alleles have been tested. But what has always been done is to look at either temperature sensitive mutants or amber mutation, growing them in different suppressors, and asking the question do these mutations have an influence on fidelity? No one has directly really gone after fidelity mutants and then attributed them to these particular proteins. There is a paper by Watanabe and Goodman (J. Virol. 25, 73, 1978) where they used different amber mutations in a variety of suppressors and they saw some very small anti-mutator effects, I believe with gene-45, and perhaps one other gene. But they're two-fold effects, three-fold effects. One doesn't know whether those are direct effects of those proteins on DNA replication, or whether it's something that's more indirect via polymerase. Nothing like synergistic reactions or anything like that has ever been looked at.

DR. WALKER: These effects on fidelity and infidelity, are they very dependent on whether the T4 DNA is modified or not?

DR. RIPLEY: It's not thought to. The argument's a little weak. The argument is that the DNA polymerase handles dCTP in the same manner as dHMCTP in vitro. The other argument is that when we do grow phage with C in their DNA rather than the modified C, there is not any big effect on spontaneous reversion frequency.

One thing you may want to think about a little bit is that there is another possibility with respect to the way antimutators act at other base pair substitution sites besides the AT sites. The fact that it fails to decrease GC base pair sites could be related to some other component of infidelity which contriubutes primarily those base pair substitutions; polymerization may not be the major cause of mutation at GC base pair sites. However, I don't have any good explanation for what huge factors are contributing to all these GC site mutations. I think that the kinds of measurements we've made in T4 with heat do not argue that deaminiation is likely to be the cause of that process.

DR. LOEB: Let me comment on one thing; the question of other proteins affecting fidelity. There is a controversy about this in the literature. In our experiments, T4 polymerase has an error rate that is close to that obtained in cells; we therefore have very little room for other proteins contributing to fidelity. Alberts has looked at the whole complex, gets an error rate close to that in a cell and claims that T4 polymerase has a high error rate. We're using different systems, but we're looking at the same place on the ØX genome. In order for him to go from an error-prone situation to 10^{-8}, he has room for a lot of interacting proteins. We claim we have only to go from 10^{-7} to 10^{-8} and single-stranded binding protein is enough.

REPLICATION AND MUTAGENESIS OF IRRADIATED SINGLE-STRAND PHAGE DNA

O.P. Doubleday, Ph. Lecomte, A. Brandenburger,
W.P. Diver and M. Radman

Université Libre de Bruxelles, Département de Biologie
Moléculaire, Laboratoire de Biophysique et Radiobiologie
rue des Chevaux 67
B - 1640 Rhode-St-Genèse (Belgium)

Mutagens may be divided into two classes on the basis of the mechanism of their mutagenic action and the requirement for cellular repair genes for their mutagenic effect. *Direct mutagens* cause subtle modifications of the bases in DNA (or its precursors) and give rise to mutations by mispairing during DNA replication. Direct mutagens do not require cellular DNA repair activity for their mutagenic effect (i.e. act independently of the bacterial *recA* and *lexA* genes) and normally have a high degree of mutagenic specificity, reflecting the specificity of mispairing. Examples of direct mutagens are deaminating agents, such as hydroxylamine and bisulfite ; some base analogs, such as 2-aminopurine ; and alkylating agents, such as ethylmethane sulfonate. *Indirect mutagens* destroy the coding properties of DNA templates, thereby blocking DNA replication, and, unless repaired, are lethal. In situations where there is a redundancy in the DNA sequence information error-free repair processes may eliminate (e.g. by excision repair) or 'tolerate' (e.g. by postreplication recombinational exchanges) these lesions. However, when the mutagen destroys a unique piece of DNA sequence information, repair by error-free processes is no longer possible, and *recA* *lexA*-dependent error-prone repair is the only alternative to lethality. Such a situation may occur when a non-coding lesion is located within an overlapping excision gap (1), or in an overlapping gap in nascent DNA (2), or in a single-stranded genome (3). Examples of indirect mutagens are ultraviolet (UV) and ionizing radiation and many chemical mutagens and carcinogens, such as aflatoxin B_1, benzo(a)pyrene, and mitomycin C.

While there is evidence that some error-prone repair is

115

constitutive (4), the observation made by Weigle (5) that both the
survival and mutagenesis of UV-irradiated phage is increased by
irradiating the host cell (Weigle reactivation or W-reactivation)
suggested that mutagenic "SOS" repair capacity is inducible (6).
The finding that UV-irradiation of host bacteria also increases the
frequency of mutation of unirradiated phage (7,8), a phenomenon
known as untargeted mutagenesis (9), indicates that the SOS repair
system causes mutations in undamaged DNA. It appears that radiation
and other treatments that interrupt DNA replication induce the error-
prone repair system (and the other recA lexA-dependent SOS functions),
but the precise molecular inducing signal(s) have not yet been
identified (10).

We have used single-stranded DNA phages to study the error-prone
repair system of E.coli. The use of phages in such an investigation
allows us to separate the effects of irradiating the 'target' DNA
from the effects of host cell irradiation (i.e. induction of the SOS
repair system). We have exploited the fact that the rate of formation
of UV-induced pyrimidine dimers is increased in the presence of Ag^+
(11,12) and decreased in the presence of Hg^{++} (12,13) to study the
role of pyrimidine dimers in the UV-induced lethality of ØX174.
When UV-irradiated ØX174 DNA was used to transfect bacteria, phage
production was increased if the irradiation had been performed in
the presence of Hg^{++} and decreased if the irradiation had been performed in the presence of Ag^+ (figure 1). Comparable effects were
observed when these DNAs were used as templates for in vitro DNA
synthesis by E.coli DNA polymerase III, i.e. relative to control
templates irradiated in buffer, more synthesis was observed when the
template had been irradiated in the presence of Hg^{++} and less synthesis was observed when the template had been irradiated in the
presence of Ag^+ (figure 2). (Assays with pyrimidine dimer-specific
T4 endonuclease V confirmed that the presence of Hg^{++} during UV-
irradiation did indeed reduce pyrimidine dimer yields, while the
presence of Ag^+ increased them, results not shown). We find that a
single pyrimidine dimer seems to be sufficient to inactivate ØX174
DNA, as indicated by the fact that the fraction of molecules not
inactivated by a given dose of UV-irradiation correlates closely
with the fraction of molecules that did not contain any pyrimidine
dimers (calculated from known yields of pyrimidine dimers in UV-
irradiated ØX174 DNA, ref.3) (figure 3).

It is apparent that induction of the SOS repair system alleviate
some of the effects of lethal lesions in DNA (figure 3). Induction
of the SOS repair system was found to increase the extent to which
UV-irradiated ØX174 DNA is converted to the double-stranded replicative form in vivo (3). The mechanism by which UV-irradiated DNA
is replicated under conditions of error-prone repair has been the
subject of considerable speculation. Villani et al. (14) found that
the blockage of DNA synthesis upon UV-irradiated templates in vitro
is accompagnied by the generation of deoxynucleoside monophosphates.

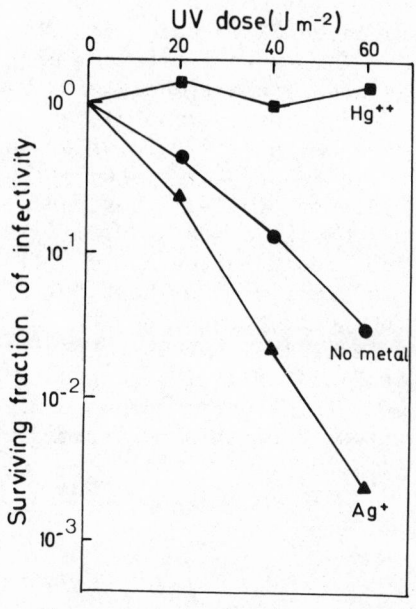

ØX174 DNA was UV-irradiated in the absence of metal or in the presence of equimolar AgNO3 or HgCl2. After removal of the metal by G-50 gel filtration the infectivity of the DNA was assayed upon bacterial spheroplasts (16).

Figure 1. *Infectivity of ØX174 DNA UV-irradiated in the presence of Ag+ and Hg++.*

[32]P-labelled Hind II fragment 2 of ØX174 RF was annealed to viral DNA and UV-irradiated in the absence of metal or in the presence of Ag+ or Hg++. After removal of the metal by gel filtration, 200 pmoles of these DNAs were replicated by 8 μl of *E.coli* DNA polymerase III fraction II (23) for 20 min at 32°C in a 30 μl reaction containing 50 mM Tris.HCl pH 7.4, 10 mM MgCl2, 10 mM DTT, 5 mM ATP, 50 μM dNTPs. The DNA was then denatured and, with size markers, electrophoresed in a 2% agarose gel which was dried and subjected to autoradiography.

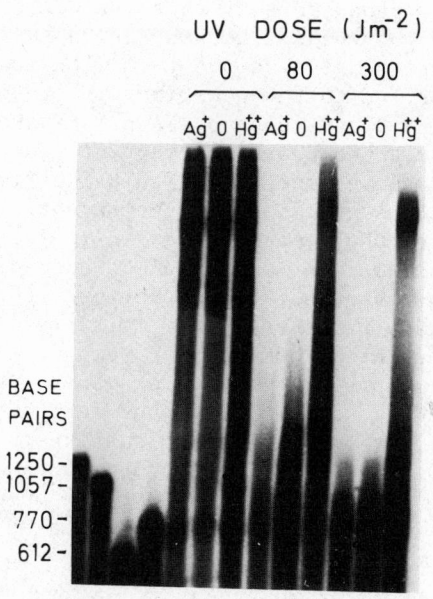

Figure 2. *Extent of DNA synthesised upon DNA templates UV-irradiated in the presence of Ag+ and Hg++.*

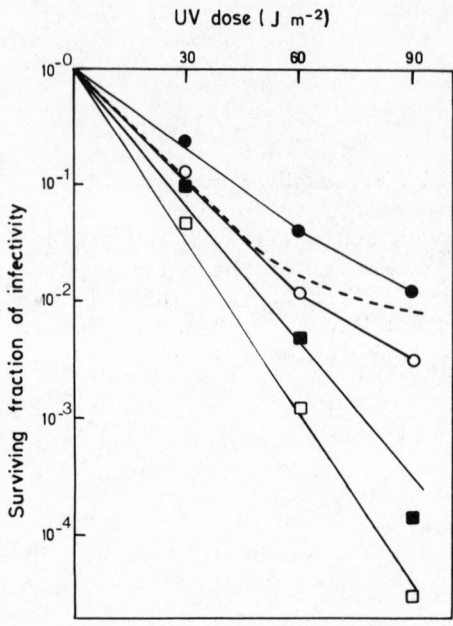

UV-irradiated ØX174 phage (□ , ■)
and DNA (O , ●) were titrated upon
unirradiated (open symbols) and UV-
irradiated (closed symbols) bacteria or spheroplasts. In order to
induce the SOS repair system a log
phase culture of HF4733 host bacteria was UV-irradiated (70 J m⁻²)
and grown for a further 20 min (15)
prior to being used as host for
ØX174 phage or being converted to
spheroplasts (16) and used to
titrate ØX174 DNA. The dotted line
indicates the fraction of molecules
calculated to contain no pyrimidine
dimer (3).

Figure 3   *Infectivity of UV-irradiated ØX174 phage and DNA in UV-
irradiated and non-irradiated hosts.*

They proposed that this was due to DNA polymerase "idling", whereby
the polymerase incorporates bases opposite pyrimidine dimers, but
subsequently excises them with its proofreading 3'→5' exonuclease.
This led to the question: is the blockage of DNA synthesis at pyrimidine dimers due to DNA polymerase proofreading activity excising the
mismatched primer nucleotide from which the polymerase might otherwise have extended, albeit inefficiently, over the non-coding lesion ?
The observation that the extent of overall synthesis upon UV-irradiated DNA templates was increased when replication was performed by
a polymerase lacking 3'→5' exonuclease activity indirectly supported
this hypothesis. Furthermore, the finding that diminished DNA polymerase proofreading activity results in increased misincorporation
upon intact homopolymer templates (17,18) is analogous to untargeted
mutagenesis. We have examined the influence of DNA polymerase proof-
reading activity upon the replication of UV-irradiated ØX174 DNA
templates by analysing the extent of elongation of radioactively-
labelled DNA primers and find that DNA polymerases such as α polymerase from *Drosophila*, that are devoid of 3'→5' exonuclease activity,
are as sensitive to inhibition as polymerases with associated 3'→5'
exonuclease activity (figure 4). These experiments (19), and those
of Moore *et al* (20), have shown that the arrest of DNA replication
by pyrimidine dimers is not due exclusively to 3'→5' exonulease
activity associated with the DNA polymerase. What are the
additional factors which determine whether a DNA polymerase may, or
may not, replicate a given lesion ? Is it possible that diminished

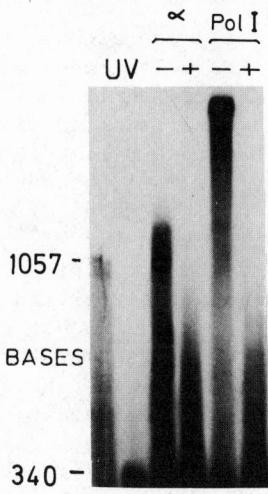

^{32}P-labelled Hind II fragment 6 of
ØX174 RF was annealed to viral DNA.
This template was UV-irradiated (500
J m^{-2}) and replicated by 0.1 units
of E.coli DNA polymerase I, large
fragment (Boehringer Mannheim) or
1.0 units of Drosophila DNA polymerase α (21) a gift of Dr.G.Villani.
The DNA was denatured and, with size
markers, electrophoresed in a 2.2%
agarose gel which was dried and
subjected to autoradiography.

Figure 4.   Replication of UV-irradiated ØX174 DNA templates by E.coli
DNA polymerase I and Drosophila DNA polymerase α.

Hind II fragment 6, annealed to intact
(i) or acid-treated (ap) (22) ØX174 DNA
was 3'-labelled by limited extension in
the presence of ^{32}P-dNTP and heated to
60°C. After G-50 gel filtration these
templates were replicated by E.coli DNA
polymerase I, large fragment in reactions containing 50 mM Tris. HCl pH 8.5,
50 μM dNTPs and either 8 mM MgCl$_2$ or
0.8 mM MnCl$_2$. Products were analysed
by electrophoresis in 0.8% alkaline
agarose gels.

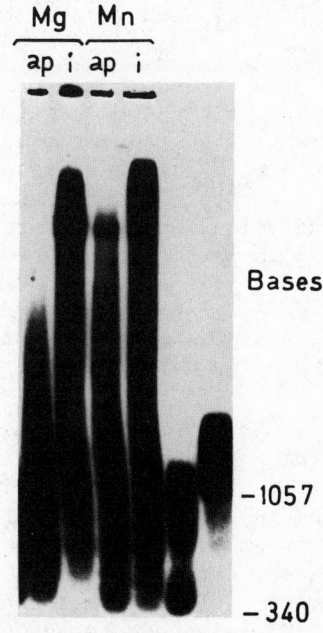

Figure 5. Replication of DNA templates containing apurinic sites in
the presence of Mg^{++} and Mn^{++}.

nucleotide discrimination by a DNA polymerase, such as has been
observed for certain mutator mutants of T4 DNA polymerase (24,25),
may allow DNA replication to proceed over non-coding lesions ? Moore
et al (20) have shown that replication of a UV-irradiated ØX174 DNA
template is inhibited one (or more) bases before a pyrimidine dimer.
However, when the mutagenic metal ion Mn^{++} (which is thought to decrease DNA polymerase fidelity by reducing nucleotide discrimination,
ref 26) was substituted in the replication reaction for Mg^{++}, synthesis was terminated upon the first base of the pyrimidine dimer.
Although we found no evidence for the replication of pyrimidine dimers
in the presence of Mn^{++}, we have observed increased extents of replication of depurinated ØX174 DNA templates in the presence of Mn^{++}
(figure 5). Since it is known that apurinic sites constitute lethal
lesions in ØX174 DNA, but that these sites may be mutagenic under conditions of SOS repair (27), these results possibly suggest a role for
decreased nucleotide discrimination in error-prone repair. Before
detailed mechanisms for UV mutagenesis are elaborated, it is pertinent
to remember that there is no evidence that pyrimidine dimers persist
in Weigle reactivated phage DNA. Thus there is no information as to
the nature of the lesion being replicated under conditions of Weigle
reactivation. While little attention has been paid to hypotheses
involving the modification or removal of non-coding lesions (as has
been shown for the miscoding 0^6-methyl guanine lesions repaired by
the adaptation response, ref 28), such hypotheses cannot be ruled out.
Perhaps the only (indirect) evidence against such models is the apparent specific requirement for DNA polymerase III in bacterial mutagenesis (29) and Weigle reactivation of phage (30).

While the mechanism of repair of pyrimidine dimers in Weigle-
reactivated phage remains obscure, we sought to determine the importance to UV mutagenesis of mutations occurring at the sites of pyrimidine dimers. We have analysed the DNA sequences of spontaneous and
radiation-induced revertants of two *amber* mutants of phage M13. Phage
M13 is particularly useful in such an investigation as its DNA has
been completely sequenced (31), and rapid DNA sequencing methods have
been developed specifically for M13 DNA (32). Furthermore, since it
is a single-stranded DNA, it is possible to identify unambiguously
the exact base or bases involved in any sequence alteration.

49 revertants of *am6H1* and 125 revertants of *am7H3* were isolated
from populations of irradiated and unirradiated phages grown in irradiated and unirradiated bacteria (Table 1), and the DNA sequence in
the region of the *amber* codon determined by the Sanger method (32,33).
The number of occurrences of each reversion sequence is shown in fig 6
Except for the TAG → GAG transversion, which was not found for either
amber, all other detectable single nucleotide changes were observed.
(TAG → TAA produces an *ochre* nonsense codon which would not be detected
as a revertant.) We are unable to assess the significance of our
failure to detect any GAG revertants, since this might be due either

to the infrequency of the $T \rightarrow G$ transversion or the inability of
glutamic acid to substitute for glutamine at these two sites.

The DNA sequence analysis revealed no obvious specificity of
mutagenesis by either UV or γ irradiation of target DNA and no
specificity of mutagenesis by the SOS repair system acting upon
either irradiated or unirradiated DNA, although induction of the
host SOS repair system increased revertant yields considerably
(Table 1). All combinations of host cell and target irradiation
induced both transitions and transversions involving each of the
three bases of the *amber* codon. UV-irradiation of target DNA did
not produce a pattern of revertant sequences suggesting that pyrimidine dimers are the primary sites of base changes, although the
potential for thymine dimer formation involving the third base of
the preceeding codon existed for both *amber* mutants. (In the case
of the *am6H1* mutation the third base of the valine codon preceeding
the *amber* codon could be replaced by any other nucleotide without
changing the sense of the codon.) Two tandem base change revertants
were observed : one from unirradiated phage and the other, involving
two purines, from γ - irradiated phage.

The small number of revertants analysed in each class makes it
difficult to assess the significance of any differences observed
between the different treatments, such as the differences observed
between irradiated and unirradiated bacteria (see references 7 and
34), or the large fraction of transversion mutants isolated from
γ - irradiated phages (see reference 35). In addition, pre-existing
mutants may make a significant contribution to spontaneous revertants
and revertants isolated after those treatments for which the reversion frequency is close to the spontaneous frequency. (Reconstruction experiments did not reveal any significant growth advantage
or disadvantage of revertants.) In an attempt to minimize the
problems of pre-existing revertants, several stocks of *amber* phages
were prepared from isolated plaques for each experiment. For those
treatments resulting in an elevated reversion frequency, the fact
that revertants were isolated from infective centres insures that
they arose independently.

UV-irradiation of phages results in an elevated mutation frequency when the phages are grown in UV-irradiated hosts (Table 1)
suggesting that UV-photoproducts in target DNA are involved in UV-
induced mutagenesis. We have presented evidence that pyrimidine
dimers are the major lethal DNA lesions induced by UV-irradiation
of single-stranded DNA phages, and that their lethal effect may be
partially alleviated by induction of the SOS repair system (figures
1 and 3). However this does not necessarily imply that pyrimidine
dimers are the major sites for UV mutagenesis. Indeed, a recent
comparison of the effects of various types of radiation upon the
survival and mutation of $\emptyset$X174 *Eam3* has shown that although

Table 1. *Survival and Mutagenesis of irradiated and unirradiated M13 grown in UV-irradiated and unirradiated bacteria*

Expt.	IRRADIATION phage	host	SURVIVAL	MUTATION FREQ. ($\times 10^{-5}$)
1. am6H1	0	0	–	11
		UV	–	77
	UV	0	1.5×10^{-4}	18
		UV	5.0×10^{-4}	190
2. am6H1	0	0	–	21
		UV	–	180
	UV	0	5.2×10^{-4}	14
		UV	6.4×10^{-4}	320
3. am7H3	0	0	–	1.3
		UV	–	2.8
	γ	0	6.6×10^{-3}	3.8
		UV	7.4×10^{-3}	14
4. am7H3	0	0	–	0.61
		UV	–	1.4
	γ	0	4.7×10^{-3}	3.8
		UV	5.3×10^{-3}	24
5. am7H3	0	0	–	0.74
		UV	–	2.1
	γ	0	2.7×10^{-3}	3.5
		UV	3.9×10^{-3}	90
6. am7H3	0	0	–	0.46
		UV	–	4.4
	γ	0	6.2×10^{-3}	1.4
		UV	1.3×10^{-2}	18
	UV	0	1.3×10^{-2}	0.73
		UV	3.1×10^{-2}	12
7. am7H3	0	0	–	0.58
		UV	–	8.2
	γ	0	3.1×10^{-3}	2.5
		UV	5.8×10^{-3}	31
	UV	0	2.0×10^{-2}	0.81
		UV	2.4×10^{-2}	17

Log phase cultures of KA787 and KA805 *SupD* grown with bubbling at 37° to an O.D.$_{560}$ = 0.3, were divided into two portions, one of which was UV-irradiated. The cultures were grown in the dark for a further 20 min, in order to maximise SOS induction (15), and were then infected with intact or irradiated M13 phage. After absorption at 20° for 3

min the infected bacteria were plated out and incubated overnight at
37°. Plaques were counted and revertants picked for DNA sequence
analysis. In experiments with *am6H1* phages, the UV dose to phages
was 90 J m^{-2} and to host bacteria was 50 J m^{-2}. In experiments with
am7H3 phages, phages were irradiated with 60 J m^{-2} or 1.5 mRad, and
host bacteria with 60 J m^{-2}. Mutation frequencies are the proportion
of revertant phages (titred on KA798) to total phages (titred on KA
805 *SupD*).

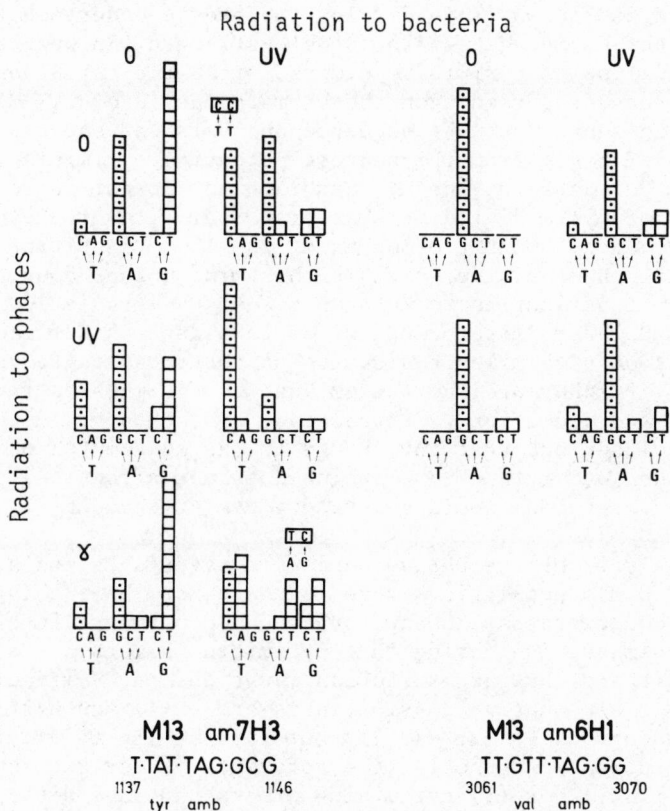

Figure 6. *DNA sequences of irradiated and unirradiated M13 am7H3 and
           am6H1 grown in UV-irradiated and unirradiated bacteria.*

Revertants from the experiments shown in table 1 were isolated and
their DNA sequences in the region of the *amber* codon determined by
the Sanger method (32,33). For *am6H1* the M13 Hha I fragment 5 (31)
was used as primer, for *am7H3* the M13 Tha I fragment 1 (31) was used
as primer. After extension by *E.coli* DNA polymerase I Klenow subfrag-
ment, the primers were removed by Hha I digestion, and the products
analysed on 12% polyacrylamide gels. Each box corresponds to an
independently isolated revertant : (⊡) transition; (□) transversion;
(⊠⊠) tandem base changes.

UV damage is the radiation damage most efficiently repaired by Weigle reactivation, it is less mutagenic (per lethal event) than other types of radiation (36). These experiments, in which virtually all mutagenesis was found to depend upon induction of the SOS repair system, afford a clear demonstration that there is no simple correlation between the lethality, reparability, and mutagenic potential of different radiation lesions in target DNA. It is difficult to assess the role of pyrimidine dimers in the UV-mutagenesis of single-stranded phages. It was already clear before we undertook this study that the UV-mutagenesis of single-stranded DNA phages is not confined to sites of pyrimidine dimers, since Bleichrodt and Verheij (7) have reported upon the UV-mutagenesis of the ØX174 Aam18 mutant, which subsequent DNA sequence analysis has shown to be separated by 2 bases from the nearest potential pyrimidine dimer site (37). Our data for the M13 am6H1 mutant do not suggest the involvement of pyrimidine dimers as principal sites for UV-mutagenesis at this codon. On the other hand, the T → C transition involving the potential thymine dimer site in the am7H3 mutant does appear to make a major contribution to mutagenesis of UV-irradiated phage at this codon. However, interpretation of this data is difficult, since this transition also makes a significant contribution to untargeted mutagenesis. Furthermore, no tandem base change mutants involving this T were induced after UV irradiation of the phage, although it should be pointed out that any change of the adjacent T of the tyrosine codon other than to C (which does not affect the sense of the tyrosine codon) would generate a *nonsense* codon.

Our analysis of the DNA sequences of revertants isolated from UV-irradiated M13 populations revealed no tandem base changes and frequently contained base changes that could not have involved pyrimidine dimers, indicating that pyrimidine dimers are not the exclusive sites of base substitution mutations in UV-irradiated M13 DNA. Two alternative possibilities are : (1) non-pyrimidine dimer UV photoproducts are significant targets for UV-induced base substitution mutagenesis (see reference 38 for a discussion of this possibility) or (2) pyrimidine dimers can have distal mutagenic effects on the same molecule. Other investigators (ref. 39 and references therein) have found that UV induces a significant proportion of tandem base change mutations. In our experiments two such mutations were detected in phages grown in UV-irradiated hosts, but could not be attributed to targeted mutagenesis at the site of pyrimidine dimers. This raises the possibility that the SOS repair system causes clusters of base changes. This idea is supported by the finding that non-tandem double base changes in the anticodon sequence of an *E.coli* tRNA gene occur at much higher frequencies after UV-irradiation than expected from the frequency of single base change mutants (40, see also 41).

We realise that the contribution of untargeted mutagenesis to UV mutagenesis may be higher for phage genomes compared to bacterial genomes, since (1) phage DNA experiences more rounds of replication in the presence of SOS mutator activity (9) and (2) post-replicative mismatch correction (42) of untargeted mutations may be less efficient for phage DNA. Our data indicate that, under the conditions employed, thymine dimers are not the principal sites of UV-induced base substitution mutagenesis of phage M13. If this is a general phenomenon, the photoreversibility of the mutagenic effects of UV on bacteria (38) and the photoreversibility of the UV-induction of the SOS repair system (5) may indicate that pyrimidine dimers are primarily inducing lesions for a process (SOS repair) necessary for targeted and leading to untargeted mutagenesis.

ACKNOWLEDGEMENTS

We are very grateful to Dr. G.N. Godson (Yale University) and Drs. B.W. Glickman and C.A. van Sluis (Leiden State University) for their kind hospitality while the DNA sequence analyses were performed. We thank them and our colleagues for useful discussions. O.P.D., Ph.L. and A.B. were recipients of Euratom grants. This work was supported by Euratom Research Contracts 224-76-1 BIO B, 156-76-1 BIO B (Brussels) and 197-76-1 BIO N (Leiden) and NIH grant A1 11633 to Dr. G.N. Godson.

REFERENCES

1. Bresler, S.E.: Theory of misrepair mutagenesis. Mutation Res., 29:467-472 (1975)

2. Sedgwick, S.G.: Misrepair of overlapping daughter strand gaps as a possible mechanism for UV induced mutagenesis in UVR strains of Escherichia coli: a general model for induced mutagenesis by misrepair (SOS repair) of closely spaced DNA lesions. Mutation Res., 41:185-200 (1976)

3. Caillet-Fauquet, P., Defais, M. and Radman, M.:Molecular mechanisms of induced mutagenesis. Replication in vivo of bacteriophage ØX174 single-stranded, ultraviolet light-irradiated DNA in intact and irradiated host cells. J. Mol. Biol., 117:95-112 (1977)

4. Bridges, B.A. and Mottershead, R.P.: Mutagenic DNA repair in Escherichia coli. VIII. Involvement of DNA polymerase III in constitutive and inducible mutagenic repair after ultraviolet and gamma irradiation. Molec. gen. Genet., 162:35-41 (1978)

5. Weigle, J.J.: Induction of mutations in a bacterial virus. Proc. Natl. Acad. Sci. USA 39:628-636 (1953)

6. Radman, M.: Phenomenology of an inducible mutagenic DNA repair pathway in *Escherichia coli* : SOS repair hypothesis. In Molecular and Environmental Aspects of Mutagenesis, Ed. by L. Prakash, F. Sherman, M.W. Miller, C.W. Lawrence and H.W. Taber, C.C. Thomas Publ. Co., Springfield, Illinois, pp.128-142 (1974)

7. Bleichrodt, J.F. and Verheij, W.S.D.: Mutagenesis by ultraviolet radiation in bacteriophage ØX174:on the mutation stimulating processes induced by ultraviolet radiation in the host bacterium. Molec. gen. Genet., 135:19-27 (1974)

8. Ichikawa-Ryo, H. and Kondo, S.: Indirect mutagenesis in phage lambda by ultraviolet preirradiation of host bacteria. J. Mol. Biol., 97:77-92 (1975)

9. Witkin, E.M. and Wermundsen, I.E.: Targeted and untargeted mutagenesis by various inducers of SOS functions in *Escherichia coli*. In DNA : Replication and Recombination, Cold Spring Harbor Symposium on Quantitative Biology, 43:881-886 (1979)

10. Gottesman, S.: Genetic control of the SOS system in *E.coli*. Cell, 23:1-2 (1981)

11. Rahn, R.O. and Landry, L.C.: Ultraviolet irradiation of nucleic acids complexed with heavy atoms - II. Phosphorescence and photodimerization of DNA complexed with Ag. Photochem. Photobiol., 18:29-38 (1973)

12. Rahn, R.O., Setlow, J.K. and Landry, L.C.: Ultraviolet irradiation of nucleic acids complexed with heavy atoms - III. Influence of Ag^+ and Hg^{2+} on the sensitivity of phage and of transforming DNA to ultraviolet radiation. Photochem. Photobiol., 18:39-41 (1973)

13. Rahn, R.O., Battista, M.D.C. and Landry, L.C.: Influence of mercuric ions on the phosphorescence and photochemistry of DNA. Proc. Natl. Acad. Sci. USA, 67:1390-1397 (1970)

14. Villani, G., Boiteux, S. and Radman, M.: Mechanism of ultraviolet-induced mutagenesis: extent and fidelity of *in vitro* DNA synthesis on irradiated templates. Proc. Natl. Acad. Sci. USA, 75:3037-3041 (1978)

15. Defais, M., Caillet-Fauquet, P., Fox, M.S. and Radman, M.: Induction kinetics of mutagenic DNA repair activity in *E.coli* following ultraviolet irradiation. Molec. gen. Genet., 148: 125-130 (1976)

16. Guthrie, G.D. and Sinsheimer, R.L.: Observations on the infection of bacterial protoplasts with the deoxyribonucleic acid of bacteriophage ØX174. Biochim. Biophys. Acta, 72:290-297 (1963)

17. Hall, Z.W. and Lehman, I.R.: An *in vitro* transversion by a mutationally altered T_4-induced DNA polymerase. J. Mol. Biol., 36:321-333 (1968)

18. Byrnes, J.J., Downey, K.M., Que, B.G., Lee, M.Y.W., Black, V.L. and So, A.G.: Selective inhibition of the 3' to 5' exonuclease activity associated with DNA polymerases: a mechanism of muta-genesis. Biochemistry, 16:3740-3746 (1977)

19. Doubleday, O.P., Michel-Maenhaut, G., Brandenburger, A., Lecomte, Ph. and Radman, M.: Inhibition or absence of DNA proofreading exonuclease is not sufficient to allow copying of pyrimidine dimers. In Chromosome Damage and Repair, Ed. by E. Seeberg and K. Kleppe, Plenum Pub. Corp., New York (in press)

20. Moore, P.D., Bose, K.K., Rabkin, S.D. and Strauss, B.S.: Sites of termination of *in vitro* DNA synthesis on ultraviolet- and N-acetylaminofluorene-treated ØX174 templates by prokaryotic and eukaryotic DNA polymerases. Proc. Natl. Acad. Sci. USA, 78:110-114 (1981)

21. Banks, G.R., Boezi, J.A. and Lehman, I.R.: A high molecular weight DNA polymerase from *Drosophila melanogaster* embryos. Purifica-tion, structure, and partial characterization. J. Biol. Chem., 254:9886-9892 (1979)

22. Lindahl, T. and Nyberg, B.: Rate of depurination of native deoxyribonucleic acid. Biochemistry, 11:3610-3618 (1972)

23. Conrad, S.E. and Campbell, J.L.: Characterization of an improved *in vitro* DNA replication system for *Escherichia coli* plasmids. Nucleic Acids Res., 6:3289-3303 (1979)

24. Hershfield, M.S.: On the role of deoxyribonucleic acid polyme-rase in determining mutation rates. Characterization of the defect in the T4 deoxyribonucleic acid polymerase caused by the TS L88 mutation. J. Biol. Chem., 248:1417-1423 (1973)

25. Reha-Krantz, L.J. and Bessman, M.J.: Studies on the biochemical basis of mutation. VI. Selection and characterization of a new bacteriophage T4 mutator DNA polymerase. J. Mol. Biol., 145: 677-695 (1981)

26. Dube, D.K. and Loeb, L.A.: Manganese as a mutagenic agent during *in vitro* DNA synthesis. Biochem. Biophys. Res. Comm., 67:1041-1046 (1975)

27. Schaaper, R.M. and Loeb, L.A.: Depurination causes mutations in
    SOS-induced cells. Proc. Natl. Acad. Sci. USA, 78:1773-1777
    (1981)

28. Olsson, M. and Lindahl, T.: Repair of alkylated DNA in *Escherichia
    coli*: methyl group transfer from O^6-methylguanine to a protein
    cysteine residue. J. Biol. Chem., 255:10569-10571 (1980)

29. Bridges, B.A., Mottershead, R.P. and Sedgwick, S.G.: Mutagenic
    DNA repair in *Escherichia coli*. III. Requirement for a function
    of DNA polymerase III in ultraviolet-light mutagenesis. Molec.
    gen. Genet., 144:53-58 (1976)

30. Fields, P.I. and Yasbin, R.E.: Involvement of deoxyribonucleic
    acid polymerase III in W-reactivation in *Bacillus subtilis*.
    J. Bacteriology, 144:473-475 (1980)

31. Van Wezenbeek, P.M.G.F., Hulsebos, T.J.M. and Schoenmakers, J.G.G.:
    Nucleotide sequence of the filamentous bacteriophage M13 DNA
    genome: comparison with phage fd. Gene, 11:129-148 (1980)

32. Sanger, F., Coulson, A.R., Barrell, B.G., Smith, A.J.H. and Roe,
    B.A.: Cloning in single-stranded bacteriophage as an aid to
    rapid DNA sequencing. J. Mol. Biol., 143:161-178 (1980)

33. Sanger, F., Nicklen, S. and Coulson, A.R.: DNA sequencing with
    chain-terminating inhibitors. Proc. Natl. Acad. Sci.USA, 74:
    5463-5467 (1977)

34. Bleichrodt, J.F. and Roos-Verheij, W.S.D.: Influence of SOS repair
    on the specificity of radiation mutagenesis in bacteriophage
    ØX174. Molec. gen. Genet., 176:155-160 (1979)

35. Glickman, B.W., Rietveld, K. and Aaron, C.S.: γ-ray induced
    mutational spectrum in the *lacI* gene of *Escherichia coli*.
    Comparison of induced and spontaneous spectra at the molecular
    level. Mutation Res., 69:1-12 (1980)

36. Yatagai, F., Kitayama, S. and Matsuyama, A.: Weigle reactivation
    and Weigle mutagenesis in phage ØX174 by various types of
    radiation. Mutation Res., 91:3-7 (1981)

37. Smith, M., Brown, N.L., Air, G.M., Barrell, B.G., Coulson, A.R.,
    Hutchison III, C.A. and Sanger, F.: DNA sequence at the C
    termini of the overlapping genes A and B in bacteriophage ØX174.
    Nature, 265:702-705 (1977)

38. Witkin, E.M.: Ultraviolet mutagenesis and inducible DNA repair in *Escherichia coli*. Bacteriological Rev., 40:869-907 (1976)

39. Coulondre, C. and Miller, J.H.: Genetic studies of the *lac* repressor. IV. Mutagenic specificity in the *lacI* gene of *Escherichia coli*. J. Mol. Biol., 117:577-606 (1977)

40. Coleman, R.D., Dunst, R.W. and Hill, C.W.: A double base change in alternate base pairs induced by ultraviolet irradiation in a glycine transfer RNA gene. Molec. gen. Genet., 177:213-222 (1980)

41. Kubitschek, H.E.: Double mutations induced in *Escherichia coli* by ultraviolet light. J. Bacteriology, 142:724-725 (1980)

42. Radman, M., Villani, G., Boiteux, S., Kinsella, A.R., Glickman, B.W. and Spadari, S.: Replicational fidelity: mechanisms of mutation avoidance and mutation fixation. In DNA: Replication and Recombination, Cold Spring Harbor Symposium on Quantitative Biology, 43:937-946 (1979)

DISCUSSION

DR. SHERMAN: How big a region did you sequence? In other words, how far away could you have detected up a double base pair change?

DR. DOUBLEDAY: Well, we were using rather high concentration gels to get that separation. I guess about 80 base pairs. We picked up two double base-pair changes in SOS-induced bacteria, and we also picked up that unselected one which must have occurred during the stock preparation because it was found in combination with two different revertant sequences.

DR. SARASIN: Did you notice any mutations due to the PydC lesions which have been suggested by W. Haseltine?

DR. DOUBLEDAY: I can't say: there's not cytosine in, or adjacent to, the TAG sequence that we're reverting, which is a shame because one would be interested in having a look at mutations at cytosine. We also have some data that the hydrates of cytosine and uracil are, at least in vitro, mutagenic. So we would have liked to have been able to see those, but we couldn't.

DR. SARASIN: Still referring to W. Haseltine' experiments, specifically his calculation of the efficiency of thymine dimer formation as a function of the two flanking nucleotides and the number of potential dimer sites, could you determine the probability of getting pyrimidine dimers at different sites in your system?

DR. DOUBLEDAY: It wouldn't be able to account for all the mutations, I'm sure. We were looking to see whether we could get the Jeffrey Miller-type results (reference 39) of tandem double base change mutations which one would have predicted one would only see under conditions of target irradiation and induction of the SOS-repair system. And even if it had been at a relatively low frequency, we would have hoped to see them, but we didn't. But we got tandem base changes in other treatments.

DR. GLICKMAN: I'd like to comment on the tandem base changes. As you just pointed out, they're predicted on the basis of pyrimidine dimers being sites for mutagenesis. In the original data, Miller quotes a fairly high percent of tandem double base changes, but in a later paper, when the sites were actually assigned to the given sequences, a lot of those presumed doubles turned out to be singles. It turns out that only something like 12 out of the 650 mutants examined were actually doubles, and of those twelve, one occurred at a non-pyrimidine dimer site and another wasn't tandem, giving an overall frequency for tandem doubles of 1.7%. I agree with your conclusion that it's the induction of SOS which is most likely to be responsible.

DR. SINGER: Did I understand you to say in the beginning that you did not know whether the pyrimidine dimer was actually being copied?

DR. DOUBLEDAY: That's right. In vivo there's no evidence for that. They have certainly not been copied, at least as yet, in vitro. I've not been able to do that; I don't think Peter Moore has.

DR. SINGER: Have you quantitated the dimers?

DR. DOUBLEDAY: Yes. We use T4 endonuclease to look at the DNA that we'd irradiated in the presence of mercury or silver. To that extent we had indeed attempted to quantify the dimer units. But more recently we've been trying simply to see whether there were primidine dimers in the Weigle reactivated RF DNA, (this was of course done in bacteria deficient in excision repair, under the yellow light and so on). We have not been able to detect them.

DR. RIPLEY: How strong is the evidence that the silver and mercury reactions really affect only pyrimidine dimers and not other things?

DR. DOUBLEDAY: They don't affect the formation of hydrates of cytosine and uracil. They're meant to inhibit triplet-excitation. Data from Setlow (reference 12) indicated that they reduce the lethality of UV irradiation by about 30%, though they still found a significant amount of residual killing, which we don't get at all. But then we're looking at single-stranded DNA and they were looking at the viability of double-stranded transforming DNA.

DR. RIPLEY: That certainly may speak to the question of inactivating events, but not necessarily to the role of those lesions in mutagenesis.

DR. DOUBLEDAY: I agree absolutely. Those data were only there to establish what is the inactivating lesion (pyrimidine dimer), and then I was trying to show a modest increase in survival for that inactivating lesion under conditions of Weigle-reactivation.

DR. RIPLEY: What effect on survival does UV-irradiation to the cells have?

DR. DOUBLEDAY: The Weigle reactivation effects are weak with M13. You tend to get a decrease in the plating efficiency, which makes life confusing. It normally works out at a factor of about 2, but not much more. Of course, the majority of the survival effects of SOS-induction in other systems would appear to be largely due to induction of uvrA and B and what not, that is to excision repair, which can't act upon single-stranded DNA like M13.

DR. LOEB: I have three sets of questions. First, I guess you're telling me that the previous reports on the ability to replicate past UV dimers have not been substantiated; is that correct? The other question is, when you did those primer experiments, did you use isolated restriction primers?

DR. DOUBLEDAY: Yes to both questions.

DR. LOEB: So you've purified a single isolated restriction fragment and measured growth. Then your ability to detect replication past apurinic sites--using manganese--would be about 10% because you're looking at the amount that goes past, on a gel. Thus, our results with manganese are similar. We find that Mn^{2+} enhances the ability to replicate past an apurinic site, using a variety of DNA polymerase.

DR. GLICKMAN: I would like to comment on why I thought the gamma experiments would be so important. In experiments that we published about two years ago we did a study of the spectrum of mutations in the lacI system, using 30 kilorads of gamma radiation. One of the questions that we later have tried to ask was, "Is there a difference in spectra at different doses?" This could be very important in thinking about chemicals and radiation in terms of thresholds and repair. We therefore did an experiment at approximately the doubling dose for mutagenesis. It appeared from this experiment that almost all of the mutants that we isolated were transitions, whereas in the 30 kilorad dose experiment we had approximately equal proportions of transitions and transversions. The state of the cell, its repair situation, will therefore affect mutational specificity. The idea for the M13 mutagenesis experiment was to take gamma irradiated phage and plate them on irradiated and non-irradiated hosts. I think your result indicates there is a substantial difference, and support the idea that the SOS-induced host cell gives a different spectrum with the gamma irradiated phage.

DR. THILLY: I'm confused by not knowing the actual frequency of thymine dimers. You spoke of it as being very low, with what I would consider very large fluences. Would it have approached the frequency that you observed for mutations and perhaps exceeded that frequency 100-fold. Were you working at, say, 5 dimers per phage?

DR. DOUBLEDAY: A little less. They're high fluences but on the other hand it's a very small target.

DR. THILLY: With a frequency of mutation at a site of 10^{-4}, and an expectation of dimers at that site of 10^{-4} or some number

close to that, shouldn't we wonder about the relationship between
the two. This is something that our discussion so far hasn't
touched.

DR. DOUBLEDAY: We were just looking to see whether mutations were
occurring at potential sites for dimers. We didn't get a
significant contribution to mutagenesis at the potential sites for
pyrimidine dimers, except possibly for the amber-7 mutation.
There we also had the increase for untargeted mutagenesis, so it's
difficult to assess.

DR. GLICKMAN: You're arguing that he is not seeing as many
mutants as you should see because, if he's got let's say 5 dimers
per phage, he has one in a thousand with a dimer at that site.
And his frequency of mutation is not 10^{-3} but 10^{-4}, or less.
But that's making a large number of assumptions.

DR. DOUBLEDAY: Not every dimer produces a mutation. Most of them
are lethal. Moreover, it is possible that only the least damaged
phages are reactivated.

DR. BRIDGES: But you haven't excluded the possibility that there
may be 80 or 90% of the dimers which are not giving rise to
mutations, yet yielding viable phage. That would have serious
implications for dimer replication models.

DR. GLICKMAN: There's another possibility, one which your data
doesn't remove. Most mutagenesis with UV is due to transition, so
you can make a model in which the SOS system puts in A's; every
time you hit a dimer you put in an A. Since this is the correct
base opposite a T, you don't see mutants.

DR. S. PRAKASH: What is the proportion, from your data and
others, of targeted versus untargeted mutations by UV? Can you
quantify it in some way?

DR. DOUBLEDAY: We had an increase of up to 10-fold by simply
untargeted mutations. And the effect of target irradiation would
perhaps double that, or a little bit more. But as I said
previously, I wouldn't want people to extrapolate too much to
cellular systems because there are differences, specifically in
the number of replication runs. And also, maybe even more
important, in the correction of untargeted mutations by the
post-replicative mismatch correction system.

DR. S. PRAKASH: Might the SOS repair system have evolved simply
to produce untargeted mutations?

DR. DOUBLEDAY: Miroslav Radman would love you for saying that.
Think teleologically.

DR. S. PRAKASH: And that seems rather bizarre. Things don't evolve that way.

DR. DOUBLEDAY: I think most of those untargeted mutations would get taken out by mismatch correction. There is evidence that the mismatch correction system does act upon untargeted mutations. And there's also evidence that phages are likely to have less efficient mismatch correction. For instance, the methylation of DNA, which instructs the process, is poor for phage lambda and its replicating all the time.

DR. LAWRENCE: You're probably underestimating the frequency of untargeted mutations, in any case, even if you want to call mutations in purine sequences untargeted; I suspect, for the reasons that you mentioned, that lesions other than dimers may also be significant. But even if you want to make that assumption, you are probably underestimating them because you have to correct for the time at which the phage replicate in relation to the time the cell is maximally induced. If you put in unirradiated phage, they replicate straight away, and a lot of that replication occurs before the cell is properly induced. But if you irradiate the phage, it replicates later.

DR. DOUBLEDAY: I gave the bacteria 20-30 minutes postirradiation incubation in an attempt to maximize SOS induction. But that's right; those are the sorts of problems you get into.

MUTATIONAL SPECIFICITY OF UV LIGHT IN E. COLI: INFLUENCE OF

EXCISION REPAIR AND THE MUTATOR PLASMID PKM101

Barry Glickman

National Institute of Environmental Health Sciences
Laboratory of Molecular Genetics
P. O. Box 12233
Research Triangle Park, North Carolina   27709

INTRODUCTION

The *lacI* system of *E. coli* provides a method for determining
UV-induced mutational specificity at a large number of sites (1,2,
3). In contrast, earlier studies in other systems have generally
relied upon the analysis of reversion at a rather limited number
of sites (4,5,6). Often, the mutants analyzed in reversion studies
were originally induced by the mutagenic treatment (7) and the
possibility therefore exists that preferentially mutable sites or
hotspots were selected and that these may have behaved atypically.
Alternatively, the original mutation may have removed a DNA
sequence target and these sites may be in fact partially immutable!
Moreover, in studies of the reversion of nonsense mutations, the
majority of "revertants" actually occur not in the structural gene
but at suppressor loci which behave unusually in their response to
UV light (4,8,9). The *lacI* system allows the examination of
forward mutagenesis at 65[†] individual sites where nonsense mutations
can arise by a single base substitution. Since both the DNA

[†]There are 72 characterized base pair substitutions which give
rise to an amber or ochre mutation by a single base change but
only 65 of these are independent sites. Seven sites correspond to
the tyrosine codons TAT and TAC which can be converted to an amber
or an ochre by an alteration at the same nucleotide.

sequence and the location of the nonsense mutations have been established (10), each mutation can be attributed to a specific transition or transversion event.

Detailed knowledge of the sites at which mutagenesis occurs and the specific base changes produced, may yield important clues about the premutagenic lesions and how they are processed. In the case of UV irradiation the mutational mechanism is poorly understood. Photoreactivation experiments suggest that mutagenesis in both excision-proficient and excision-deficient strains is largely or exclusively due to pyrimidine dimers (11,12). Nevertheless, considerable evidence exists that not all UV-induced mutagenesis is targeted (13). For example, undamaged bacteriophage λ plated on UV-irradiated hosts shows high levels of a $recA^+$-dependent mutagenesis ("in-direct mutagenesis") (14). According to a current model, UV-induced mutagenesis in $E.$ $coli$ proceeds through a $recA^+$-dependent, inducible, error-prone repair pathway called "SOS" repair [reviewed by Witkin (13)]. It is postulated that where alternative repair pathways are unable to function, e.g., at gaps resulting from the blockage of DNA replication past pyrimidine dimers, "SOS" repair allows the "bypass" of the "non-templating" damage, but does so at the cost of replicational fidelity (15,16). This model suggests that pyrimidine dimers are the pre-mutagenic lesions but makes no clear prediction as to whether most mutation would be "untargeted" (perhaps the result of reduced replicational fidelity) and occur at any site.

The situation is, however, more complex. Evidence has accumulated suggesting that the mutational pathways for UV-induced mutagenesis differ in excision-proficient and excision-deficient strains (11,17,18). In the case of repair-proficient strains it is hypothesized that mutations arise at sites within structural genes during excision repair gap filling. This process is $lexA^+$-$recA^+$-dependent, but protein synthesis is not required for mutation fixation and the process may be considered constitutive (18). However, mutagenesis in excision-deficient cells, and at sites such as in suppressor t-RNA loci refractory to excision repair in excision-proficient strains, involves error-prone post-replication repair which also requires the $lexA^+$-$recA^+$ genotype but in this case appears to be an inducible process requiring protein synthesis. Moreover, due to the low level of UV-induced mutagenesis in Uvr^+ bacteria, it has been postulated that excision-proficient bacteria have an error-free post-replication excision-dependent repair process which is absent in excision-deficient cells (19). Another difference between the mutational response of excision-proficient and excision-deficient strains is the inability of excision-deficient strains to demonstrate a phenomenon known as mutation frequency decline or MFD (20-23). MFD is the rapid reduction of UV-induced mutagenesis under conditions unfavorable to protein synthesis. This process is specific for suppressor t-RNA loci (9, 20) and differences in the level of MFD at specific sites in the t-RNA loci have been interpreted to mean that the local configuration of the DNA affects their susceptibility to excision repair (24).

An investigation of the mutational specificity of UV irradiation and the influence of DNA repair processes upon this specificity would be informative in identifying the critical premutational lesions (if mutagenesis is targeted) and improving our understanding of the mutational process. In this study we have used the *lacI* system of *E. coli* to analyze the mutational specificity of UV light in excision repair proficient and excision repair deficient strains of *E. coli*.

We have also investigated the effect of the mutator plasmid pKM101 on mutational specificity. The mechanism by which these plasmids enhance mutagenesis and provide UV protection is unknown. However, because they too depend upon functional $recA^+$ and $lexA^+$ genes it has been suggested that pKM101 provides the host with enhanced error-prone repair capacity (25-30). The synergistic interaction between a *tif-sfiA* double mutant and pKM101 (27,31) further suggests that pKM101 may interact with the cellular "SOS" system. More recently, the demonstration that pKM101 can reverse the non-mutability of a *umuC* strain (32) lends support to this hypothesis. Our basic premise in undertaking these experiments is that, if pKM101 exerts its action through simple enhancement of the "SOS" repair pathway, then the mutational specificity in the presence of pKM101 will be similar to that in its absence. On the other hand, a significantly altered specificity might indicate that a different mutational mechanism was responsible. Moreover, since pKM101 makes the detection of some mutagens possible, information on how it affects specificity would have practical implications.

In these studies, the *lacI* system has been supplemented by including the frameshift *trpE9777* to monitor frameshift mutagenesis.

MATERIALS AND METHODS

Strains, Media and the *LacI* System

Unless otherwise stated, the materials and techniques for the *lacI* forward mutation system were the same as those used by Miller *et al*. (1) and Coulondre and Miller (2,3). The wild-type strain NR3835 (F' *pro-lac/ara*, Δ*(prolac)*, *thi*, *trpE9777*) has been previously described (33). Strain NR3951 [same as NR3835 except *(bioFCD, uvrB, chlA)*] was constructed by P1*vir*-mediated transduction using C261 [*str*, *his*, Δ*(bioFCD, uvrB, chlA)*] as donor (34); chlorate-resistant transductants of NR3835 were selected and

screened for UV sensitivity. Biotin was supplemented at 0.5 μg/ml.
Strains NR3835/pKM101 and NR3951/pkM101 are derivatives of NR3835
and NR3951 into which pKM101 has been introduced as described
previously (35).

UV-Irradiation and Mutant Selection

UV irradiation was carried out as described previously (36)
and mutational spectra were prepared from 10-15 independent cultures
grown to 4-6 x 10^8 cells/ml. LacI mutants were selected by plating
10 μl of the irradiated cell suspension in 3 ml soft agar on plates
containing phenyl-β-D-galactoside. Sufficient growth occurred on
the selective plates to ensure full expression of all mutants since
the addition of various limiting amounts of glucose or glycerol as
alternative carbon sources did not increase the observed mutation
frequency. The spontaneous mutational spectra obtained by this
method did not differ from that found previously (2; B. W. Glickman,
unpublished). At the UV fluences used to produce the spectra (100
Jm^{-2} for the Uvr^+ and 5 Jm^{-2} for the ΔUvrB strains), the contribution of spontaneously arising mutants is less than 1%. Since
the mutants are selected directly following irradiation by plating
in top agar, the induced mutants are of independent origin. Each
independent culture used to produce the UV spectra was screened
for spontaneous jackpots which would have interfered with the
induced spectra.

In order to arrive at the UV-induced mutation frequencies, a
correction is necessary for the number of spontaneous mutants
arising on each plate. Since the number of spontaneous mutants
arising per plate was proportional to the number of cells plated
within the range of cell densities employed, the number of induced
mutants per plate could be calculated by correcting directly for
the survival of the UV-irradiated cell suspension.

Experimental Modifications for pKM101-Bearing Strains

Slight modifications were required to obtain the UV spectra
of the pKM101-bearing strains. Since such strains grow more
slowly on minimal medium (28,37), mutant selection plates were
incubated 3 rather than 2 days. In the final analysis of the
mutants the presence of pKM101 was found to interfere with deletion
mapping. This required elimination of the plasmid, which was
accomplished by transferring the F' pro-lacI-factor X7026r by
liquid crosses rather than by replica mating. Transfer was
accomplished in 30 minutes at 37° without agitation in a mating
mixture containing 4 x 10^7 donors/ml and 4 x 10^8 acceptors/ml as
described by Miller (38). The resultant sexductants were screened
for pKM101 by testing for ampicillin resistance. Most F'-sexduc-
tants (>95%) had not received pKM101, and were therefore satisfac-
tory for the mapping experiments.

RESULTS

UV Responses of NR3835 and NR3951 and the Effect of pKM101

Cell survival and mutagenesis following UV irradiation of NR3835 (Uvr[+]) and NR3951 (ΔUvrB) and their pKM101-bearing derivatives are given in Figure 1. As expected, pKM101 provides UV protection to both the Uvr[+] and ΔUvrB strains and greatly enhances *lacI* mutagenesis. Reversion of the *trpE9777* frameshift mutant is greatly enhanced by pKM101 in the Uvr[+] strain but only slightly in the ΔUvrB strain.

The *lacI* mutational responses of these strains are summarized in Table I. At higher UV fluences the mutation frequency of pKM101-bearing strains is up to 10-fold higher than in the absence of the plasmid. At low doses, however, the plasmid effect is much larger. For example, in the Uvr[+] strain the plasmid effect at 5 Jm^{-2} is 400-fold (data not shown), and at 10 Jm^{-2} is greater than 500-fold.

Mutational Spectra of Uvr[+] and ΔUvrB Strains

The mutational spectra for the Uvr[+] strain (Figures 2 and 3) were obtained following a UV fluence of 100 Jm^{-2}. This dose resulted in cell survival similar to that obtained by Coulondre and Miller (3), while the mutation frequency was slightly lower than they reported. This may reflect the use of exponentially growing rather than stationary cells. Nonetheless, the UV spectra obtained by direct plating of the Uvr[+] strain are very similar to those of Coulondre and Miller (2,3). Mutational hotspots are observed at sites A23, O24 and O27, and in similar ratios as reported by Coulondre and Miller (2,3). The 5 Jm^{-2} dose used to produce the UV spectrum for the ΔUvrB strain was chosen to generate a frequency of induced amber and ochre mutants similar to that produced by 100 Jm^{-2} in the Uvr[+] strain. The spectra for the ΔUvrB strain are given alongside those of the Uvr[+] strain in Figures 2 and 3.

A comparison of the spectra obtained with the Uvr[+] and ΔUvrB strains reveals a new hotspot at site A24 in the ΔUvrB strain. This site accounted for almost 19% of all mutants arising among the 36 potential amber sites. The frequency with which A24 mutants were induced in the ΔUvrB strain was 149×10^{-8} compared with 2.9×10^{-8} in the wild type. A24 mutants arise as a consequence of a $G \cdot C \rightarrow A \cdot T$ transition at the site of a potential cytosine-cytosine dimer.

Table II describes the mutational specificities in terms of the base pair substitution required to generate the mutants, while Table III summarizes the effects of the potential target on mutation frequencies. Table IV reports the results in terms of whether a

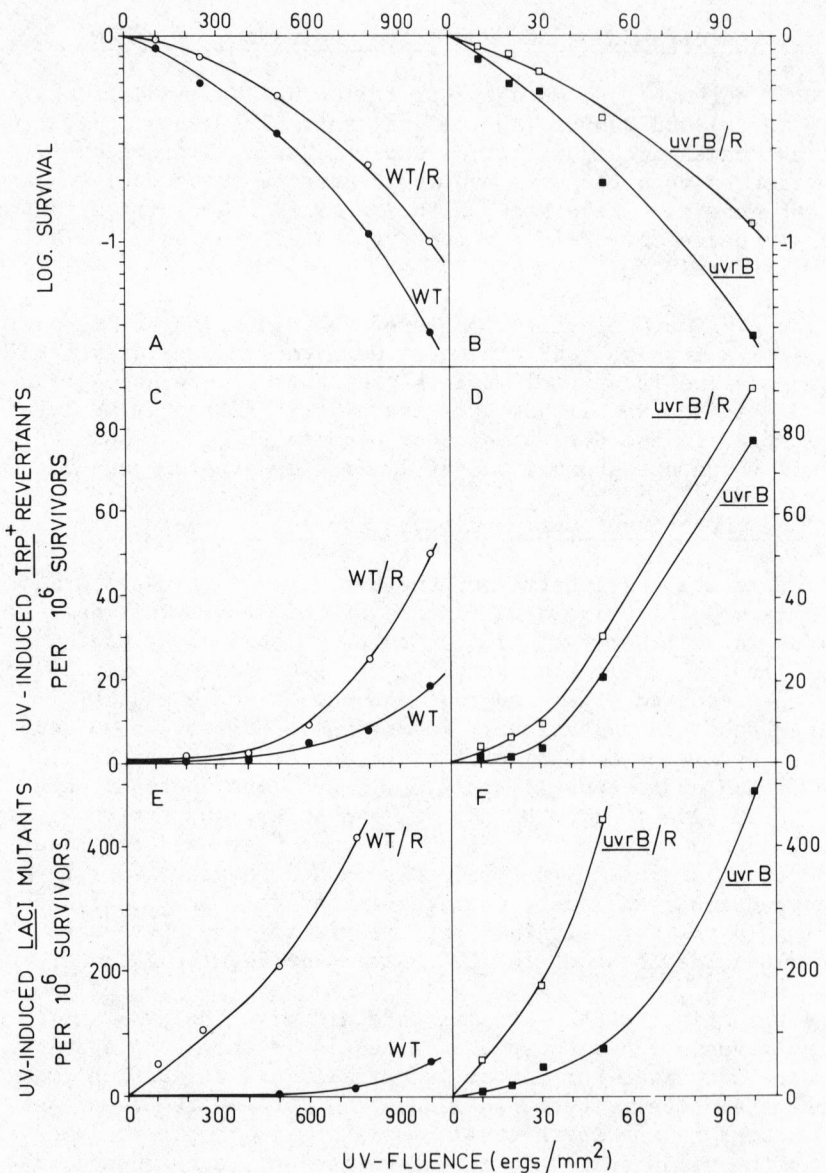

Figure 1. Effect of UV fluence on survival (Panels A and B), *trpE9777* reversion (Panels C and D) and *lacI⁻* mutagenesis (Panels E and F) in the wild-type and ΔUvrB strains with and without pKM101. Symbols: ●, NR 3835; ○, NR3835/pKM101; ■, NR 3951; □, NR3851/pkm101.

TABLE 1. UV-induced mutation frequencies for Uvr$^+$ and ΔUvB strains with and without the mutator plasmid pKM101.

UV FLUENCE J/M2	UV-INDUCED LacI FREQUENCY	PER CENT AMBER MUTANTS	UV-INDUCED LacI AMBER FREQUENCY	PER CENT OCHRE MUTANTS	UV-INDUCED LacI OCHRE FREQUENCY
NR3835 (Uvr$^+$)					
0	0.0	0.96	0.000	0.61	0.000
10	0.3	1.7	0.005	1.2	0.004
25	0.6	2.3	0.014	2.7	0.018
50	2.5	8.5	0.20	4.5	0.12
75	22	11.4	2.5	5.6	1.20
100	56	8.3	4.6	6.4	3.5
NR3835/pKM101					
0	0.0	1.9	0.00	1.8	0.00
5	25	8.2	2.1	4.3	1.1
10	60	12.6	7.6	7.6	4.6
25	100	11.2	12	9.4	10
50	210	14.6	31	10.7	22
75	430	13.8	59	8.0	34
NR3951 (ΔUvrB)					
0	0.0	1.4	0.00	0.8	0.00
0.5	3.4	2.1	0.07	1.0	0.04
1.0	6.7	2.3	0.16	1.3	0.09
2.5	53	4.4	2.3	5.0	2.7
5.0	75	10.4	8.0	6.7	5.0
10.0	490	7.3	36	4.2	21
NR3951/pKM101					
0	0.0	3.0	0.00	2.0	0.00
1.0	65	9.5	6.2	12.5	8.1
2.5	190	13.1	24	14.1	26
5.0	450	11.4	51	11.7	53
10.0	950	10.7	102	12.8	120

Mutation frequencies are per 10^6 survivors. The results are averages of 3-4 independent experiments.

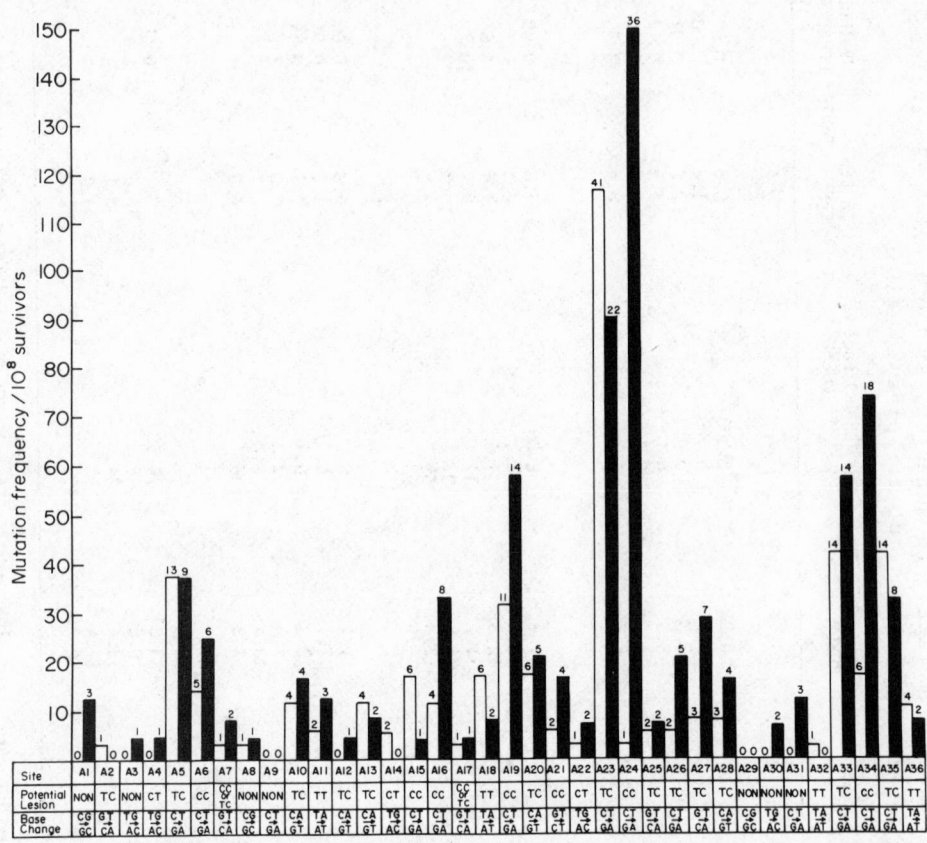

Figure 2. Frequency distribution of *lacI⁻* amber mutants. The UV fluences were 100 and 5 Jm⁻² for the wild-type (NR3835) and ΔUvrB strain (NR3951) respectively. The open bars reflect the wild-type data and the solid bars the ΔUvrB data. 160 and 193 amber mutants were analyzed in the wild-type and ΔUvrB strains respectively.

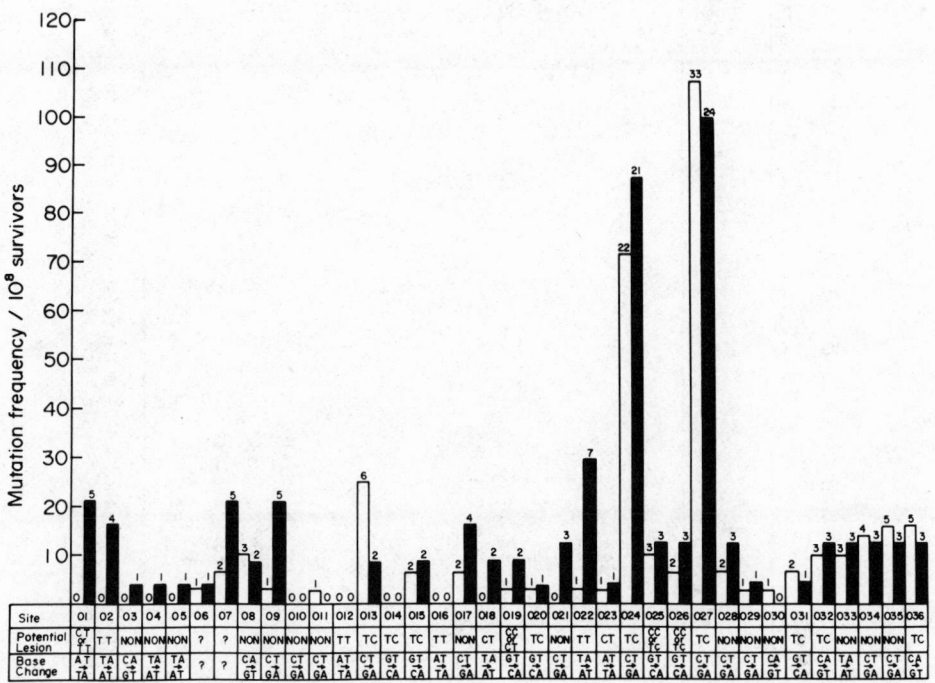

Figure 3. Frequency distribution of *lacI*⁻ ochre mutants in the Uvr⁺ and ΔUvrB strains. Symbols and UV fluences as described in Figure 2. 108 and 120 ochre mutants were analyzed in the wild-type and ΔUvrB strains respectively.

TABLE II Base substitutions detected following UV-irradiation.

BASE SUBSTITUTION	AMBER MUTANTS			OCHRE MUTANTS		
	SITES AVAILABLE	SITES FOUND	FREQUENCY[a]	SITES AVAILABLE	SITES FOUND	FREQUENCY[a]
Uvr+ (NR3835)						
GC → AT	14	12	340 (28.3)	12	11	250 (21.0)
hotspots removed	13	11	220 (17.0)	10	9	72 (7.2)
GC → TA	9	8	69 (7.7)	12	10	75 (6.3)
AT → TA	5	5	37 (7.5)	10	3b	16 (1.64)
AT → CG	5	2b	8.6 (1.7)	--	--	--
GC → CG	3	1b	2.8 (0.9)	--	--	--
ΔUvrB (NR3951)						
GC → AT	14	13	610 (43.7)	12	10	280 (23.3)
hotspots removed	12	11	370 (31.1)	10	8	90 (9.2)
GC → TA	9	8	110 (12.4)	12	10	90 (7.3)
AT → TA	5	4	33 (3.6)	10	7	97 (9.7)
AT → CG	5	4	25 (5.0)	--	--	--
GC → CG	3	2b	17 (5.5)	--	--	--

Dose to Uvr+ was 100 Jm^{-2} and ΔUvrB 5 Jm^{-2}. [a] Frequency x 10^8 mutants/survivor; number in parenthesis is site average (frequency divided by number of sites). [b] Sample size is 5 isolates or fewer.

transition or transversion event occured at a potential dimer or non-dimer site. Interpretation of the data requires understanding an important characteristic of the *lacI* system: A·T → G·C transitions cannot be detected, so that all transitions involve G·C → A·T. It must also be appreciated that the assignment of base substitutions in Table II does not take into account the nature of the potential DNA target, while the data in Table III do not discriminate among substitutions and Table IV defines events only as transitions or transversions at potential dimer and non-dimer sites. Awareness of these differences in the display of the data will facilitate interpretation.

In both the Uvr[+] and ΔUvrB strains, G·C → A·T transitions predominate (Table II). The predominance of transitions is best seen in Table IV from which it can also be concluded that transitions occur at potential dimer sites for more frequently than at non-dimer sites. When the hotspots are eliminated from consideration, transition events still predominate among the amber but not the ochre mutants (Table II). The reason for this difference is that of the 10 transition sites available among the ochre collection only one, 013, provides a dimer site.

From Table III it can be seen that thymine-thymine dimers do not constitute target sites in either the Uvr[+] or ΔUvrB strain. However, transitions cannot be measured at these potential dimer sites. The thymine-cytosine targets include the hotspots for G·C → A·T transitions at sites A23, 024 and 027. The average frequency of mutagenesis at these three hotspots is 99×10^{-8} in the Uvr[+] strain and 93×10^{-8} in the ΔUvrB strain compared with site averages of 9.6×10^{-8} and 13×10^{-8} at all potential thymine-cytosine dimer sites.

Mutations at potential cytosine-cytosine dimer sites account for the majority of the higher mutation frequencies seen in the ΔUvrB strain at the doses used. Mutation frequencies at these sites in the ΔUvrB strain, even excluding the A24 hotspot seen only in that strain, are 2- to 4-fold higher than in the Uvr[+] strain. This is true for both transitions and transversions. The site average for transitions is 16×10^{-8} and 57×10^{-8} at these sites for the Uvr[+] and ΔUvrB strains, respectively. Removing the A24 site from consideration reduces the average mutation frequency to 39×10^{-8}.

The excision repair deficient strain shows a consistently 2-fold greater mutation frequency at non-dimer sites than does the Uvr[+] strain. (Table III and IV). The numbers of isolates involved are large enough (20 in the Uvr[+] strain and 37 in the ΔUvrB strain) not to be influenced by spontaneous mutation. When these data are partioned into transitions and transversions (Table IV), it can be seen that transitions at non-dimer sites are several-fold rarer

Table III Influence of target site on mutation frequency.

POTENTIAL TARGET	SITES[a] AVAILABLE	FREQUENCY OF AMBERS[b] Urv+	ΔUvrB	SITES[a] AVAILABLE	FREQUENCY OF OCHRES[b] Urv+	ΔUvrB
TT	4	35 (8.6)	29 (7.3)	5	3.3 (0.65)	70 (13)
TC	15	310 (21)	360 (21)	12	260 (22)	270 (23)
TCC	17	200 (12)	280 (16)	13	85 (6.5)	120 (9.0)
CC	9	110 (12)	370 (41)	3	20 (6.5)	33 (11)
CC[d]			220 (28)			
CT	3	8.9(3.0)	120 (4.1)	3	3.3 (1.1)	33 (11)
TOTAL[e]	31	460 (15)	760 (24)	26	290 (12.5)	410 (18)
TOTAL[f]		340 (11)	520 (18)		110 (5.1)	220 (10)
NON-DIMER	7	2.9(0.4)[g]	41 (5.9)	15	75.0 (5.0)	120 (8.1)

a Sites forming potential dimers of more than one class; e.g., sequences such as CCT or TTC are counted in both potential classes. There are 2 such amber sites and 4 ochre sites. b Frequency of mutants x 10^8. The number in parenthesis is the per-site average. c Corrected for hotspots. Sites A23, 024 and 027 are excluded from the data of both the Uvr+ and ΔUvrB strains. d Corrected for the A24 site in the ΔUvrB data. e Sum of all mutations arising at potential dimer sites. f Sum of all mutations arising at potential dimer sites excluding all hotspots. g Single event recorded.

TABLE IV Frequency of transitions and transversions at potential dimer and non dimer sites.

POTENTIAL SITES	STRAINS	TRANSITION SITES AVAILABLE	TRANSITION FREQUENCIES		TRANSVERSION SITES AVAILABLE	TRANSVERSION FREQUENCIES	
Pyr-Pyr	UvrB$^+$	14	540	(38.2)	36	200	(5.4)
	ΔUvrB	14	780	(55.7)	36	340	(9.5)
hotspots removed	UvrB$^+$	11	240	(21.6)	--	--	
	ΔUvrB	10	350	(35.2)	--	--	
hotspots only	UvrB$^+$	3	300	(99.2)	--	--	
	ΔUvrB	4	440	(109.6)	--	--	
non dimer	UvrB$^+$	11	52	(4.8)	11	25	(2.3)
	ΔUvrB	11	100	(9.5)	11	52	(4.8)

The number in parenthesis is the site average.

than at potential dimer sites. In the case of transversions, this
difference is only about two-fold.

Since the mutational responses at individual sites might show
different dose-response kinetics, UV spectra were determined at
lower UV fluences (10 Jm^{-2} and 0.5 Jm^{-2} in the case of the Uvr^{+} and
ΔUvrB strains, respectively). At these doses, cell survival is not
significantly reduced and the mutation frequency is less than
doubled. The spectra therefore largely reflect spontaneous
occurrences, but the induction of mutants at specific sites
(particularly hotspots) might be discernable. The spectra for the
amber and ochre mutants produced in the Uvr^{+} strain did not show
any indication of excess mutagenesis at the A23, O24 or O27 hotspots
(data not shown). However, in the case of the ΔUvrB strain (Figure
4), the hotspot A24 accounts for almost 20% of the total mutants.
This is at least ten-fold greater than expected from the spontaneous
spectra and is even higher than the spontaneous deamination hotspots
at A6, A15 and A34 (39).

Mutational Spectra in the Presence of pKM101

Details of the mutational responses of the Uvr^{+} and ΔUvrB
strains carrying pKM101 have been summarized in Figure 1 and Table
I. The mutational spectra for the Uvr^{+} strain bearing pKM101 after
treatment with 75 Jm^{-2} are presented in frequency histograms alongside the spectra obtained at 100 Jm^{-2} in the absence of the mutator
plasmid (Figures 5 and 6). In this comparison, the induced nonsense
mutation frequency is about 6-fold higher in the presence of pKM101
than in its absence. In Figures 7 and 8 the mutational spectra
obtained in the ΔUvrB strain in the presence of the plasmid are
compared to the spectra in its absence. These spectra are from
cells treated with 5 Jm^{-2}. The nonsense mutation frequency was
enhanced 6- to 10-fold by the introduction of the plasmid.

A summary of the base substitutions required to generate the
recovered mutants is given in Table V. Although the absolute
frequencies obtained in the presence of the plasmid are considerably
greater than in its absence, the general characteristics of the
mutant distribution are similar. The G·C → A·T transition predominates as in earlier studies (Table II). At the UV dose
administered, the spectra in the presence of the plasmid seem
qualitatively similar to those obtained in its absence. The only
notable difference in any of the pKM101 spectra is found in the
ΔUvrB amber spectrum, where A19, a "warm spot" without the plasmid,
is the most common amber mutation (Figure 7).

Table VI considers the effect of the mutator plasmid on the
distribution of base substitutions with respect to the nature of
the potential DNA target. Within each strain, pKM101 has little

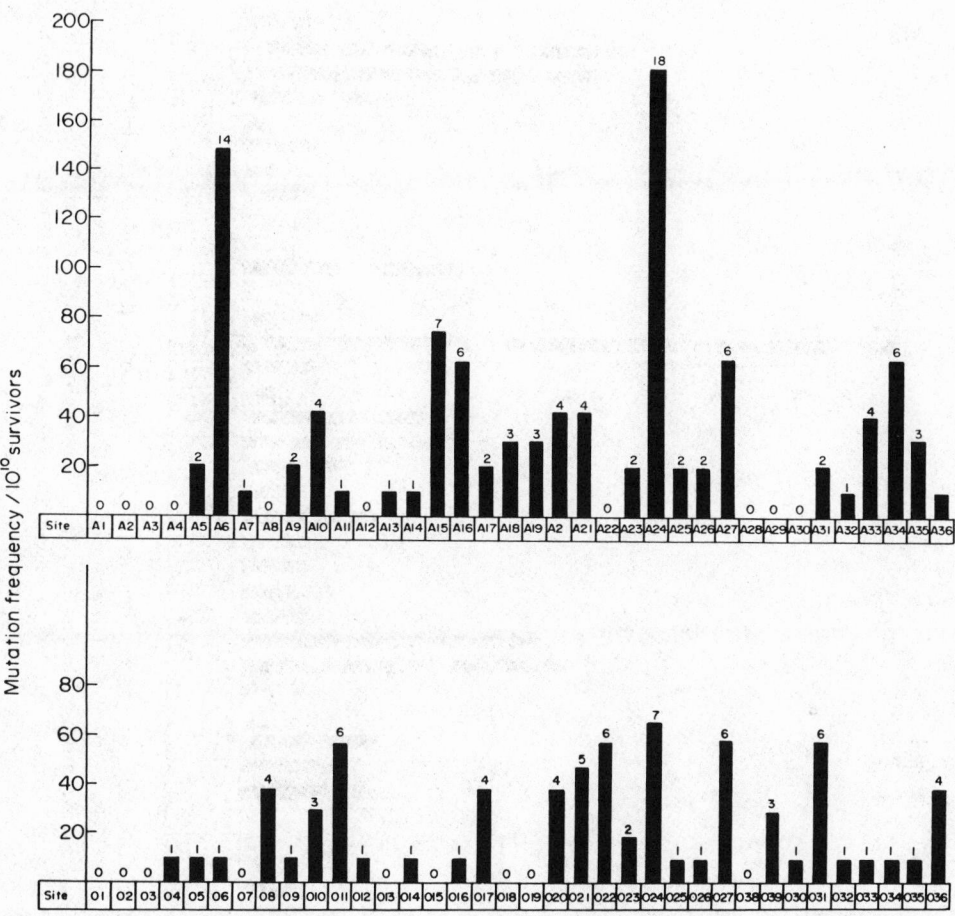

Figure 4.   The distribution of $lacI^-$ amber and ochre mutants in the
ΔUvrB strain following a UV dose of 0.5 Jm^{-2}.   101 amber and 74 ochre
mutants were analyzed.

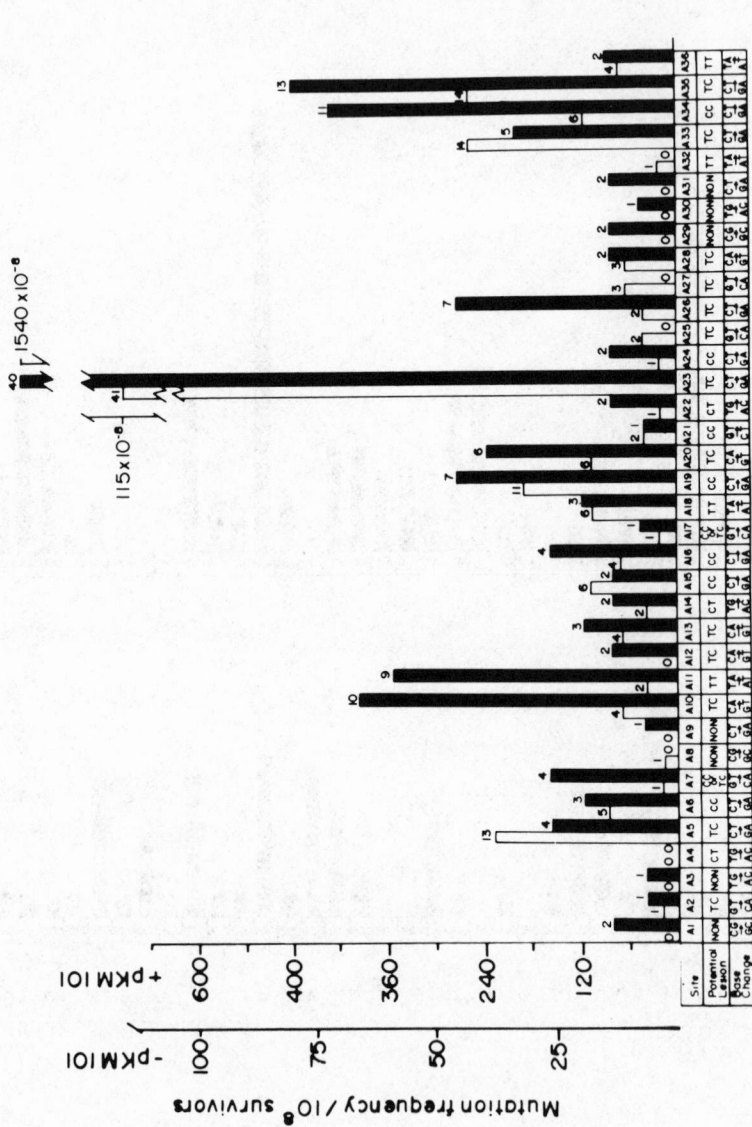

Figure 5. Frequency distribution of *lacI*⁻ amber mutants obtained in the wild-type strain in the presence and absence of pKM101. The UV-fluence was 100 and 75 Jm⁻² in the presence and absence of pKM101 respectively. 160 mutants obtained in the absence of pKM101 and 155 mutants obtained in the presence of pKM101 were analyzed. The open bars give the mutational spectrum without the plasmid; the solid bars, with pKM101.

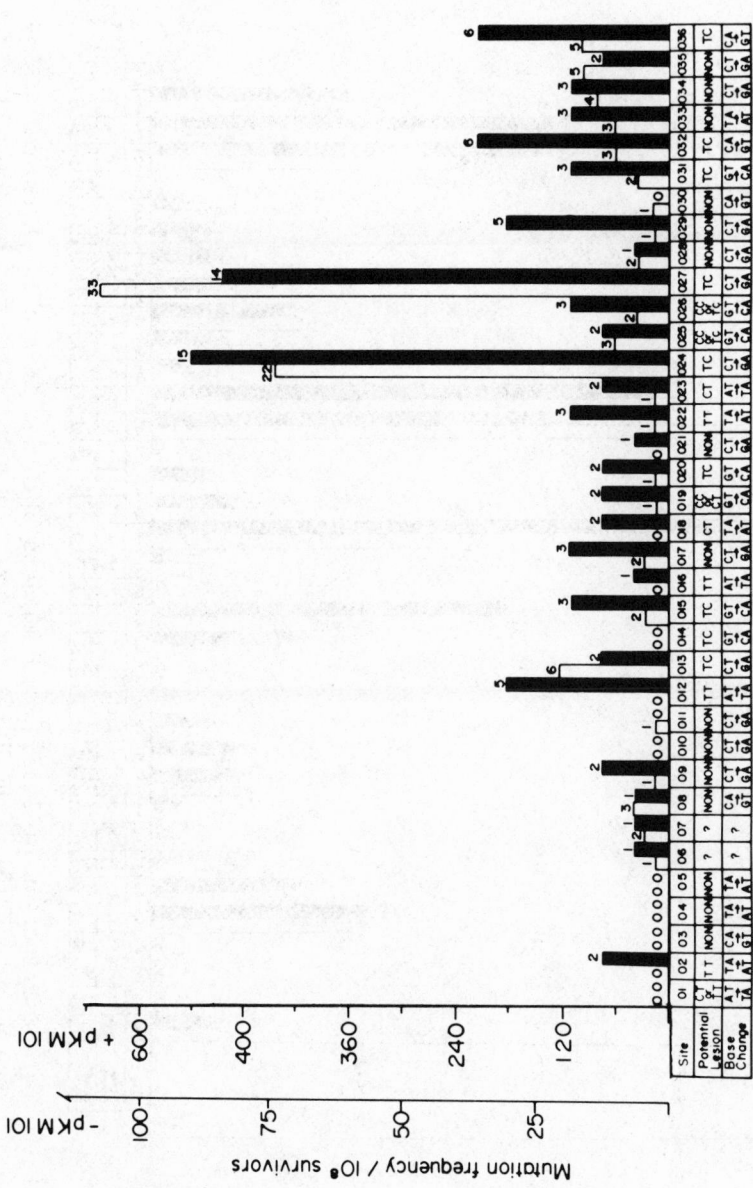

Figure 6.   Frequency distribution of *lacI⁻* ochre mutants obtained in the wild-type strain in the presence and absence of pKM101. The UV-fluence was 100 and 75 Jm⁻² in the presence and absence of pKM101 respectively. The open bars give the mutational spectrum without the plasmid; the solid bars with pKM101. 108 and 96 mutants were analyzed in the presence and absence of pKM101.

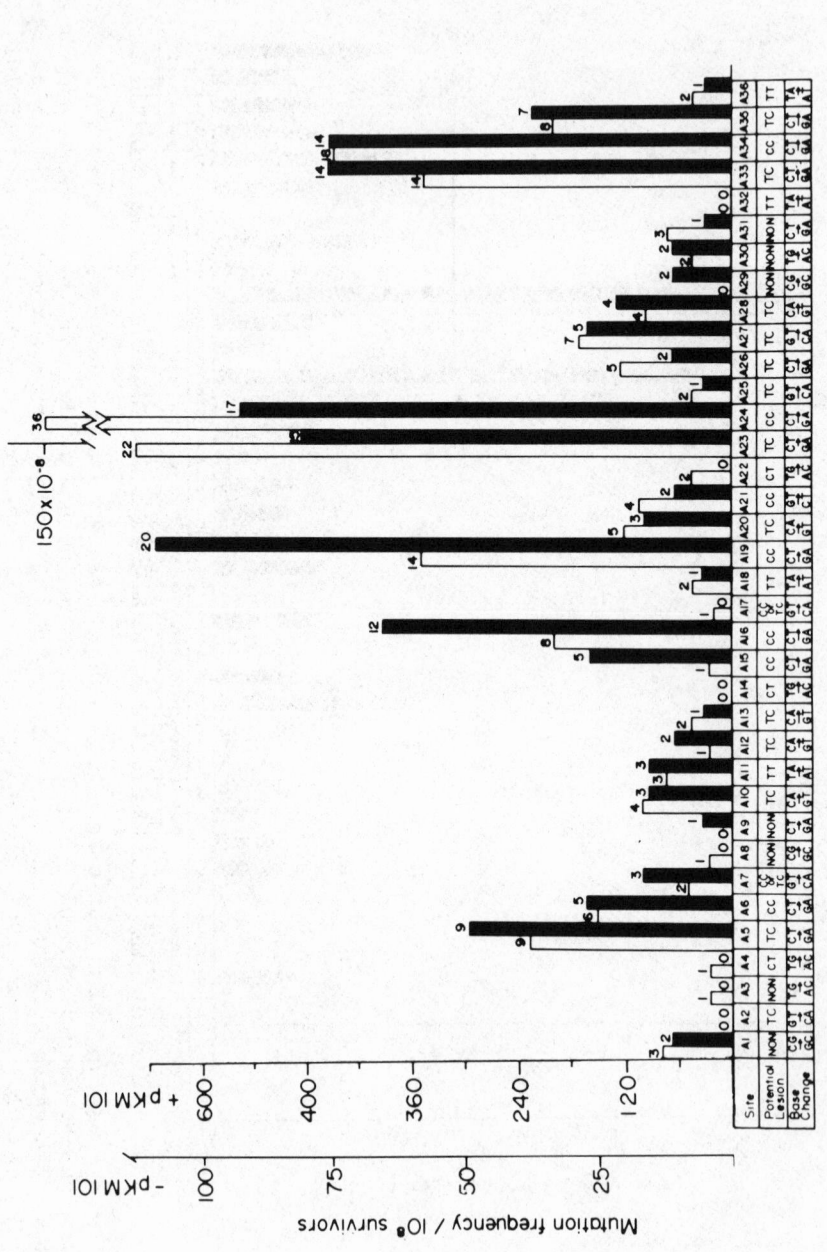

Figure 7. Frequency distribution of *lacI*⁻ amber mutants obtained in the ΔUvrB strain (NR3591) in the presence and absence of pKM101. The UV-fluence was 5.0 Jm⁻². The open bars give the mutational spectrum without the plasmid; the solid bars with pKM101. 193 and 157 mutants were analyzed in the presence and absence of pKM101 respectively.

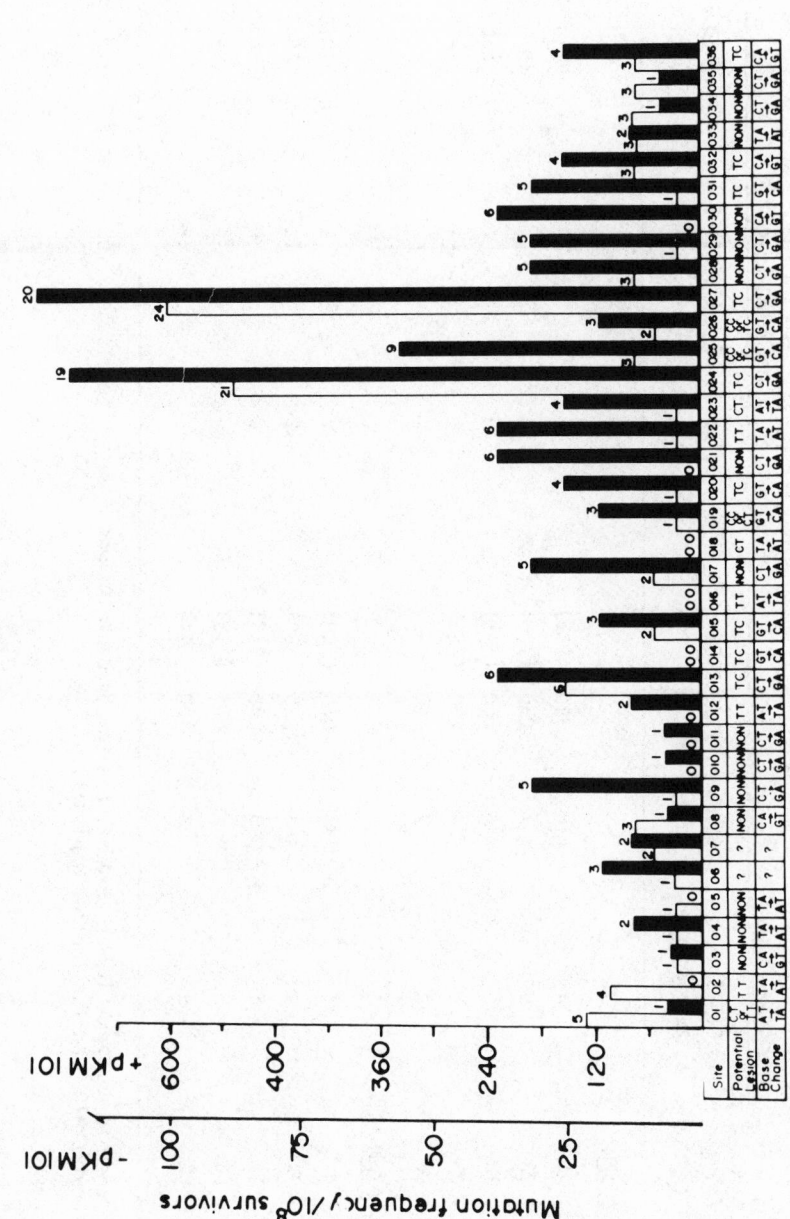

Figure 8. Frequency distribution of *lacI*⁻ ochre mutants obtained in the ΔUvrB strain (NR3591) in the presence and absence of pKM101. The UV-fluence was $5.0~Jm^{-2}$. The open bars give the mutational spectrum without the plasmid; the solid bars with pKM101. 120 and 140 mutants were analyzed in the presence and absence of pKM101 respectively.

TABLE V Distribution of base substitution following UV irradiation of pKM101 bearing strains.

BASE SUBSTITUTION	AMBER MUTANTS			OCHRE MUTANTS		
	SITES AVAILABLE	SITES FOUND	FREQUENCY	SITES AVAILABLE	SITES FOUND	FREQUENCY
Uvr⁺/pKM101						
GC → AT	14	14	3940 (281)	12	10	1730 (144)
GC → TA	9	7	1080 (120)	12	9	1010 (84)
AT → TA	5	4	660 (131)	10	7	650 (65)
AT → CG	5	4	230 (46)	--	--	---
GC → CG	3	2	150 (51)	--	--	---
ΔUvrB/pKM101						
GC → AT	14	14	4040 (290)	12	12	2810 (234)
CG → TA	9	8	720 (80)	12	12	1580 (131)
AT → TA	5	3	160 (133)	10	6	638 (64)
AT → CG	5	1	65 (13)	--	--	---
GC → CG	3	2	130 (44)	--	--	---

The dose to Uvr⁺/pKM101 was 75 Jm^{-2}, and to ΔUvrB/pKM101 was 5 Jm^{-2}. See Table 2 for comparison with strains without pKM101 and for further explanation.

TABLE VI Per cent distribution of UV-induced lacI mutations on the potential target sites.

POTENTIAL TARGET	SITES AVAILABLE	Uvr^+ -pKM101 %	Uvr^+ $^+$pKM101 %	ΔUvrB -pKM101 %	ΔUvrB $^+$pKM101 %
TT	9	4.6	8.9	7.2	4.6
TC	27	69	61	47	48
CT	6	1.5	3.0	3.4	1.6
CC	12	15	16	30	30
NONDIMER	22	9.4	11.2	12.5	16

Doses are those given in Figures 2,3 and 5-8.

effect upon this distribution. The high frequency of mutagenesis
seen in the ΔUvrB strain at potential cytosine-cytosine dimers is
also found in the.pKM101 derivative. In general, the spectra
obtained in the presence of the mutator plasmid are very similar to
those generated in its absence.

DISCUSSION

The Nature of Mutational Hotspots

A striking difference between the mutational spectra obtained
for the Uvr[+] and ΔUvrB strains is the A24 hotspot exclusive to the
excision repair defective strain (Figure 2). This site is further
unique in that the induction of A24 mutants was detected in the
ΔUvrB strain at UV fluences at which mutagenesis at other hotspots
was not observed (Figure 4). The A24 mutation is produced by a
transition at a potential cytosine-cytosine dimer. Five other
transitions arise at potential cytosine-cytosine dimer sites, yet
A24 is clearly different. Upon examining the nucleotide sequence
surrounding the A24 site we found, as indicated in Figure 9, that
site A24 occurs within a quasipalendromic sequence for which a
secondary structure can be predicted. In this structure, site A24
(position "7" in Figure 4) is located in the unpaired terminal
loop of a hairpin structure (labelled "B"). The location of the
A24 hotspot within a configuration of potential secondary structure
is intriguing because of the suggested relationship between the
high UV mutability of suppressor loci and their potential secondary
structure (24).

Secondary structure has been predicted to play a role in a
number of processes such as the initiation of DNA replication and
the control of transcription. Cruciform structures have recently
been demonstrated in plasmid DNA (40,41). The energetic costs of
forming such structures are large, especially in double-stranded
DNA. However, upon the addition of DNA gyrase to relaxed and
unbroken DNA, palindrome formation is favored in the negatively
supercoiled DNA (42). The stability of the proposed secondary
structures can be inferred from values calculated for tRNA (43).
The ΔG (compared with single-stranded DNA) for the proposed
secondary structure containing the A24 site is -7.4 kcal/mole,
making this hairpin structure potentially stable in single-stranded
DNA.

Obviously we do not yet know the molecular events resulting
in the A24 hotspot. The high mutability of this site could be
related either to a predisposition to form the premutagenic lesion
in locally denatured DNA in the unpaired loop, or to differences
in processing premutagenic lesions. While we cannot exclude the

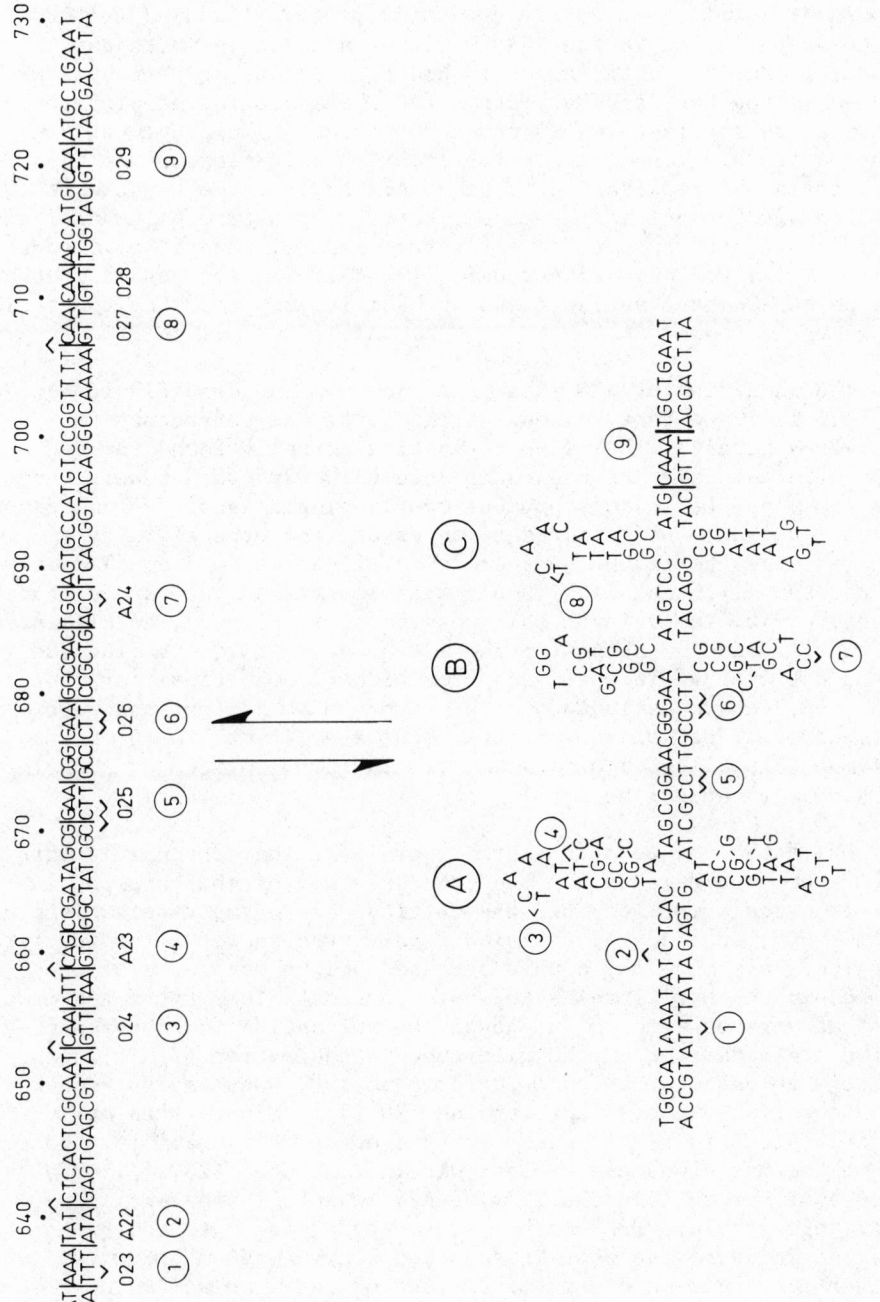

Figure 9. The proposed secondary structure for the 630–730 base pair region of the *lacI* gene of *Escherichia coli*.

possibility that an excisable lesion is preferentially (50-fold) formed at this site in the DNA, we favor a model in which the A24 hotspot is due to differences in how the lesions at this site are handled in the Uvr$^+$ and Uvr$^-$ strains. There are indeed differences in how these strains process their DNA damage. In a Uvr$^+$ strain mutation fixation can occur prior to DNA replication, while in a Uvr$^-$ strain DNA replication is required (13,18). We hypothesize that the A24 hotspot is due to the refractory nature of that site's specific DNA structure to an error-free mode of repair, e.g., post-replicational recombination repair (13,44). The net result is that lesions at A24 have an increased probability of producing a mutation in the Uvr$^-$ strain.

The induction of A24 mutants at low UV fluences (Figure 4) may also indicate that lesions at this site are refractory to error-free repair. The two-hit kinetics normally found for UV-induced mutagenesis has been suggested to be due to a requirement that two separate lesions produce overlapping daughter strand gaps (45) which cannot be repaired by an error-free process. The occurrence of the A24 hotspot at low UV fluences (0.5 Jm^{-2}) does not necessarily mean that two-hit kinetics are absent, and a study is underway to investigate this possibility. However, an alternate hypothesis which accounts for the A24 hotspot at low UV fluences on the basis of an independence from either "SOS" repair or its induction, can be excluded on the grounds that A24 mutants were not recovered following UV treatment of a *recA*$^-$ strain (B. W. Glickman, unpublished) or a *umuC*$^-$ strain (B. W. Glickman, A. Kolman and R. Dunn, unpublished).

While the nature of the premutagenic lesion responsible for the A24 hotspot remains unknown, the presence of this hotspot in the Uvr$^+$ strain implies that the critical lesion must be excisable. Furthermore, because the excision repair process (at least in vitro) functions only on fully double-stranded DNA, the absence of the hotspot in the Uvr$^-$ strains suggests that both isomeric forms must exist at various times. One possible explanation for the absence of the A24 hotspot from the wild-type is the action of MFD which, although thought to be specific for nonsense, suppressors, occurs only in excision-proficient strains (20-23). MFD is seen at specific sites in t-RNA genes (24) and occurs very rapidly: up to 50% of the mutations can be lost within 2 minutes (20,24). MFD could easily result in the loss of A24 mutants in the excision-proficient strain. MFD can be blocked by adding nutrient broth or casamino acids to the post-irradiation growth medium. However, neither the addition of nutrient broth or casamino acids to selection plates nor up to 18 hours incubation in media containing these components increased the recovery of A24 mutants in the wild type strain (results not shown). It therefore seems unlikely that MFD is responsible for the absence of this hotspot in the Uvr$^+$ strain.

The mutational hotspots 024, 027 and A23 are seen in both the
Uvr[+] and ΔUvr strains and occur within the same region of the gene;
in fact, almost 50% of the UV-induced nonsense mutants arose in
only about 6% of the *lacI* gene. Intriguingly, each of these hotspots is also located at an unpaired site within potential secondary
structures. These three sites all involve G·C → A·T transitions
at DNA sequences where thymine-cytosine dimers could form, as well
as the Py-C* lesion described by Hazeltine et al. (personal
communication). The possible importance of the Py-C* lesion will
be discussed later.

The apparent relationship between mutational hotspots and
secondary structure prompted a computer-assisted search for potential secondary structure throughout the *lacI* gene. Several
sequences were identified where potential secondary structure could
be drawn. In only one such sequence, however, was a nonsense site
(A35) available. The sequence surrounding A35 is very similar to
that around the hotspot 024 (10,46). The potential secondary
structure has a ΔG of -12.4 kcal/mole and A35 mutants arise via a
G·C → A·T transition at a potential thymine-cytosine dimer site,
just as do 024, 027 and A23 mutants. Since we are largely ignorant
of the factors involved in determining the existence of secondary
structure, e.g., the sequence specificity of local supercoiling,
the sequence and site specificities of DNA binding proteins which
might facilitate or hinder the formation of secondary structure,
the proximity of initiation or termination sites of Okazaki fragments, and the effect of local secondary structure, we are unable
to assess the reasons why, if secondary structure were to be a
determinate of mutational hotspots, A35 is less mutable by UV than,
e.g., A23 or 024.

Nonetheless, the localization of the hotspots at sites within
potential secondary structures is exciting. Such isomers are most
likely to form during DNA replication just when mutations may be
fixed by error-prone repair processes (13,44,45). They may also
exist as recombinational intermediates during a process which
normally provides an opportunity for error-free repair (13). We
therefore propose that secondary structures can modulate the extent
of both error-free and error-prone repair and, as a result, can
contribute to the determination of mutational hotspots.

The Specificity of UV Mutagenesis: General Comments

The *lacI* mutational specificity system allows the identification of G·C → A·T transitions and all four transversions;
A·T → G·C transitions do not give rise to nonsense codons and
therefore remain undetected. Bearing this limitation in mind,
mutational specificity can be compared in the Uvr[+] and ΔUvrB
strains (Table II). All detectable base substitutions were

recovered. Even the least frequent events occurred at frequencies
significantly higher than witnessed spontaneously, e.g., $G \cdot C \rightarrow C \cdot G$
and $A \cdot T \rightarrow C \cdot G$ transversions were recovered at frequencies 50- to
200-fold higher among the induced mutants than occurred spontane-
ously (2,3,33). The $G \cdot C \rightarrow A \cdot T$ transition, however, predominates in
both the Uvr$^+$ and ΔUvrB strains, and these occurred primarily at
potential dimer sites. This is true for the amber collection even
when the $G \cdot C \rightarrow A \cdot T$ hotspots are removed from consideration. The
predominance of $G \cdot C \rightarrow A \cdot T$ transition events is consistent with
earlier observations made in bacteriophage T4 (47,48) and *E. coli*
(2-6). The predominance of transitions over transversions has also
been noted in yeast (49,50).

Mutational Specificity: A Role for Pyrimidine Dimers?

Evidence for a role for cyclobutane pyrimidine dimers as
premutational lesions is mainly based upon the reversal of mutagenesis seen in excision repair deficient strains (c.f. 13,18).
As a result, the role of pyrimidine dimers as premutagenic lesions
is far from clear. UV light produces a multiplicity of products
(51) and photoreactivation may not be exclusive to pyrimidine
dimers. Moreover, the Uvr$^-$ strains of *E. coli* fail to excise a
varietyof DNA damages including 4NQO and mitomycin benomyl adducts,
and O^6-ethylguanine (52).

Studies on the mutational specificity of UV light have not
directly supported a role for pyrimidine dimers in UV-induced
mutagenesis. Meistrich and Drake (48) examined the mutational
specificity of 313nm light on bacteriophage T4 in the presence of
a derivative of acetophenone (a treatment thought to result in the
preferential induction of thymine-thymine dimers), but recovered
primarily $G \cdot C \rightarrow A \cdot T$ transitions. In a recent study of UV-induced
reversion of bacteriophage M13 amber mutants, Schaaper and Glickman
(53) found that pyrimidines preceeded by a purine (at which no dimer
could form) mutated at frequencies comparable to pyrimidines
preceeded by a pyrimidine (where dimer formation is possible).
Furthermore, the DNA sequences of UV-induced revertants of M13 amber
mutants containing a potential dimer site do not indicate a strong
preference for base substitutions at the site of the potential dimer
(54,55).

The occurrence of most UV-induced mutagenesis in the *lacI* gene
at potential dimer sites, as shown by Coulondre and Miller (2,3,39)
and confirmed here for both the Uvr$^+$ and ΔUvrB strain (Tables III
and IV), is consistent with a role for pyrimidine dimers as
important determinants of UV mutagenesis. However, the mutational
profile for mutagenesis in the *lacI* gene does not reflect the
expected distribution of pyrimidine dimers in the DNA (56). While
this may indicate the modulating role of DNA repair in producing

the final mutational spectrum, it may also indicate that other DNA damages are involved in the production of mutation at some of the sites. A recently suggested candidate for such a premutational lesion is a modified cytosine residue 3' to a pyrimidine, i.e., the Py-C* lesion of Hazeltine et al. (personal communication). Indeed, most mutagenesis seen after UV irradiation appears to be the G·C → A·T transition (Tables II and IV). However, the inability to measure A·T → G·C mutagenesis in the *lacI* system prevents the direct comparison which might resolve this question.

A number of possibilities can be envisioned. Possibly Py-C* actually is the premutagenic lesion and the altered cytosine directly codes the incorporation of adenine, producing G·C → A·T transitions. Alternatively, "SOS" repair may preferentially insert adenine across from pyrimidine dimers. This later explanation would account for both the high specificity for G·C → A·T transitions and the low level of transversions detected at potential dimer sites (Table IV). If pyrimidine dimers represented the major mutational lesion in *E. coli*, then the high degree of mutational specificity would be in contrast to the hypothesized mechanism of random insertion of bases at "non-coding" lesions (13,15,16). In this context it should be pointed out that the origin of the idea the pyrimidine dimers are non-coding is the observation that these lesions block DNA synthesis in vitro and at the same time result in high turnover of dNTPs. To date no studies have adequately determined the specificity of this turnover. Thus not only do we lack proof that the blockage of synthesis by a dimer is related to its informational defect, but we also lack insights into the characteristics of that informational content. Moreover, while a pyrimidine making up part of a dimer may not pair properly, it seems likely that it would appear more "pyrimidine-like" to DNA polymerase than "purine-like" and would therefore direct the insertion of a purine. Mispairing would then result in the G·C → A·T transition at potential pyrimidine-cytosine dimers as is observed in the *lacI* system. The poor induction of transversions at potential dimer sites (see Table III) is also consistent with this model since most transversions are thought to occur through purine: purine intermediates (where several pairing schemes are possible), but not through pyrimidine:pyrimidine intermediates because of steric considerations (57).

An observation interpreted as indicating a role for pyrimidine dimers in UV-induced mutagenesis is the recovery of *lacI*⁻ nonsense mutants involving tandem base pair substitutions (3,4). The occurrence of such tandem doubles was predicted by the model involving the random insertion of bases across from pyrimidine dimers (15,16). In our study, however, the number of uncharacterized nonsense mutants, and therefore potential tandem double base changes, was very limited and amounted to less than 1% of the mutants examined

in the Uvr[+] and ΔUvrB strains. This low frequency, particularly in
the ΔUvrB strain where dimers are not excised, does not favor the
quantitative predictions made by the random insertion hypothesis.
It is our view that the occasional occurrence of clustered mutations
reflects the intense local mutator activity of "SOS" repair rather
than the presence of a dimer. Tandem double mutations have been
detected in yeast following ionizing radiation (58), and in sequenced
bacteriophage M13 revertants among both unirradiated phage plated
upon UV-induced bacteria and γ-irradiated phage plated upon an induced host (54). Moreover, multiple mutational events have been
shown to occur more frequently in *E. coli* after UV than would have
been predicted on the basis of random damage distribution (59). A
significant number of non-tandem double mutants was also reported
by Coulonde and Miller (3) following UV irradiation.

The recent assignment within the DNA sequence (10) of many of
the suspected tandem double mutations isolated by Coulondre and
Miller (2,3) allows a re-examination of these data. Only 12 of the
653 mutants (1.8%) induced by UV were in fact double base pair
substitutions. Among these, one site (X1) is a tandem double
mutation converting the ACG codon of amino acid position 5 to a UAG
codon and therefore involves a non-dimer site. Another, X14,
probably involves the conversion of the GAU codon at amino acid
position 247 to a UAG codon (i.e., a close but not tandem double).
Mutagenesis at these sites is consistent with a model in which the
reduced replicational fidelity of "SOS" repair allows for multiple,
non-targeted mutagenesis and is furthermore consistent with the
mutator phenotype of "SOS" induced cells as exemplified by studies
on the *tif* mutant (for a discussion of *tif* see reference 13).

The contribution and specificity of indirect mutagenesis still
remains to be determined and such a study is in progress in our
laboratory. However, as long as we are unable to ascertain the
contribution of indirect mutagenesis to mutational spectra, we
cannot adequately evaluate the contribution of any lesion, known or
unknown. In this respect, however, it seems likely that the A24
hotspot seen in the ΔUvrB strain is related to a specific lesion,
albeit a cytosine-cytosine dimer, the Py-C* product or a yet-to-be
identified photoproduct, since it seems unlikely that the
specificity of indirect mutagenesis would be greatly influenced by
the operation of an excision repair system.

UV Specificity in the Presence of pKM101

The introduction of pKM101 into the Uvr[+] and ΔUvrB strains
considerably enhanced the mutagenic response (see Figure 1 and
Table I). The mutational spectra in the presence and absence of
pKM101 (Figures 5-8) are very similar. Even the spectral differences between the Uvr[+] and ΔUvrB strains are observed (i.e., the

increased yield of mutations at potential cytosine-cytosines sites
and the A24 hotspot unique to the excision-defective strain). The
only clear affect of pKM101 on the mutational spectrum was the
increased proportion of A19 mutants relative to A24 mutants in
the ΔUvrB strain. Otherwise, the mutational spectra, when converted
to the frequencies of each class of base substitution without
regard to the target site, were very similar (compare Tables II and
V). Also, comparing the plasmid effects at potential dimer and
non-dimer sites (Table VI) showed no differential effects.

It should be noted, however, that the spectral comparisons
were done at greatly differing levels of mutagenesis. In fact, at
different UV doses producing similar levels of mutagenesis, spectral
differences might be expected. Whether these would then reflect
differences in the distribution of DNA damage, altered levels of
interaction between pKM101 and incompletely induced cellular "SOS"
response, or actual differences in the mechanism of plasmid-mediated
mutagenesis, would be difficult to determine.

We have made preliminary comparisons of UV spectra at very low
UV fluences. Experiments at 0.5 Jm^{-2} with the ΔUvrB strains (results not shown) indicate that the mutational spectra contain
representatives of all the hotspots recovered at 5 Jm^{-2} in the
absence of the plasmid even though only the A24 hotspot is recovered
after 0.5 Jm^{-2} in the ΔUvrB strain (Figure 4). The differences in
the spectra obtained in this comparison reveal the quantitative
difficulties of establishing an "induced" spectrum at doses producing negligible increases in mutagenesis above the spontaneous
background.

The Mechanism of Action of pKM101

As outlined in the introduction, our basic premise was that
any large differences in mutational spectra in the presence of
pKM101 would indicate a difference in the mechanisms of error-prone
repair supplied by the plasmid and the host. The lack of any
detected effect of pKM101 on the UV mutational spectra may mean
that the plasmid operates by an error-prone pathway very similar
to that of its host, though other interpretations are possible. A
number of further observations, however, support the suggestion that
the two error-prone pathways are inter-related. Like "SOS" repair,
pKM101 depends upon functional $recA^+$ and $lexA^+$ host alleles for its
mutator effect. Similarly, pKM101, like its host, requires a
functional $lexC^+$ gene for UV mutagenesis, though not for its
protective effect (B. W. Glickman, P. A. Todd and A. Dunn,
unpublished). Another observation which indicates a relationship
between pKM101 and "SOS" repair is the synergistic effect on
mutagenesis of pKM101 in tif^--$sfiA$ double mutants at low UV fluences
(31). Our interpretation of these experiments is that at low UV

fluences, the turning on of "SOS" by *tif* at 42° provides increased
levels of *recA* protein (13) and this in turn enhances the mutator
effect of pKM101. The nearly linear kinetics for UV mutagenesis
in the presence of pKM101 (see Figure 1 and Table I) are consistent
with this interpretation.

One interpretation of these results is that pKM101 supplies
a constitutive component which channels DNA lesions preferentially
through error-prone rather than error-free pathways. The "SOS"
model accounts for the dose-squared mutational response by requiring
two DNA lesions to produce overlapping single-strand gaps (45)
unrepairable by error-free repair. The removal of a requirement
for a second DNA lesion to produce UV mutagenesis is indicated by
the linear rather than dose-squared mutational response seen in the
presence of pKM101, at least in the case of *lacI⁻* mutagenesis in
Figure 1. Lesions which normally would have been repaired in an
error-free manner are now causing mutation in the presence of
pKM101. If dimers are the cause of the hotspot at A24, their
mutagenic efficiency in the ΔUvrB strain is 10-30% (calculations
not shown). In the presence of pKM101 however, their efficiency
approaches 100%. This may explain why the A24 hotspot is
relatively reduced in the ΔUvrB/pKM101 strain (Figure 7). If
pKM101 were to reduce the relative contribution of error-free
repair, then the level of mutagenesis at site A24 which we believe
is refractory to error-free repair, should be more similar to other
sites having similar DNA sequences as is indeed observed (e.g.,
site A19 in Figure 7).

One objection raised against models involving the interaction
between pKM101 and host "SOS" repair is that of Goze and Devoret
(60) who found pKM101-mediated UV-protection to depend upon
functional *uvrD* and *uvrE* genes, even though host cell mutagenesis
is independent of these functions. In subsequent experiments we
(35) demonstrated that pKM101 required these genes only for its
protective effect and not its mutator effect. It therefore appears
possible to retain the mutator effect but lose the protective
effect of pKM101 in *uvrD* and *uvrE* mutants, whereas just the
reverse is observed in the *lexC⁻* strain.

Restoration of the mutability of a *umuC* strain by pKM101 [and
restoration of the mutability of a *uvm* strain by R46 (61)] may be
indicative of plasmid independence from this "SOS" component or may
reflect an ability of a pKM101 product to replace the *umuC* product
(62). Our data do not distinguish between these hypotheses.
Instead, we prefer the more general suggestion that pKM101 provides
a shunt promoting error-prone rather than error-free processes.
This shunt may indeed be achieved by the constitutive production
of a "*umuC*-like" product. The mechanism of pKM101 action is
currently under investigation in a number of laboratories.

Our results support potential pKM101 mechanisms which produce enhanced mutagenesis with specificities similar or identical to those of the host's error-prone repair pathway.

ACKNOWLEDGMENTS

The assistance of D. L. Parker, B. Dumas, A. R. Kersey and J. C. Glickman in the preparation of this manuscript is greatfully acknowledged. The work presented was carried out both at the University of Leiden and N.I.E.H.S. and involves contributions of P. A. Todd, J. Brouwer, N. Guijt, G. Moessen, C. van Teijlingen, and R. Dunn. The manuscript benefited greatly from extensive discussions with Drs. C. A. van Sluis, L. Ripley and J. Drake.

REFERENCES

1. Miller, J. H., Ganem, D., Lu, P., Schmitz, A.: Genetic studies of the lac repressor. I. Correlation of mutational sites with specific amino acid residues: Construction of a colinear gene-protein map. J. Mol. Biol., 109:275 (1977)

2. Coulondre, C., Miller, J. H.: Genetic studies of the lac repressor. III. Additional correlation of mutational sites with specific amino acid residues. J. Mol. Biol., 117:525 (1977)

3. Coulondre, C., Miller, J. H.: Genetic studies of the lac repressor. IV. Mutagenic specificity of the lacI gene of Escherichia coli. J. Mol. Biol., 117:577 (1977)

4. Osborn, M., Person, S., Phillips, S., and Funk, F.: A determination of mutagen specificity in bacteria using nonsense mutants of bacteriophage T4. J. Mol. Biol., 26:437 (1967)

5. Person, S., and Osborn, M.: The conversion of amber suppressors to ochre suppressors. Biochem., 60:1030 (1968)

6. Person, S., McCloskey, J. A., Snipes, W., and Bockrath, R. C.: Ultraviolet mutagenesis and its repair in an Escherichia coli strain containing a nonsense codon. Genetics, 78:1035 (1974)

7. Yanofsky, C., Ito, J. and Horn, V. D.: Amino acid replacements and the genetic code. Cold Spring Harbor Symp. Quant. Biol., 31:151 (1966)

8. Bridges, B. A., Dennis, R. E. and Munson, R. J.: Mutation in E. coli B/r WP2 trp⁻ by reversion or suppression of a chain terminating codon. Mutation Res., 4:502 (1967)

9. Bridges, B. A., Dennis, R. E., and Munson, R. J.: Differential induction and repair of ultraviolet damage leading to true reversions and external suppressor mutations of an ochre codon in Escherichia coli B/r WP2. Genetics, 57:892 (1967)

10. Miller, J. H., Coulondre, C., Farabaugh, P. J.: Correlation of nonsense sites in the lacI gene with specific codons in the nucleotide sequence. Nature (Lond.), 274:770 (1978)

11. Nishioka, H. and Doudney, C. O.: Different modes of loss of photoreversibility of mutation and lethal damage in ultraviolet-light resistant and sensitive bacteria. Mutation Res., 8:215 (1969)

12. Jagger, J.: Ultraviolet inactivation of biological systems. In Photochemistry and Photobiology of Nucleic Acids, Ed. by S. Y. Wang. Academic Press, New York, Vol. II, p. 147 (1976)

13. Witkin, E. M.: Ultraviolet mutagenesis and inducible DNA repair in Escherichia coli. Bacteriol. Rev., 40:869 (1976)

14. Defais, M., Fauquet, P., Radman, M. and Errera, M.: Ultraviolet reactivation and ultraviolet mutagenesis of λ in different genetic systems. Virology, 43:495 (1971)

15. Radman, M.: Phenomenology of an inducible mutagenic DNA repair pathway in Escherichia coli: SOS repair hypothesis. In Molecular and Environmental Aspects of Mutagenesis, Ed. by L. Prakash, F. Sherman, M. W. Miller, C. M. Lawrence, and H. W. Taber. Thomas, Springfield, Ill., p. 128 (1974)

16. Radman, M.: SOS repair hypothesis: Phenomenology of an inducible DNA repair which is accompanied by mutagenesis. In Molecular Mechanisms for Repair of DNA, Ed. by P. C. Hanawalt and R. B. Setlow. Plenum Press, New York, p. 355 (1975)

17. Nishioka, H. and Doudney, C. O.: Different modes of loss of photoreversibility of ultraviolet light-induced true and suppressor mutations to tryptophan independence in an auxotrophic strain of Escherichia coli. Mutation Res., 9:349 (1970)

18. Bridges, B. A. and Mottershead, R. P.: Mutagenic DNA repair in Escherichia coli. VII. Constitutive and inducible manifestations. Mutation Res., 52:151 (1978a)

19. Green, M. H. L., Bridges, B. A., Eyfjord, J. E. and Muriel,
    W. J.: Mutagenic DNA repair in Escherichia coli. V.
    Mutation frequency decline and error-free post-replication
    repair in an excision-proficient strain. Mutation Research,
    42:33 (1977)

20. Witkin, E. M.: Radiation induced mutations and their repair.
    Science, 152:1345 (1966)

21. Clarke, C. H.: Caffeine and amino acid effects upon try$^+$
    revertant yield in UV-irradiated hcr$^+$ and hcr$^-$ mutants of
    E. coli B/r. Mol. Gen. Genet., 99:97 (1967)

22. Clarke, C. H.: Mutation frequency decline and its relation-
    ship to excision repair. Stud. Biophys., 36/37:277 (1973)

23. Munson, R. J. and Bridges, B. A.: Non-photoreactivating repair
    of mutational lesions induced by ultraviolet and ionizing
    radiations in Escherichia coli. Mutation Res., 3:461 (1966)

24. Bockrath, R. C. and Palmer, J. E.: Differential repair of
    premutational UV-lesions at tRNA genes in E. coli. Molecular
    Gen. Genet., 156:133 (1977)

25. Tweats, D. J., Thompson, M. L., Pinney, R. J., Smith, J. T.:
    R factor-mediated resistance to ultraviolet light in strains
    of Escherichia coli deficient in known repair functions. J.
    Gen. Microbiol., 93:103 (1976)

26. Monti-Bragadin, C., Babudri, N., and Samer, L.: Expression
    of the plasmid pKM101-dtermined DNA repair system in recA$^-$
    and lex strains of Escherichia coli. Molec. gen. Genet.,
    145:303 (1976)

27. Walker, G. C.: Plasmid (pKM101)-mediated enhancement of
    repair and mutagenesis: Dependence on chromosomal genes in
    Escherichia coli K12. Molec. gen. Genet., 152:93 (1977)

28. Waleh, N. S. and Stocker, B. A. D.: Effect of host lex, recA,
    recF and uvrD genotypes on the ultraviolet light-protecting
    and related properties of plasmid R46 in Escherichia coli.
    J. Bacteriol., 137:830 (1979)

29. Mortelmans, K. E. and Stocker, B. A. D.: Ultraviolet light
    protection, enhancement of ultraviolet light mutagenesis,
    and mutator effect of plasmid R46 in Salmonella typhimurium.
    J. Bacteriol., 128:271 (1976)

30. Mortelmans, K. E. and Stocker, B. A. D.: Segregation of the
     mutator property of plasmid R46 from its ultraviolet-
     protecting property. Molec. gen. Genet., 167:317 (1979)

31. Doubleday, O. P., Green, M. H. L. and Bridges, B. A.:
     Spontaneous and ultraviolet-induced mutation in Escherichia
     coli: Interaction between plasmid and tif-1 mutator effects.
     J. Gen. Microbiol., 101:163 (1977)

32. Walker, G. C. and Dobson, P. P.: Mutagenesis and repair
     deficiencies of Escherichia coli umuC mutants are suppressed
     by the plasmid pKM101. Molec. gen. Genet., 172:17 (1979)

33. Glickman, B. W.: Spontaneous mutagenesis in Escherichia coli
     strains lacking 6-methyladenine residues in their DNA: An
     altered mutational spectrum in dam⁻ mutants. Mutat. Res.,
     61:153 (1979)

34. Cleary, P. P., Campbell, A. and Chang, R.: Location of promoter
     and operator sites in biotin gene cluster in Escherichia coli.
     Proc. Natl. Acad. Sci. U.S.A., 69:2219 (1972)

35. Todd, P. A. and Glickman, B. W.: UV protection and mutagenesis
     in uvrD, uvrE and recL strains of Escherichia coli carrying the
     pKM101 plasmid. Mutat. Res., 62:451 (1979)

36. Glickman, B. W., Zwenk, H., van Sluis, C. A. and Rorsch, A.:
     Isolation and characterization of an X-ray sensitive UV
     resistant mutant of Escherichia coli K12. Biochim. Biophys.
     Acta, 254:114 (1971)

37. Fowler, R. G., McGinty, L. and Mortelmans, K. E.: Spontaneous
     mutational specificity of drug resistance plasmid pKM101 in
     Escherichia coli. J. Bacteriol., 140:929 (1979)

38. Miller, J. H.: Experiments in molecular genetics. Cold Spring
     Harbor Laboratory, Cold Spring Harbor, New York, (1972)

39. Miller, J. H., Coulondre, C. and Farabaugh, P. J.: Molecular
     basis of base substitution hotspots in Escherichia coli.
     Nature, 274:775 (1978)

40. Lilley, D. M. J.: The inverted repeat as a recognizable
     structural feature in supercoiled DNA molecules. Proc. Natl.
     Acad. Sci., 77:6488 (1980)

41. Panayatos, N. and Wells, R. D.: Cruciform structures in
     supercoiled DNA. Nature, 289:466 (1981)

42. Gellert, M., Mizuchi, K., O'Dea, M. H., Ohmori, H. and
     Tomizawa, J.: DNA gyrase and DNA supercoiling. Cold Spr.
     Harb. Symp. Quant. Biol., 43:35 (1978)

43. Tinoco, I., Borer, P. N., Dengler, B., Levine, M. D., Uhlenbock,
     O. C., Crothers, D. M. and Gralla, J.: Improved estimation
     of secondary structure in ribonucleic acids. Nature New
     Biology, (1973)

44. Kimball, R. F.: The relation of repair phenomena to mutation
     induction in bacteria. Mut. Res., 55:85 (1978)

45. Sedgwick, S. G.: Misrepair of overlapping daughter strand
     gaps as a possible mechnism for UV-induced mutagenesis in
     uvr strains of Escherichia coli. A general model for induced
     mutagenesis by misrepair (SOS repair) of closely spaced DNA
     lesions. Mutation Res., 41:185 (1976)

46. Farabaugh, P. J.: Sequence of the lacI gene. Nature, 274:765
     (1978)

47. Drake, J. W.: Properties of ultraviolet-induced rII mutants
     of bacteriophage T4. J. Mol. Biol., 6:268 (1963)

48. Meistrich, M. L. and Drake, J. W.: Mutagenic effects of
     thymine dimers in bacteriophage T4. J. Mol. Biol., 66:107
     (1972)

49. Prakash, L. and Sherman, F.: Mutagenic specificity: reversion
     of iso-1-cytochrome c mutants of yeast. J. Mol. Biol.,
     79:65 (1973)

50. Lawrence, C. W. and Christensen, R. B.: Absence of relation-
     ship between UV induced reversion frequency and nucleotide
     sequence at the cyc-1 locus of yeast. Molec. Gen. Genet.,
     177:31 (1979)

51. Wang, S. Y.: Photochemistry and photobiology of nucleic acids.
     Academic Press, New York, Vol. I (1976)

52. Todd, P. A., Brouwer, J. and Glickman, B. W.: Influence of
     DNA. Repair deficiencies on MMS and EMS induced mutagenesis.
     Mut. Res., (in press)

53. Schaaper, R. M. and Glickman, B. W.: Mutability of bacterio-
     phage M13 by Ultraviolet Light: Role of pyrimidine dimers.
     (submitted)

54. Brandenburger, A., Godson, G. N., Radman, M., Glickman, B. W.,
     van Sluis, C. A. and Doubleday, O. P.: Radiation-induced
     base substitution mutagenesis in single stranded DNA phage
     M13. Nature, (in press)

55. Doubleday, O. P., Brandenburger, A. and Radman, M.: This
     volume

56. Haseltine, W. A., Gordon, L. K., Lindan, C. P., Grafstrom, R.
     H., Shaper, N. L. and Grossmon, L.: Cleavage of pyrimidine
     dimers in specific DNA sequences by a pyrimidine DNA-glyco-
     sylase of M. luteus. Nature, 285:634 (1980)

57. Topal, M. D. and Fresco, J. R.: Complementary base pairing
     and the origin of substitution mutations. Nature (Lond.)
     263:285 (1976)

58. Stewart, J. W. and Sherman, F.: Demonstration of UAG as a
     nonsense codon in Bakers' yeast by amino acid replacements
     in Iso-1-cytochrome c. J. Mol. Biol., 68:429 (1972)

59. Coleman, R. D., Dunst, R. W. and Hill, C. W.: A double base
     change in alternate base pairs induced by ultraviolet
     irradiation in a glycine transfer RNA gene. Molec. Gen.
     Genet., 177:213 (1980)

60. Goze, A. and Devoret, R.: Repair promoted by plasmid pKM101
     is different from SOS repair. Mutat. Res., 61:163 (1979)

61. Steinborn, G.: Uvm mutants of Escherichia coli K12 deficient
     in UV mutagenesis. Molec. Gen. Genet., 175:203 (1979)

62. Walker, G. C. and Dobson, P. P.: Mutagenesis and repair
     deficiencies of Escherichia coli umuC mutants are suppressed
     by the plasmid pKM101. Molec. Gen. Genet., 172:17 (1979)

DISCUSSION

DR. THILLY: One part to the formalism I think you were trying to stress, is that in any competition between repair pathways, the pre-existing condition can be altered by switching to less error-free repair as well as to more error-prone. In trying to read and understand the literature from a formalistic or geometrical approach, I often find the assumption is that there is an increase in some repair capacity as opposed to a decrease which causes these shifts in mutability. Would you like to comment?

DR. GLICKMAN: I think that's very valuable question. When I use the word shunt, I don't really mean to imply that one is up versus one down. If the system works by supplying a product, constitutively, that allows error prone repair to occur, it may be strictly competitive so that error-free repair is avoided. I don't really see why PKM101 would replace umuC as a product if it simply blocked an error-free pathway. But things are undoubtedly going to be more complex than that answer would suggest.

DR. WALKER: At least one thing you can argue for PKM101 is it increases both survival and mutagenesis. If it were blocking an error-free pathway, it would have to decrease the amount of error-free pathway in such way that inspite of that, survival is increased, which gets you into a bit of a circle.

DR. GLICKMAN: One of the most commonly used arguments against pKM101 working by SOS-type pathway has to do with the observation of Goze and Devoret which shows that pKM101 is dependent upon uvrD and uvrE with respect to enhancement of survival. Since the host cell is not dependent upon uvrD and uvrE for SOS repair in the absence of pMK101, the idea is, from that paper, that pKM101 doesn't work by the SOS pathway. However, if one looks at uvrD and uvrE strains for mutagenesis with pKM101, one finds there is no dependence for mutagenesis. Alternatively, if one turns the situation around with lexC, the host doesn't mutate if it's a lexC mutant, and pKM101 does not restore mutability. However, it does give some protection. So apparently you can separate protective and mutational effects, although I'm not sure I understand the mechanism.

DR. WALKER: There is one very bizarre construction that we have come across. Every point mutant save one and every insertion mutant ever gotten in pKM101 abolished both the protective effect and the enhanced mutagenic effect. As shown in my paper, there's a good chance there may be two proteins coded for by pKM101 that are involved in this, although I'm not sure whether it's one transcription unit or two. We know there is one and we know the direction of transcription; if there are two proteins, they're transcribed in the same direction. A while ago we got one in vivo-derived deletion mutant that took out most of pKM101 and made a fusion somewhere around the beginning of this 2,000 base-pair region. That particular derivative of pKM101 makes the cell even more mutable than does pKM101, at least under several conditions; there are a couple of constraints on that. But strikingly it had the property of making cells more sensitive rather than more resistant to killing by UV. It's the one rather clear situation where we've been able to uncouple, to some extent, the mutagenesis and survival.

DR. GLICKMAN: And there are different protective plasmids which have different ratios of mutation enhancement and protection.

DR. WALKER: If you're thinking about how these things might work, another fact is that both umuC and at least one pKM101 protein are induced by DNA damage and are under lexA control, in a manner which is consistent with lexA acting as a repressor. So if you're thinking of models and how they work, they have to be something in which you turn the process on by DNA damage. That puts some constraints on how you build your model.

DR. BRIDGES: Could I address this question of secondary structure? We've done some studies with some DNA gyrase mutants which should be relevant here. To get cruciform structures in vivo, you've got to provide free energy, and the maintaining of negative superhelicity is one way of doing this in a cell. Indeed, the E. coli chromosome has a high degree of negative superhelicity which is largely maintained by DNA gyrase. Atte von Wright and I have done some experiments with gyrase mutants. We did them for a quite different reason to start with because we were looking to see whether superhelicity affected photoproduct yield with psoralen plus UVA.

In doing that, we had to construct a series of repair-deficient strains, uvr plus and minus, rec plus and minus, double mutants--all with and without gyrase. We found that the reduced gyrase activity in our mutants did reduce the superhelicity in the E. coli chromosome. We measured this in sucrose gradients by doing a titration with varying concentrations of ethidium bromide. We found that when superhelicity was reduced, recombination repair was also reduced. This is expected because superhelicity is necessary for recA to catalyze the invasion of a duplex by single-stranded ends. What we didn't expect to see was that excision repair activity was enhanced when there was less superhelicity. We have no explanation for that.

Now, having done that we then thought that the systems we had would be of relevance to the sort of thing that Barry was talking about because if you reduce the superhelicity, then quite a small reduction in superhelicity is going to have quite a large effect on the amount of hair-pinning and cruciform formation inside the cell, and on the amount of secondary structure. So, Maurice Southworth and I have looked at mutation induction. We've looked at various loci but the most interesting ones are ochre suppressor loci where we know we're dealing with anti-codon mutations in one of these loops of a potential hairpin. The first thing is that in uvr-minus background we found that there was absolutely no effect of the gyrase-B mutation on the frequency of UV-induced mutation. The curves were identical. And this was true even though the bacteria were more sensitive because there was less recombination repair going on. This didn't have the effect of shunting repair from an error-free mode into an error-prone mode. The mutation frequencies in the survivors were the same.

In an uvr-plus background, there was a small decrease, a very small decrease, in mutation frequency. And we were able to study this with a temperture-sensitive mutant where we could arrange for the gyrase deficiency to occur either before or after UV. When it was before, so that you had, as it were, parental DNA that was less superhelical, those were the conditions in which you got more excision repair, and the amount of extra excision repair you got was enough to account for the slightly lower mutation frequency because less damage was going through to be replicated. So, all in all, we found no evidence from this study to suggest that cruciform structure in itself, at least at this site, had any effect on UV mutagenesis.

Now, of course there is an awful lot one could do. One really needs to do this in the lacI system since one can make lots of arguments as to why things might be different in different systems. Nevertheless, the fact remains that we didn't get even a small hint of an effect of superhelicity on UV mutagenesis.

DR. GLICKMAN: Two comments might be made. One is that the suppressor loci do not behave in the same way as this site, even though we believe the secondary structure might be important. For example, mutation frequency decline is specific for some suppressor loci in an Uvr+ background. We've looked for mutation frequency decline in the lacI system and we do not see it. Mutation frequency decline is very rapid and it's prevented by the addition of nutrient broth or casimino acids. The addition of these, to the media or to the plates directly following irradiation, doesn't make a A24 appear in the Uvr+ strain. In that sense we have some differences.

The second thing I should mention is that we hope to address this kind of a problem with some topoisomerase mutants of Salmonella. Paul Margolin has a very strange result with this mutant, one that I think is not entirely explainable at the

moment. It's that the toposiomerase mutant, supX, is non-mutable
by UV, and a few other agents.

DR. BRIDGES: As I understand it, the most logical explanation is
the one that Paul Margolin himself suggests; that it's for some
reason switching off the SOS system as a whole. In which case you
can't use it to test the superhelicity argument.

DR. GLICKMAN: But we can use it in E. coli if we can get the gene
over. There might be some differences but the restrictions would
not be as large. For example, Mike Volkert has a recA
operator-constitutive mutant--one can't imagine being able to do
such experiments. It's in the thinking stage, but I believe there
are two approaches.

   One is the approach that Ripley described earlier that is to
alter the site so that the secondary structure cannot be formed,
or to create a site where it can be formed. That's one way to
go. And the other way to go would be the genetic analysis, but
one still needs to understand a great deal more of the isomerases
involved and influence upon other processes.

DR. WALKER: A suggestion for your results, built on several acts
of faith. Paul Margolin's observation is that the supX
(topoisomerase minus) mutant of Salmonella is nonmutable. It can
be restored by putting in pKM101, at least to some extent. But
that doesn't fix the whole supX phenotype; it only fixes the
nonmutability. Perhaps for some reason the Salmonella equivalent
of umuC can't be read in a supX mutant, maybe because its promoter
is only read at low superhelical density, and in a supX mutant,
DNA is not sufficiently relaxed. It would make sense for a gene,
one turned on in response to damage, to have a promoter that's
read once a domain is relaxed.

   If other genes are induced by DNA damage, such as uvrA and uvrB
and so on, and were subject to the same kind of control, it would
really be a control that's overlaid on top of lexA control; you
could take lexA on or off and still have a second kind of control
which would be dependent on how well RNA polymerase recognizes the
exposed promoter. In a gyrase minus mutant the amount of
supercoiling would be reduced, and uvrA, B and C, which are known
to be inducible and under the same kind of control, would perhaps
be read at a higher level, giving you the otherwise paradoxical
increase in excision repair. It's a rationalization which could
explain the observations.

DR. GLICKMAN: I think it's a nice idea. One of the few
situations--the only situation perhaps-- where we can say we
perhaps have targeted mutagenesis, is the CC site at A24. It's
hard to imagine the specificity of mutagenesis caused by indirect

mutagenesis, untargeted mutagenesis, being different in a Uvr⁺
an uvr⁻ strains. So I think the uvr⁻ hot spot A24 is a
potential site for a CC dimer. And I think that the uvr
dependence there might also be related to this kind of fine
control.

DR. BRIDGES: I wasn't clear why you couldn't distinguish between
PydC and pyrimidine dimer sites in your system. Why can't you
distinguish them if you do a well-designed photoreactivation study?

DR. GLICKMAN: We are in the middle of photoreactivation studies.
The basic problem is that the PydC lesion, if the same as that
described in the past by Wang, is one which is highly
photoreversible.

DR. BRIDGES: But not enzymatically.

DR. GLICKMAN: But not enzymatically. So we are doing
photoreactivation experiments at 4 and 37 , and we're looking at
the spectrum. If there is some kind of difference, one could
think of doing phr mutants, but the temperature difference should
be informative.

DR. BRIDGES: It might be better doing it with what is effectively
a dose-rate difference, because in enzymatic PR the time taken for
an enzyme molecule to find another dimer is minutes. So if you
gave your PR at the same intensity, either in a minute or as a
series of high flashes at 5 minute intervals, as has been used by
Harm and others, you could get a very fair distinction, I would
have thought better than with temperature.

DR. GLICKMAN: There is another observation and I should mention,
to do with Bockrath's work. Bockrath is the man who has really
worked out mutation frequency decline. He has made the very
interesting observation that if you UV-irradiate, then briefly
heat, and then photoactivate, you get enormous mutagenesis in an
ung- strain and not ung+ strain. It happens that he's looking at
a PydC site again. This particular damage, the C damage, is one
which is very sensitive to deamination on heating. It goes to a
U, you now photoactivate whatever it is, and in an ung- strain
you get high mutagenesis at this site.

We have the ung strains in the lacI system; hopefully we'll
soon get some data on this kind of an effect, but I don't think it
necessarily helps us to find the lesion. It might help us look at
a hot spot versus non-hot spots, we may find something. But I
don't think it will tell us wheter PydC is involved.

DR. WALKER: It occurred to me that in calculating the stability
of those cruciforms, since a dimer is thought to cause local

denaturation within about a 5 base pair region, your energy
difference between the cruciform and non-cruciform structure, is
actually bigger because that dimer may be more easily accommodated
in a loop than trying to jam it into a single-strand structure.
It's hard to know what number to put on that, but it goes in the
right direction.

DR. GLICKMAN: There is a problem here: dimers might stabilize in
this way but if one looks at hot spots in the yeast the good
reverters are not found in the unpaired regions; they are in the
stems. That may be quite different to what we see in the two E.
coli examples, but remember that different structural isomers can
be drawn at many sites.

    I wonder what happens when the mutation arises, forward
mutation rather than reversion. I wonder if there isn't some
effect on mutation fixation by creating such structures that are
more stable. I think it's a difficult problem to approach. I
think it has to be stressed, we don't really know what's going on
when one talks about DNA metabolism, we don't know how much we're
dealing with recombinational structures, whether all mutation is
fixed after UV by the same mechanism or by a variety of mechanism,
and the contribution of the different mechanisms to mutations in
different sites. I find this kind of thinking about such a
specific site intriguing and I'd like to know more about the
appearance of such sites, in which one such structure could be
created or destroyed.

DR. RIPLEY: There has been an awful lot of discussion about
stability of secondary structures without much talk about the
frequency of their formation. The stability is relatively
unimportant if they never form. Formation might be influenced by
nicks or gaps in the DNA. For example, if such a potential site
of secondary structure is at the end of an Okazaki piece, one
might imagine breathing at that end creating a structure which
would then find itself in this secondary form with a great deal
larger frequency than some other site in the DNA which wasn't near
such a frequently-occurring nick.

DR. GLICKMAN: But you've made some calculations on frameshifts at
or near potential Okazaki initiation sites. Do they fit?

DR. RIPLEY: They fit. What's known in T4 is that Okazaki pieces
in vivo begin with the sequence A,C,X,Y,Z, XYZ being any three
bases. So that doesn't tell you very much about where Okazaki
pieces start because there are a lot of potential AC sites in any
given DNA sequence. But what I have done is look at where AC
sequences are, relative to frameshift hot spots that we've been
studying in the T4 system.

We've looked at three sites each of which are 6 A's in a row in the wild type DNA sequence. They are all spontaneous hot spots but they have very differet frequencies of generating frameshift mutation. Two of them are quite high, one is considerably lower. The two that are high contain within the run of 6A's the potential for coding an Okazaki primer at that site. The one site which is much cooler does not code for such a primer within about 30 or 40 base pairs on either side. It's only a correlation and it's only one site.

DR. SHERMAN: But stretches of GC are also very hot spots?

DR. RIPLEY: They may be very hot but they may also have AC sites nearby. The point would be to compare two GC stretches, one of which was adjacent to an AC and one which was not.

DR. LOEB: If I look at half a lacI gene, do you get secondary structure if I run it through a computer. If I just take a single-strand of the lacI gene?

DR. GLICKMAN: We've been very curious about how much potential secondary structure there is anywhere. And it's non-random. There are regions of genes where there is a lot of potential secondary structure and there might be 300 nucleotides where there is very little that we can see, and then another bunch. In the lacI system there are 5 or 6 nice potential secondary structures. And I know there are regions where there is very little to be seen.

DR. LOEB: Can you draw a picture of how you can get excision repair in a clover-leaf structure?

DR. GLICKMAN: In a Uvr⁻ strain where the hot spot exists that we're really quite confident in, the A24, there is no excision repair. The lesions that are the hot spots at A23, O24 and O27, they would be single-stranded in such a secondary structure, and then probably for that short period of time, maybe during what's called postreplication excision repair, refractory to such a process. It's a possible reason why even if it's PydC which may be excized by uvrA, B, C dependent enzymes, they might be hotter than other PydC sites. We don't know because we don't know yet their formation levels really. We will soon.

SESSION III

GENETIC ANALYSIS OF MUTAGENESIS

MODERATOR: HELEN I. EBERLE

REGULATION AND FUNCTION OF CELLULAR GENE PRODUCTS INVOLVED IN UV

AND CHEMICAL MUTAGENESIS IN E. COLI

Graham C. Walker, Stephen J. Elledge, Karen L. Perry
Anne Bagg, and Cynthia J. Kenyon

Biology Department
Massachusetts Institute of Technology
Cambridge, MA 02139

The past few years have seen a remarkable increase in our understanding of the strategies employed by cells in dealing with damage to their genetic material. Of particular interest has been the recognition that mutagenesis by a variety of agents including UV, methyl methanesulfonate (MMS), and 4-nitroquinoline-1-oxide (NQO) is not a passive process. Rather it requires the intervention of a cellular system which processes damaged DNA in such a way that mutations result. Mutagenesis is not a necessary consequence of DNA damage since, if this system is inactivated, no mutations result. In this paper we would like to summarize some of the recent work of our lab on the cellular molecular mechanisms involved in UV and chemical mutagenesis.

The Concept of Error-Prone Repair System

This "mutagenesis system" is inducible. Its activity is observed in wild type cells only following treatments that either damage DNA or block replication (1). This particular property of the system is probably best illustrated by the fact that UV-irradiated bacteriophage λ are not mutated unless the E. coli strains they infect have been exposed to such an inducing treatment (2,3). In addition to increasing the mutation frequency of UV-irradiated bacteriophage, treatment of host cells with low levels of DNA damaging agents also increases the fraction of surviving phage (2,3). These inducible mutagenesis and reactivation activities have been called Weigle or W-reactivation respectively (4).

The ability of E. coli or its bacteriophage to be mutated by
UV and chemical agents can be blocked by mutations at three
bacterial loci, recA, lexA, or umuC (1,5,6). These mutations
simultaneously reduce or eliminate W-reactivation. Because of
the association of an inducible mutagenesis activity with what
appears to be an inducible repair activity it has been proposed
that mutations result from the operation of an "error-prone
repair" system (1,4). To date, the biochemical mechanism of
"error-prone repair" has not been established nor have the
effects on mutagenesis and repair even been rigorously shown to
result from the same process (7,8).

Two of the mutations inactivating the "error-prone repair"
system, recA⁻ and lexA⁻, have other effects on the cell. They
also prevent the induction of the various SOS responses such as λ
induction and filamentous growth which are seen in cells treated
with various DNA damaging agents (1).

Phenotypes of umuC Mutants

In contrast to the pleiotropic effects of recA⁻ and lexA⁻
mutations, umuC⁻ mutations seem to specifically affect
"error-prone repair". UmuC mutants were first isolated by Kato
and Shinoura (5) who screened a set of EMS-mutagenized bacteria
searching for nonmutable phenotypes. In addition to the expected
mutations mapping at the recA and lexA loci they obtained a third
class termed umuC which mapped at 25 min on the E. coli map.
Cells containing a umuC mutation i) are nonmutable with UV, MMS,
NQO, and a number of other agents (5) ii) lack the W-reactivation
response (5,9) and iii) are slightly UV sensitive
(5). Unlike recA or lexA mutants, umuC mutants still show other
SOS responses such as λ induction (5), filamentous growth (5),
and induction of the recA protein (9). The uvm mutants
independently isolated by Steinborn (6,10) appear to be identical
to umuC mutants.

We have recently been able to isolate mutants of umuC which
were generated by the insertion of either the bacteriophage
Mud(Ap, lac) (11) or the transposon Tn5 into the umuC gene (12;
Elledge and Walker, unpublished results). They were identified
by screening the appropriate insertion mutants of E. coli for
nonmutable phenotypes. The umuC::Mud(Ap, lac) mutation has been
characterized in detail (12) and it causes the same phenotype as
the EMS-derived umuC mutants identified by Kato and Shinoura (5).
The Mud(Ap, lac) insertion maps to the position of the umuC gene
and thus the nonmutability of the strain is presumably due to an
insertion of the Mud(Ap, lac) within the umuC gene or proximal to
it within the same transcriptional unit. This insertion mutation
of umuC is likely to be a null allele; the mutation is recessive

and was detected after screening a relatively modest number (17,000) of random Mud(Ap, lac) insertion mutants (12). The simplest interpretation of such a null allele is that the umuC gene product is an active participant in the processing of DNA damage which results in mutations. In other words, it seems likely that umuC is a key gene for "error-prone repair".

Induction and Regulation of umuC Expression

The insertion of the Mud(Ap, lac) phage (12) into the umuC gene created an operon fusion which placed the structural gene for β-galactosidase (lacZ) under the control of the regulatory region of the umuC gene. We found that β-galactosidase synthesis was induced by treatment of the umuC-lac fusion strain with DNA-damaging agents such as UV, MMS, NQO, and N-methyl-N'-nitro-N-nitrosoguanidine (MNNG) (12). Since the Mud(Ap, lac) appears to be inserted with the umuC transciptional unit this provides direct evidence that the expression of the umuC gene is induced by DNA damage

By transducing various mutations into a strain carrying the umuC::Mud(Ap, lac) fusion we have been able to genetically analyze the regulation of the umuC gene. As shown in Table 1 the introduction of a recA⁻ or lexA⁻ mutation prevented the UV induction of β-galactosidase expression in a umuC::Mud(Ap, lac) strain. In contrast, the introduction of a putative null allele of lexA, termed spr, into the strain resulted in the constitutive high level expression of β-galactosidase (Table 1). Thus the lexA protein seems to play a negative regulatory role in the

Table 1. UV Induction of β-Galactosidase in the umuC::Mud(Ap, lac) Fusion Strain and Its Derivatives[a]

Strain	Relevant Genotype	β-Galactosidase (Units/A$_{600}$) After UV Irradiation	
		0 min	90 min
GW1104	lexA⁺tif-1	4.6	76
GW1105	lexA3 tif-1	1.4	1.7
GW1106	lexA⁺recA56	1.8	1.4
GW1107	spr-51 tif-1	262	245
GW1108	spr-51 recA56	187	193

[a] Cells were grown in a supplemented minimal medium at 30°C and then were UV irradiated (10 J/m²). β-Galactosidase activity was measured as described previously (12).

control of the umuC gene. Once lexA activity had been eliminated
from a cell there did not seem to be a requirement for recA
function since a spr recA derivative also showed high level
constitutive expression of β-galactosidase (Table 1). These
observations suggest that the lexA protein is the direct
repressor of the umuC gene and that induction occurs when, in
response to DNA damage, the lexA protein is cleaved by the recA
protease (12,13). It should be noted that these experiments
analyze only the transcription regulation of umuC and do not
address the issue of whether there is any translation regulation
of umuC or whether any posttranslational modification of the umuC
gene product is required for its activity.

Regulation of din (Damage Inducible) Genes

We and others have used the Mud(Ap, lac) operon fusion
vector to identify other genes besides umuC whose expression is
induced by DNA damage. Mud(Ap, lac) fusions have been obtained
to uvrA (14,15), uvrB (15,16), sfiA (17), dinA, dinB, dinD, and
dinF (14). Both recA (18) and lexA (19) have been shown to be
inducible by other means. Genetic (14-17; Kenyon and Walker,
unpublished results) and biochemical (Brent and Ptashne, in
press; Little and Mount, in press) analysis has indicated that
the regulation of expression of each of these other genes is
similar to that of umuC with lexA being the direct repressor of
each gene with induction occurring when lexA is proteolytically
cleaved by the recA protease (Fig. 1). Thus umuC is one member
of a regulatory network in E. coli which consists of at least ten
genes with probably more still to be discovered (20,21).

Cloning of the umuC Gene

We have recently suceeded in cloning the umuC gene of E.
coli K12 (Elledge and Walker, unpublished results). We had
initially attempted to do this by preparing a bank of random E.
coli fragments in the multicopy vector pBR322 and then screening
these hybrid plasmids for their ability to complement a umuC
strain. We did not identify the umuC gene by this procedure
perhaps because the increased gene dosage resulted in
overproduction of the product of umuC or some closely linked
gene; this overproduction may have been deleterious to cell
survival or have resulted in an alternative phenotype. In an
alternative approach we first obtained a Tn5 insertion in the
umuC gene and then cloned out a restriction fragment containing
part of Tn5 (including the neomycin phosphotransferase) as well
as a small piece of the adjacent chromosomal DNA. This
restriction fragment was then used to screen λ banks containing
random fragments of E. coli DNA. A λ phage was identified whose

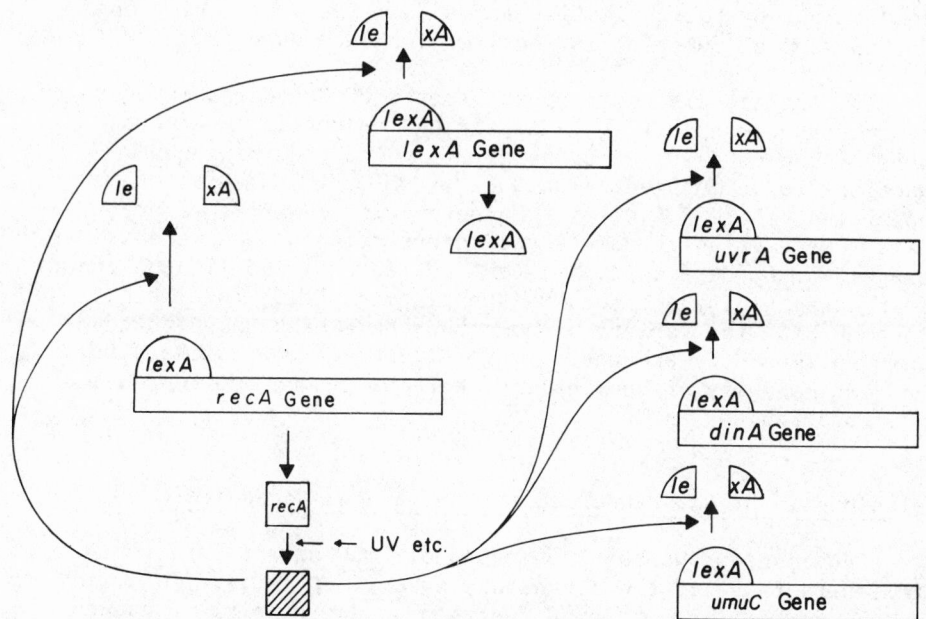

Fig. 1. Model for the regulation of E. coli genes that are
turned on by DNA damage.

DNA hybridized to DNA from the umuC region of the chromosome.
Since this particular λ phage also complemented the nonmutability
of a umuC mutant (21) we conclude that it carries the umuC gene.
We anticipate that this work will be of considerable help in
attempting to identify the role of the umuC gene product in
mutagenesis.

Nature of umuC Requirement for Mutagenesis

Chemical mutagens seem to fall into two classes, those which
do not require umuC function to be mutagenic and those which do.
Two examples of umuC-independent mutagens are MNNG and ethyl
methanesulfonate (EMS) (6,12,22). In general mutagens in the
first class introduce lesions such as O^6-methylguanine which
alter nucleic acid bases in such a way that they can directly
mispair with incorrect bases. Two recent pieces of evidence
particularly support the contention that mispairing base
derivatives such as O^6-methylguanine are the premutagenic lesions
for this class of chemical mutagens. Firstly, almost all of the
base pair changes caused by these agents are GC AT transition
(23) and secondly, the direct removal of the methyl group from
O^6-methylguanine as part of the adaptive response (24) prevents
mutagenesis by such agents. It would appear that mutations can

occur simply by polymerization across a mispairing lesion and
that no special function is required for such an event.

In contrast, it would appear that umuC-dependent mutagens do
not introduce many, if any, mispairing lesions into DNA but
rather cause other types of damage. Agents requiring umuC
functions to be mutagenic include UV, MMS, NQO, X-rays, and
neocarcinostatin (5,6,9,25). Moreover, in so far as we are
aware, there is an extremely good correlation between agents
being i) umuC$^+$-dependent, ii) recA$^+$-dependent, and iii) having
their mutagenicity enhanced by the plasmid pKM101 (see below).
Since such a wide variety of compounds have been shown to have
their mutagenicity enhanced by pKM101 (26) it seems likely that
the vast majority of chemical mutagens will turn out to require
umuC function.

Relationship of umuC Function to Spontaneous Mutagenesis

The spontaneous mutation rates of umuC mutants are very
similar to those of the corresponding umuC$^+$ strains. This lack
of an important role of umuC function in determining spontaneous
mutation rates is consistent with the low level of expression of
umuC in uninduced cells discussed above. Moreover we have found
that umuC mutations have negligible effects on the high
spontaneous mutation rates seen in mutH, mutL, and mutS strains
(H. Lust and G.C. Walker, unpublished results). The products of
the mutH, mutL, and mutS genes are required for the
methyl-directed mismatch repair system (27,28) which is thought
to play a major role in accurately correcting base mismatches
resulting from polymerase errors during DNA replication. Thus it
seems the umuC gene product does not influence fidelity of DNA
replication in uninduced cells.

Growth of a tif sfiA mutant at 42°, a condition which
results in the induction of SOS functions, has been reported to
cause very substantial increases in the rate of reversion of his
and arg ochre mutations in E. coli K-12 (29) and of a trp ochre
mutation in E. coli B/r (30). Since the umuC gene product is
expressed under SOS-inducing conditions and this increase in
spontaneous mutation rate is not seen in a tif sfiA umuC mutant
it appears that the umuC product is playing a role in the
increase in spontaneous mutation rate (Kenyon and Walker,
unpublished results). It could do so either by processing
otherwise cryptic lesions in an error-prone fashion or by
decreasing the fidelity of DNA replication. However, it does not
appear that expression of umuC leads to a generalized decrease in
replication fidelity since the reversion of a number of other
point mutants is not increased or is increased to only a small
extent by the growth of a tif mutant at 42° (A. Schauer and G.C.

Walker, unpublished results). Moreover, R. Brent and M. Ptashne
(personal communication) found that growth of a tif sfiA strain
at 42° caused only a relatively small increase in the spontaneous
frequency of lacI⁻ mutants. Thus, whatever its basis, the
potential for the umuC gene product to influence the spontaneous
mutation rate seems to be quite limited.

Analogs to umuC on Naturally Occurring Plasmids?

    The plasmids pKM101 and R46 are members of a subset of
naturally occurring plasmids which increase the susceptibility of
cells to UV and chemical mutagenesis and also increase their
resistance to killing by UV (31,32). The plasmid pKM101 was
derived from the clinically-isolated plasmid R46 by a series of
in vivo manipulations which resulted in the loss of a single 14
kb region of DNA containing several drug resistances (33-35).
Because of its ability to enhance chemical mutagenesis pKM101 was
incorporated into the Ames Salmonella tester strains for the
detection of carcinogens as mutagens (36) and it has played a
major role in the success of the system (37-39). The plasmid not
only increases the susceptibility of cells to base substitution
mutagenesis but also to frameshift mutagenesis with the
appropriate compounds.

    It seems likely that pKM101 may exert its effects by coding
for an analog of the chromosomal umuC gene. The ability of
pKM101 to enhance mutagenesis and resistance to UV killing is
recA⁺lexA⁺-dependent as are the SOS functions (40). However the
introduction of pKM101 into a umuC mutant i) restores the ability
of the strain to be mutated by UV, MMS, NQO, and other agents ii)
restores the ability of the strain to carry out W-reactivation of
UV-irradiated bacteriophage λ and iii) makes the cells more
resistant to UV killing (9,12). It does not affect other SOS
functions such as recA induction or λ induction (9,41). The
simplest interpretation of this result is that pKM101 carries an
analog of the umuC gene which we have termed muc (mutagenesis-UV
and chemical). We have discussed the other formal possibilities
for the relationship of muc to umuC elsewhere (42).

The muc Function of pKM101

    We have been able to obtain both MNNG-derived (43) and Tn5
insertion mutants (42) of pKM101 which were no longer able to
increase the susceptibility of cells to base substitution
mutagenesis. These same plasmid mutants were unable to increase
frameshift mutagenesis, make cells more resistant to UV killing,
or restore the ability to carry out W-reactivation to umuC
mutants (42,43). Thus it appears all these processes are closely

related. The position of Tn5 in twenty muc⁻ insertion mutants
was determined and all mapped within an approximately 1900 bp
region of pKM101 (42). The muc region has been recently cloned
into the vector pBR322 and we have found that the hybrid plasmid,
which contains a 2200 bp fragment of pKM101 DNA, is capable of
increasing both the susceptibility of cells to chemical
mutagenesis and their resistance to UV killing (Perry and Walker,
unpublished results). Thus the muc region of pKM101 is both
necessary and sufficient for plasmid-mediated mutagenesis and
UV-resistance. If pKM101 does in fact work by providing an
analog of the umuC gene product then the muc region codes for
that umuC analog.

The observation that pKM101 increased the capacity of cells
to carry out W-reactivation of UV-irradiated bacteriophage (44)
and that this phenomenon was recA⁺lexA⁺-dependent (45) raised the
possibility that the pKM101 muc function was inducible. We have
recently used in vitro technology (46) to construct a gene fusion
of the amino terminus end of the muc gene to the β-galactosidase
gene (Elledge and Walker, unpublished results). A fusion protein
is produced which has β-galactosidase activity and its expression
is stimulated by DNA damaging agents in a recA⁺lexA⁺-dependent
fashion. This indicates that the gene product of the muc gene is
a protein. Moreover it shows that there is a further similarity
between the plasmid-borne muc gene and the chromosomally-carried
umuC gene in that the expression of each is stimulated by DNA
damaging agents in a manner which is controlled by the recA and
lexA proteins.

How Do the umuC and muc Gene Products Act?

The mechanism by which damaged DNA is processed to give rise
to mutations is not yet clear. The umuC and muc gene products
could act at either a regulatory or a mechanistic level in the
process. Certainly it is simplest to assume that they act at a
mechanistic level. Both umuC⁻ and muc⁻ mutants can be generated
by insertions and the expression of both of these genes is
induced by DNA damage. These observations place severe
constraints on any models which view the umuC or muc products as
being further elements in a regulatory cascade that ultimately
governs the expression of some as yet unidentified gene needed
for "error-prone repair".

Even if the umuC and/or muc gene products do play a
mechanistic role in "error-prone repair" other proteins may well
be involved. In particular, it is still an open question as to
whether the recA and lexA proteins play mechanistic roles in
"error-prone repair" in addition to their roles in regulating the
expression of umuC. For example, the observation that spr recA

strains are nonmutable even though we have shown they express the
umuC gene product at high level (12) could be due to: i) the recA
and/or lexA gene products playing mechanistic roles in
"error-prone repair" ii) the need to induce some additional
function which is not expressed in spr recA strains or iii) the
requirement for some activity of the recA or lexA proteins to
activate the umuC gene product. In addition, essential proteins,
for example the products of some of the dna genes, could
participate in "error-prone repair". The screenings which have
been carried out searching for nonmutable phenotypes (5, 6, 12)
might not have detected alleles of such genes since no attempts
were made to look for conditional mutants.

Any model to explain the mechanism of action of umuC or muc
must acccount for the following observations: i) both base
substitution and some frameshift mutations can result from umuC
or muc dependent processes (5,6,33,42,47) ii) loss of umuC or muc
function results in an increase in sensitivity to killing by UV
and some agents (42, 43) implying that these functions can help
repair lethal lesions iii) $umuC^+$ or muc^+-dependent mutagenesis
can be seen with sublethal doses of some mutagens (40) implying
that DNA containing otherwise cryptic lesions can be processed to
give mutations iv) the ability of umuC or muc dependent
processes to introduce mutations into DNA is far greater if the
DNA template contains lesions (40,48).

The most popular model to explain "error-prone repair" has
been that polymerization occurs across a noncoding lesion such as
a pyrimidine dimer with the resulting incorporation of incorrect
nucleotides (7). Limited evidence has been presented that such
events occur in vitro for UV-irradiated single strand templates
in SOS-induced cells (49). However, attempts to demonstrate
transdimer synthesis in vitro have been unsuccessful to date (7,
8) although the recent results of Lippke et al. (50) suggests
that pyrimidine dimers may not be the primary premutagenic lesion
introduced by UV irradiation. Schaaper et al. (51) have
presented evidence that apurinic sites may be processed by this
type of mechanism. Polymerization across apurinic sites has been
demonstrated in vitro and apurinic sites have been shown to act
as premutagenic lesions in SOS-induced cells but not in uninduced
cells (51). Bridges and his colleagues have argued that induced
mutagenesis has a requirement for polymerase III function (52)
and recently Fields and Yasbin (53) have shown that in B.
subtilis W-reactivation can be prevented by an inhibitor of
polymerase III.

It seems clear that mutagenesis does not result from the
action of either of two inducible $recA^+lexA^+$-dependent repair
pathways which have been studied - long patch excision repair
(54) and postreplicational repair (55). Long patch excision

repair occurs normally in umuC⁻ mutants (P.K. Cooper, personal
communication) and a uvrA umuC⁻ strain performed
postreplicational repair as efficiently as a uvrA⁻umuC⁺ strain
(55). At this stage it might be wise to keep an open mind as to
the possible mechanism of "error-prone repair". For example, the
in vivo results referred to above could formally be explained by
a model in which an oligonucleotide having short regions of
homology to either side of a noncoding lesion could be "grafted"
into a DNA so that incorrect bases would be placed opposite the
lesion. We hope that our progress to date on cloning the umuC
and muc genes and obtaining relevant mutants will allow a
systematic approach towards determining the molecular basis of
"error-prone repair" whatever its mechanism.

Are Some Bacteria Naturally Deficient in umuC Function?

In the course of looking at the effects of pKM101 in E. coli
K12 and S. typhimurium LT2 some differences have become apparent.
The nature of these differences leads us to raise the speculative
possibility that S. typhimurium is somewhat deficient in umuC
function relative to E. coli K12. For example, S. typhimurium

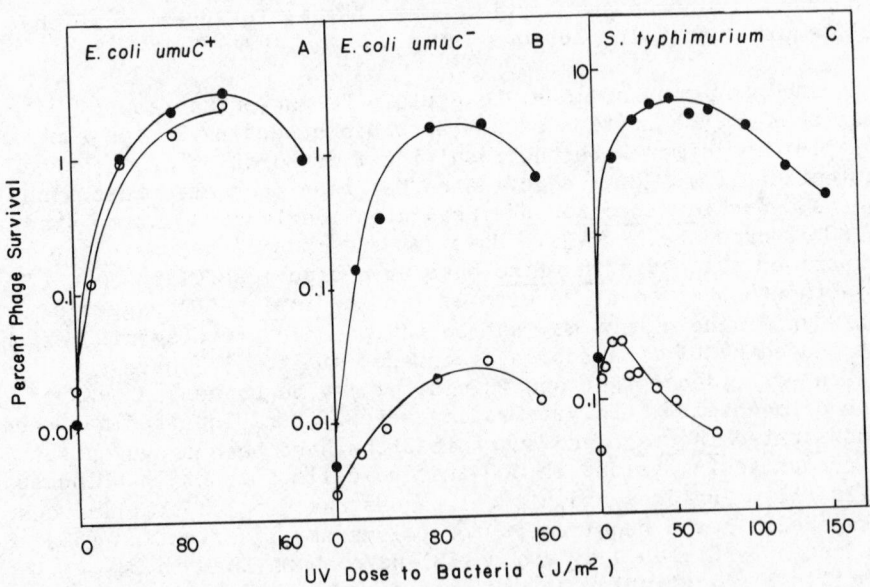

Fig. 2. Phenotypic similarity of S. typhimurium LT2 and E. coli
K12 umuC mutants. A,B. Weigle-reactivation of UV-irradiated
bacteriophage λ in E. coli K12 uvrA⁺umuC⁺ and uvrA⁺umuC⁻ strains
respectively, with (●) and without pKM101 (0)(9). C. Weigle
reactivation of UV-irradiated bacteriophage P22 in uvr⁺ S.
typhimurium with (●) and without (0) pKM101 (44).

shows only a relatively weak W-reactivation response relative to
E. coli K12 (44). Similarly an E. coli umuC mutant has only a
very weak W-reactivation response (Fig. 2) (5, 9, 39). The
introduction of pKM101 substantially increases the capacity of S.
typhimurium and E. coli umuC mutants to W-reactivate
UV-irradiated phage yet makes a relatively smaller difference in
an E. coli umuC strain (Fig. 2) (9, 44). In addition the
introduction of pKM101 increases the susceptibility of S.
typhimurium and E. coli umuC mutants to MMS or UV mutagenesis to
a much greater extent than it increases the susceptibility of E.
coli umuC$^+$ strains (39). A final piece of circumstantial
evidence comes from a comparison of the genetic maps of the two
organisms. The arrangement of genes on the respective bacterial
chromosomes is almost identical except that they apparently
differ by an inverison of a ten minute segment of the chromosome
in one area (Fig. 3) (56). We have previously noted that the
umuC gene in E. coli maps very close to one of the endpoints of
this inversion (39). Thus umuC function could have been affected
at some point in the evolutionary divergence of these two
organisms.

     Certain other bacteria such as Haemophilus influenzae are
naturally nonmutable by UV (57) and thus resemble E. coli umuC$^-$
mutants. Moreover H. influenzae shows an inducible accurate

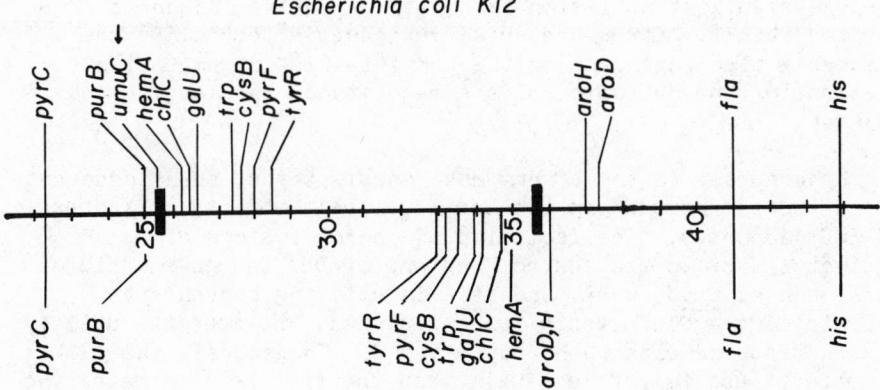

Fig. 3. Comparison of a portion of the S. typhimurium LT2
genetic map with a portion of the E. coli K12 genetic map
[adapted from Sanderson and Hartman (56)]. The portion of the
map from 25.5 to 35.5 U inside the heavy bars is inverted between
the two genera. Only those genes mapped in both genera are shown
except for umuC whose approximate position on the E. coli map (5)
is indicated.

reactivation of UV-irradiated bacteriophage (58) which resembles the inducible uvrA-dependent reactivation of UV-irradiated bacteriophage which is seen in umuC mutants of E. coli (9).

The fact that the umuC-suppressing muc gene is found not only on naturally occurring plasmids but also possibly on a transposable element [the muc gene of pKM101 is surrounded by inverted repeats (35)] suggests that it may be sometimes advantageous for a bacterium to have umuC-like function even if it normally is deficient in this function.

Implications for Genetic Toxicity Testing

The introduction of the plasmid pKM101 into the Ames Salmonella tester strains (36) greatly increased the susceptibility of the strains to mutagenesis by a wide variety of chemical mutagens (26). As we have discussed above, it seems that pKM101 is acting by increasing the amount of umuC-like function in the cells. This may be of particular significance since S. typhimurium LT2, which was chosen as the test organism for historical reasons, may be naturally deficient in umuC function. In addition, it seems that the basal level of expression of the muc gene product is probably higher than that of the umuC gene product (12, 40) so that a compound can be detected as a mutagen if it introduces a premutagenic lesion into the DNA, even if that lesion is not an effective inducer of umuC function itself. Other bacterial mutagenesis tests (59) are presumably also dependent on the participation of umuC-like function for the detection of the majority of carcinogens and mutagens.

In contrast to short term mutagenesis tests, the "Inductest" (60) detects compounds on the basis of their ability to induce the SOS responses. The isolation of operon fusions of β-galactosidase to the control regions of DNA damage-inducible genes such as umuC, uvrA, and sfiA is allowing convenient variants of the "Inductest" to be designed. However the utility of such tests remains to be established. In general, the "Inductest" has faired less well than the Ames test in detecting known carcinogens (61). This may be because the "Inductest" reacts to lesions which act as inducing signals for the SOS responses or can initiate the formation of such inducing signals. In contrast the Ames tester strains, especially the pKM101-containing strains, primarily measure whether a lesion can give rise to a mutation either by direct mispairing or by being processed by an "error-prone repair" system. Possibly the ability of a compound to introduce a premutagenic lesion into DNA may be a better predictor of its carcinogenicity than its ability to generate an inducing signal for a particular bacterial regulatory system.

Implications for Higher Organisms

A nonmutable phenotype can apparently result from a loss of cellular function. This suggests that the DNA damage resulting from UV and many chemical mutagens is not intrinsically mutagenic. Since the cells of many higher organisms including humans, can be mutated by UV and many of these same chemicals, it seems reasonable to argue that these organisms possess analogous processing systems.

Both "error-free" and "error-prone" repair systems are inducible in E. coli. A considerable amount of the phage reactivation which is observed in a W-reactivation experiment is apparently due to the induction of the uvrA and uvrB gene products (15) and is not related to the induction of "error-prone repair". The "adaptive" response of E. coli is inducible and prevents the introduction of mutations as a consequence of DNA damage (62). Thus experiments in mammalian cells simply showing inducible reactivation of a virus without at the same time following the frequency of viral mutagenesis do not necessarily demonstrate the existence of error-prone repair. Moreover, an "error-prone repair" system could exist in a mammalian cell yet, at least in principle, be constitutively expressed. "Inducible repair" does not imply "error-prone repair" and "error-prone repair" does not necessarily imply "inducible error-prone repair".

Finally the demonstration in E. coli that DNA damage elicits the expression of a set of genes suggests at least three ways that DNA damage could have long term biological consequences such as cancer in higher organisms. 1. The expression of new gene functions could cause epigenetic changes. 2. A function, such as a transposase, could be induced by DNA damage. This function could then cause permanent genetic alterations irrespective of whether the DNA was damaged or not. 3. A function such as umuC could be induced by DNA damage. This function could then operate on the damaged DNA template to cause permanent genetic changes.

Acknowledgements

We thank J. Geiger, J. Krueger, B. Mitchell, P. Pang, W. Shanabruch, and S. Winans for their many useful discussions and L. Withers for help in preparing the manuscript.

This work was supported in part by a Rita Allen Scholar Award to G.C.W. and Public Health Service grants 5-R01-CA21615-04 from the National Cancer Institute and 1-R01-GM28988-01 from the National Institute of General Medical Sciences. C.J.K. was a National Science Foundation Predoctoral Fellow and now is

supported by a Johnson and Johnson Associated Industries Fund
Fellowship. K.L.P. and S.J.E. were supported by an NIH
Predoctoral Training Grant Number GM07287. A.B. was supported in
part by the Undergraduate Opportunities Program at the
Massachusetts Institute of Technology.

References

1. Witkin, E.M.: Ultraviolet mutagenesis and inducible DNA
      repair in Escherichia coli. Bacteriol. Rev., 40: 869 (1976)
2. Weigle, J.J.: Induction of mutations in a bacterial virus.
      Proc. Nat. Acad. Sci. U.S.A., 39: 628 (1953)
3. Defais, M., Fauquet, P., Radman, M., Errera, M.: Ultraviolet
      reactivation and ultraviolet mutagenesis of λ in different
      genetic systems. Virology, 43: 495 (1971)
4. Radman, M.: SOS repair hypothesis: phenomenology of an
      inducible DNA repair which is accompanied by mutagenesis.
      In Molecular Mechanisms for Repair of DNA, Ed. by P.C.
      Hanawalt, and R.B. Setlow. Plenum Publishing Corp, New
      York, part A, p. 355 (1978)
5. Kato, T., Shinoura, Y.: Isolation and characterization of
      mutants of Escherichia coli deficient in induction of
      mutations by ultraviolet light. Molec. Gen. Genet., 156:
      121 (1977)
6. Steinborn, G.: Uvm mutants of Escherichia coli K12 deficient
      in UV mutagenesis. Molec. Gen. Genet., 165: 87 (1978)
7. Boiteux, S., Villani, G., Spadari, S., Zambrano, F., Radman,
      M.: Making and correcting errors in DNA synthesis: in vitro
      studies of mutagenesis. In DNA Repair Mechanisms, Ed. by
      P.C. Hanawalt, E.C. Friedberg, C.F. Fox. Academic Press,
      New York, p. 73 (1978)
8. Villani, G., Boiteux, S., Radman, M.: Mechanisms of
      ultraviolet-induced mutagenesis: extent and fidelity of in
      vitro DNA synthesis on irradiated templates. Proc. Natl.
      Acad. Sci. U.S.A., 75: 3037 (1978)
9. Walker, G.C., Dobson, P.P.: Mutagenesis and repair
      deficiencies of Escherichia coli umuC mutants are
      suppressed by the plasmid pKM101. Molec. Gen. Genet., 172:
      17 (1979)
10. Steinborn, G. Uvm Mutants of Escherichia coli K12 Deficient
      in UV Mutagenesis. Molec. Gen. Genet., 175: 203 (1979)
11. Casadaban, M.J. and Cohen, S.N.: Lactose genes fused to
      exogenous promoters in one step using a Mu-lac
      bacteriophage: In vivo probe for transcriptional control.
      Proc. Natl. Acad. Sci. U.S.A., 76: 4530 (1979)
12. Bagg, A., Kenyon, C.J. and Walker, G.C.: Inducibility of a
      gene product required for UV and chemical mutagenesis in
      Escherichia coli. Proc. Natl. Acad. Sci. U.S.A., in press
      (1981)

13. Little, J.W., Harper, J.E.: Identification of the lexA gene
    product of Escherichia coli K-12. Proc. Nat. Acad. Sci.
    U.S.A., 76: 6147 (1979)
14. Kenyon, C.J. and Walker, G.C.: DNA-damaging agents stimulate
    gene expression at specific loci in Escherichia coli. Proc.
    Natl. Acad. Sci. U.S.A., 77: 2819 (1980)
15. Kenyon, C.J. and Walker, G.C.: Expression of the uvrA gene of
    Escherichia coli is inducible. Nature, 289: 810 (1981)
16. Fogliano, M. and Schendel, P.F.: Evidence for the
    inducibility of the uvrB operon. Nature, 289: 196 (1981)
17. Huisman, O. and D'Ari, R.: An inducible DNA replication-cell
    division coupling mechanism in Escherichia coli. Nature,
    290: 797 (1981)
18. McEntee, K., Hesse, J.E. and Epstein, W.: Identification and
    radiochemical purification of the recA protein of
    Escherichia coli. Proc. Natl. Acad. Sci. U.S.A., 73: 3979
    (1976)
19. Brent, R. and Ptashne, M.: The lexA gene product represses
    its own promoter. Proc. Natl. Acad. Sci. U.S.A., 77: 1932
    (1977)
20. Walker, G.C., Kenyon, C.J., Bagg, A., Langer, P.J. and
    Shanabruch, W.G.: Mutagenesis and cellular responses to DNA
    damage. J. Nat. Cancer Institute Monograph, in press (1981)
21. Walker, G.C., Kenyon, C.J., Bagg, A., Elledge, S.J., Perry,
    K.L. and Shanabruch, W.G.: Regulation and functions of
    Escherichia coli genes induced by DNA damage. In Molecular
    and Cellular Mechanisms of Mutagenesis, Ed. by J.F. Lemontt
    and W.M. Generoso. Plenum Press, New York, in press (1981)
22. Schendel, P.F. and Defais, M.: The role of the umuC gene
    product in mutagenesis by simple alkylating agents. Molec.
    Gen. Genet., 177: 661 (1980)
23. Coulondre, C. and Miller, J.H.: Genetic studies of the lac
    repressor IV. J. Molec. Biol., 117: 577 (1977)
24. Olsson, M. and Lindahl, T.: Repair of alkylated DNA in
    Escherichia coli: methyl group transfer from
    O^6-methylguanine to a protein cysteine residue. J. Biol.
    Chem., 255: 10569 (1980)
25. Eisenstadt, E., Wolf, M. and Goldberg, I.H.: Mutagenesis by
    neocarcinostatin in Escherichia coli and Salmonella
    typhimurium: requirement for $umuC^+$ or plasmid pKM101. J.
    Bacteriol., 144: 656 (1980)
26. McCann, J., Choi, E., Yamasaki, E. and Ames, B.N.: Detection
    of carcinogens as mutagens in the Salmonella/-microsome
    test: assay of 300 chemicals. Proc. Natl. Acad. Sci.
    U.S.A., 72: 5135 (1975)
27. Glickman, B.W. and Radman, M.: Escherichia coli mutators
    deficient in methylation-instructed mismatch repair
    correction. Proc. Natl. Acad. Sci. U.S.A., 77: 1063 (1980)

28. Radman, M., Wagner, R.E., Glickman, B.W. and Meselson, M.:
    DNA methylation, mismatch correction and genetic stability.
    In Progress in Environmental Mutagenesis, Ed. by M.
    Alecevic. Elsevier/North Holland Biomedical Press,
    Amsterdam, p. 121 (1980)
29. George, ᵀ., Castellazzi, M. and Butlin, G.: Prophage
    induction and cell division in E. coli III. Mutations sfiA
    and sfiB restore division in tif and lon strains and permit
    the expression of mutator properties of tif. Molec. Gen.
    Genet., 140: 309 (1975)
30. Witkin, E.M.: Thermal enhancement of ultraviolet mutability
    in a tif-1 uvrA derivative of Escherichia coli B/r:
    evidence that ultraviolet mutagenesis depends upon an
    inducible function. Proc. Natl. Acad. Sci. U.S.A., 71: 1930
    (1974)
31. Drabble, W.T. and Stocker, B.A.D.: R (transmissible
    drug-resistance) factors in S. typhimurium: pattern of
    transduction by phage P22 and ultraviolet protection
    effect. J. Gen. Microbiol., 53: 109 (1968)
32. Mortelmans, K.E. and Stocker, B.A.D.: Ultraviolet light
    protection, enhancement of ultraviolet light mutagenesis,
    and mutator effect of plasmid R46 in Salmonella
    typhimurium. J. Bacteriol., 128: 271 (1976)
33. Mortelmans, K. and Stocker, B.A.D.: Segregation of the
    mutator property of plasmid R46 from its
    ultraviolet-protecting property. Molec. Gen. Genet. 167:
    317 (1979)
34. Langer, P.J. and Walker, G.C.: Restriction endonuclease
    cleavage map of pKM101: relationship to parental plasmid
    R46. Molec. Gen. Genet., in press (1981)
35. Langer, P.J., Shanabruch, W.G. and Walker, G.C.: Functional
    organization of the plasmid pKM101. J. Bacteriol., 145:
    1310 (1981)
36. McCann, J., Spingarn, N.E., Kobori, J. and Ames, B.N.:
    Detection of carcinogens as mutagens: bacterial tester
    strains with R factor plasmids. Proc. Natl. Acad. Sci.
    U.S.A., 72: 979 (1975)
37. McCann, J. and Ames, B.N.: Detection of carcinogens as
    mutagens in the Salmonella/microsome test: assay of 300
    chemicals: discussion. Proc. Natl. Acad. Sci. U.S.A., 73:
    950 (1976)
38. Walker, G.C.: Theory and design of short term bacterial tests
    for mutagenesis. In Assessing Chemical Mutagens: the Risk
    to Humans, Ed. by V.K. McElheny and S. Abrahamson. Cold
    Spring Harbor Laboratory, Cold Spring Harbor, New York, p.
    63 (1979).
39. Walker, G.C.: Molecular principles underlying the Ames
    Salmonella/microsome test: elements and design of short
    term mutagenesis test. Ed. by A.R. Kolber and T.K. Wang.
    Plenum Press, New York, in press (1981)

40. Walker, G.C.: Plasmid(pKM101)-mediated enhancement of repair
    and mutagenesis: dependence on chromosomal genes in
    Escherichia coli K-12. Molec. Gen. Genet., 152: 93 (1977)
41. Goze, A. and Devoret, R.: Repair promoted by plasmid pKM101
    is different from SOS repair. Mutat. Res., 61: 163 (1979)
42. Shanabruch, W.G. and Walker, G.C.: Localization of the
    plasmid (pKM101) gene(s) involved in recA$^+$lexA$^+$-dependent
    mutagenesis. Molec. Gen. Genet., 179: 289 (1980)
43. Walker, G.C.: Isolation and characterization of mutants of
    the plasmid pKM101 deficient in their ability to enhance
    mutagenesis and repair. J. Bacteriol., 133: 1203 (1978b)
44. Walker, G.C.: Inducible reactivation and mutagenesis of
    UV-irradiated bacteriophage P22 in Salmonella typhimurium
    LT2 containing the plasmid pKM101. J. Bacteriol., 135: 415
    (1978a)
45. Dobson, P.P. and Walker, G.C.: Plasmid (pKM101)-mediated
    Weigle reactivation in Escherichia coli K-12 and Salmonella
    typhimurium LT2: Genetic dependence, kinetics of induction,
    and effect of chloramphenicol. Mutat. Res., 71: 25 (1980)
46. Casadaban, M.J., Chou, J. and Cohen, S.N.: In vitro gene
    fusions that join an enzymatically active β-galactosidase
    segment to amino-terminal fragments of exogenous proteins:
    Escherichia coli plasmid vectors for the detection and
    cloning of translation initiation signals, J. Bacteriol.,
    143: 971 (1980)
47. Kato, T. and Nakano, E.: The effects of the umuC36 mutation
    on ultraviolet-induced base changes and frameshift
    mutations in Escherichia coli. Mutat. Res., in press
    (1981)
48. Ichikawa-Rho, H. and Kondo, S.: Indirect mutagenesis in phage
    lambda by ultraviolet preirradiation of host bacteria. J.
    Molec. Biol., 97: 77 (1975)
49. Caillet-Fauquet, P., Defais, M. and Radman, M.: Molecular
    mechanisms of induced mutagenesis: Replication in vivo of
    bacteriophage φX174 single-stranded, ultraviolet
    light-irradiated DNA in intact and irradiated host cells.
    J. Molec. Biol., 117: 95 (1977)
50. Lippke, J.A., Gordon, L.K., Brash, D.E. and Haseltine, W.A.:
    The distribution of ultraviolet light induced damage in a
    defined sequence of human DNA: Detection of alkaline
    sensitive lesions at pyrimidine-cytosine sequences. Proc.
    Natl. Acad. Sci. U.S.A., in press (1981)
51. Schaaper, R.M. and Loeb, L.A.: Depurination causes mutations
    in SOS-induced cells. Proc. Natl. Acad. Sci. U.S.A., 78:
    1773 (1981)
52. Bridges, B.A., Mottershead, R.P. and Sedgwick, S.G.:
    Mutagenic DNA repair in Escherichia coli III. Requirement
    for a function of DNA polymerase III in ultraviolet light
    mutagenesis. Molec. Gen. Genet., 144: 53 (1976)

53. Fields, P.I. and Yasbin, R.E.: Involvement of
    deoxyribonucleic acid polymerase III in W-reactivation in
    Bacillus subtilis. J. Bacteriol., 144: 473 (1980)
54. Cooper, P.K. and Hanawalt, P.C.: Role of DNA polymerase I and
    the rec system in excision repair in Escherichia coli.
    Proc. Natl. Acad. Sci. U.S.A., 69: 1156 (1972)
55. Kato, T.: Effects of chloramphenicol and caffeine on
    postreplication repair in uvrA⁻umuC⁻ and uvrA⁻recF⁻ strains
    of Escherichia coli. Molec. Gen. Genet., 156: 115-120
    (1977)
56. Sanderson, K.E. and Hartman, P.E.: Linkage map of Salmonella
    typhimurium. edition V., Microbiol. Rev., 42: 471 (1978)
57. Kimball, R.F., Boling, M.E., and Perdue, S.W.: Evidence that
    UV-inducible error-prone repair is absent in Haemophilus
    influenzae Rd., with a discussion of the relation to
    error-prone repair of alkylating-agent damage. Mutat. Res.,
    44: 183 (1977)
58. Notani, N.K. and Setlow, J.E.: Inducible repair system in
    Haemophilus influenzae unaccompanied by mutation. J.
    Bacteriol., 143: 516 (1980)
59. Hollstein, M., McCann, J., Angelosanto, F.A. and Nichols,
    W.W.: Short term tests for carcinogens and mutagens. Mutat.
    Res., 65: 133 (1979)
60. Moreau, P., Bailone, A. and Devoret, R.: Prophage λ induction
    in Escherichia coli K12 envA uvrB: A highly sensitive test
    for potential carcinogens. Proc. Natl. Acad. Sci. U.S.A.,
    73: 3700 (1976)
61. Speck, W., Santella, R.M. and Rosenkranz, H.S.: An evaluation
    of the prophage λ (Inductest) for the detection of
    potential carcinogens. Mutat. Res., 54: 101 (1978)
62. Samson, L. and Cairns, J.: A new pathway for DNA repair in
    Escherichia coli. Nature, 267: 281 (1977)

DISCUSSION

DR. SARASIN: You showed that a umuC⁻ strain seems to possess a
residual Weigle reactivation activity. Is it a significant one?
If yes, what is your explanation for it?

DR. WALKER: My guess is that when you look at Weigle reactivation
you look at a composite of umuC and a uvrA dependent process.
There may be some synergistic interaction between them, that I
can't rule out. But certainly a umuC mutant has residual Weigle
reactivation. In fact, that's what Salmonella looks like;
residual reactivation in Salmonella can be abolished by uvrA.
Conversely, pKM101 introduced into Salmonella or into a umuC⁻
mutant gives more or less normal W-reactivation, even in uvrA, B
or C strains (Fig. 2).

DR. SARASIN: So, what is your working hypothesis for the decrease
of Weigle reactivation in umuC⁻ strains?

DR. WALKER: I would say that at least some of the increase in
phage survival that one sees due to umuC dependent processes would
be carried out by the same protein that causes the increase in
mutagenesis. I think it's important to be cautious about this.
We don't yet have a rigorous demonstration that the process that
causes the increase (even in the umuC dependent part of the
increase) in survival is the same process that causes the increase
in mutagenesis. They must have something in common. They
certainly appear to have umuC function in common. However, let me
throw in some hedges now.

   pKM101 is suggesting to us that there may be two proteins
involved. So far there has been no effort to complement the
various umuC mutants. They all map at one locus. And we're just
about going to try and do that with the cloned umuC gene. As I
alluded to in previous discussion, we have one rather bizarre
derivative of pKM101. In one of our earlier attempts to clone, I
thought we'd practice by cloning out the β lactamase gene of
pKM101. We cut with EcoRI, but it turned out there was only one
EcoRI site in the plasmid, in the middle of the β lactamase
gene, so that didn't work out so well. But we did get a vector
inserted into the β-lactamase gene of this plasmid and that
doesn't seem a very stable situation. What ultimately happened
was that much of pKM101 was deleted, but not the pKM101 origin of
replication. If we took out the vector we regained the ampicillin
resistance and the plasmid still replicated. We now have a
peculiar deletion mutant; the location of the exact ends of the
deletion are a little bit fuzzy, but one is close to the muc
region that is important for mutagenesis. This particular
derivative increases mutagenesis, in many situations to a greater
extent than pKM101. But it either doesn't increase the resistance

of the cells to killing, or it in some cases even makes them UV sensitive.

Now, we've been able to complement this particular mutant. I think the easiest and cleanest experiment is if one looks at UV to phage and phage survival; this is not a Weigle reactivation experiment, but simply looking at phage survival on unirradiated host. In a cell with no plasmid at all, one sees less survival of irradiated phage than in cells with pKM101. This peculiar deletion mutant did not show any increase in phage survival. Our standard muc:Tn5 insertion mutants (of pKM101) also give no increase. However, when the deletion mutant and any of the insertion mutants were put together, we now saw a recovery of the ability to reactivate phage, suggesting there may be some sort of complementation going on, although there are a couple of more complicated explanations.

The key point is that this peculiar deletion mutant is still able to increase mutagenesis, but couldn't seem to increase survival. So I think we have to be very careful still in thinking about whether the increase in survival that one sees is necessarily due to the same process that gives us the increase in mutagenesis.

The hope is that having these things cloned and getting the products and what-not, your're finally in a position to carry out a systematic attack on figuring out what these particular proteins do. However, this may not be trivial. It seemed to take forever to clone umuC, and it took us almost two years to figure out how to see the products of pKM101 being expressed. We didn't really figure it out until we finally made a lac fusion to the pKM101 muc region, and analyzed it; then we were able to figure out how to manipulate the system so we could see the protein being expressed.

DR. GLICKMAN: I have a number of questions, the first being you have a number of din genes which haven't been identified as belonging to a given class. Because you didn't mention properties I presume they don't have any specific properties. Could you go into that?

DR. WALKER: That's essentially it. They turn on, they are not deficient in the obvious things we've looked at, they're very limited; they're recombination efficient, they're not UV-sensitive, they still carry out mutagenesis. DinD is not recF or uvrD if that's what you're thinking about. The one which is very near to leu is not sfiB; it's also very near to polymerase II which has been shown to be inducible. The joke around the lab is there's got to be a Pol II mutant because it has a Pol II phenotype, which is no phenotype at all. We're in the process of looking; it seems to be taking us forever to get around to doing that experiment. The other two I don't even have a clue about.

DR. GLICKMAN: The other thing I was curious about, you were able to look at kinetics of induction at each of these sites because of your insertion. There's one difference published between uvrA and B and that has to do with the expression time of colicin after induction with UV. The uvrB⁻ mutant responds nearly immediately and the uvrA⁻ mutant only several hours later. Do you see a difference in uvrA and uvrB in the expression time?

DR. WALKER: We haven't done very much with uvrB; by the time we got our fusions Dean Rupp was well on his way to isolating the product. The beginning of the recA, lexA and uvrA genes have been sequenced by Aziz Sancar in Dean Rupp's lab, and one finds a homologous sequence in front of all of them, which by footprinting and chemical protection experiments has been shown to be the binding site for the lexA protein. So there is what would appear to be an operator site for lexA in front of recA, lexA and uvrA. In fact Roger Brent has shown that lexA has two of those boxes in front of it.

Now, uvrB is very interesting; Dean Rupp's lab has found that it has two promoters. One of the promoters has a lexA box, an apparent lexA homology site, on top of it and the other one doesn't. So my guess is there is probably a difference between uvrA and uvrB in that uvrA has probably got one promoter that's controlled by lexA, whereas uvrB seems to have two promoters, one that's driven constitutively and a second one which is regulated by recA and lexA. However, we found some funny things about our uvrB fusions as opposed to the recA ones which were very nicely behaved. And so there may be a more complicated business going on with uvrB.

DR. S. PRAKASH: How do the levels of induction for beta galactosidase compare, for example, with uvrA, uvrB, umuC versus recA? In other words, what fold difference is there in the induction of recA on the one hand and the others?

DR. WALKER: We haven't looked at a recA fusion. You have to do it in a diploid situation because you need recA function for turning on. I think we in fact have some. I think we picked them up as UV-sensitives but we haven't bothered to characterize them in detail because of the fact it's pretty much known. The fold inductions that we see range widely; the most modest ones are in the range of 3-5 fold. UmuC is certainly 25 fold and my guess, from looking at spr recA, is that recA can be turned on even more, like 50 to 100 fold.

DR. SARASIN: But by how many times is the <u>recA</u> protein itself induced?

DR. WALKER: It's turned on at very large amounts. I don't think any of these things are turned on to the extent that <u>recA</u> is. We haven't found one anyway.

DIFFERENTIAL SENSITIVITY TO CERTAIN DNA-DAMAGING TREATMENTS,
INCLUDING 8-METHOXYPSORALEN PLUS 340 nm ULTRAVIOLET IRRADIATION,
BETWEEN K-12 AND B/r STRAINS OF *ESCHERICHIA COLI*

B. A. Bridges*, A. von Wright*, M. Stannard* and
E. Moustacchi**

*   MRC Cell Mutation Unit, University of Sussex
    Falmer, Brighton BN1 9QG, England

** Institute Curie, Centre Universitaire
    91405 Orsay-Cedex, France

INTRODUCTION

For want of information to the contrary, it is usually assumed
that when a molecule of a mutagen is near to a molecule of DNA,
reaction will occur with a probability that is essentially the same
whether or not the DNA is in a bacterial or human chromosome, whether
or not it is complexed with proteins, and whatever its degree of
superhelicity.   A moment's thought suggests that this assumption
is unlikely to be strictly true although it would be a bold man who
would predict to what extent chromosomal structure might influence
the probability of interaction with mutagens.   The present paper
records a simple bacterial model where the simplest interpretation
of the data suggests that the structure rather than sequence of the
DNA seems to affect the probability of interaction with intercalating agents such as 8-methoxypsoralen (8-MOP).

The work began with an observation that K-12 strains of
Escherichia coli were more resistant to photosensitization by 8-MOP
and ultraviolet light of around 340 nm (UVA) than strains of the
B/r WP2 family (1).   This was true even when strains (*recA uvrA*)
were compared that are deficient in all known pathways for repair of
8-MOP plus UVA damage.   In these early experiments the difference
in sensitivity was six to seven-fold as judged by loss of colony-
forming ability, and five-fold as judged by inhibition of DNA
synthesis.   (More recently, for reasons which we do not understand,
we see a sensitivity difference of two to four-fold).   Although
there was some evidence of a slow residual repair ability in the

uvrA recA K-12 strains it seemed unlikely to be responsible for the differential sensitivity since this was fully expressed within a few minutes of treatment.   Thus although the possibility could not be excluded that a hitherto unknown repair system might operate extremely rapidly to remove psoralen adducts in K-12 bacteria, a number of different lines of evidence suggested that the main cause was a lower photoproduct yield in K-12 compared with B/r WP2 bacteria.

The most obvious explanation to be considered, and eliminated, was that the intracellular concentration of 8-MOP was lower in K-12 than in B/r WP2 bacteria.   Fluorimetric measurements did not support this hypothesis and T3 phage DNA was equally sensitive to 8-MOP plus UVA, whether treated inside K-12 bacteria, inside B/r WP2 bacteria, or suspended in buffer (1).   If the different sensitivities of these two families of bacteria are indeed due to different photoproduct yields, then this must presumably be the result of some difference in chromosomal structure that renders the DNA of K-12 bacteria less accessible to 8-MOP than that of B/r WP2.   We take the term chromo-somal structure to include the presence of molecules associated with the DNA, e.g. DNA binding proteins.

EFFECT OF OTHER INTERCALATING AGENTS AND FRAMESHIFT MUTAGENS

8-MOP not only interacts photochemically but also intercalates with DNA and, without involvement of UVA, can give rise to frame-shift mutations (2,3).   We have therefore examined the cytotoxic effect of a number of other intercalating and frameshift mutagens and compared them with non-intercalating mutagens in the K-12 strain AB2480 (4) and the B/r strain WP100 (5).   Both of the strains carry the DNA repair-deficient genes *uvrA* and *recA* and are equally hypersensitive to far ultraviolet light (1).   Bacteria were grown to exponential phase in Oxoid Nutrient Broth and viability was measured on Oxoid Nutrient Agar No.2 at 37°.

Figure 1 shows clearly the differential cytotoxic effect of four frameshift mutagens known to react covalently with DNA. Exposure to hydroxyacetylaminofluorene and hycanthonemethane sulphonate was in phage buffer (6) at room temperature.   Hydroxy-acetylaminofluorene would be expected to react covalently in an intercalated state.   Benz(a)pyrene was incubated with bacteria plus S9 rat liver microsomal fraction as described by Hubbard et al. (7).   Although benz(a)pyrene itself is known to intercalate with DNA the adduct formed by the activated form benz(a)pyrene diolepoxide is not intercalated.   This metabolite may not, however, predominate under our reaction conditions where the K-region epoxide would be expected to be the major active metabolite.   Nitro-o-phenylene diamine treatment was carried out in nutrient broth at

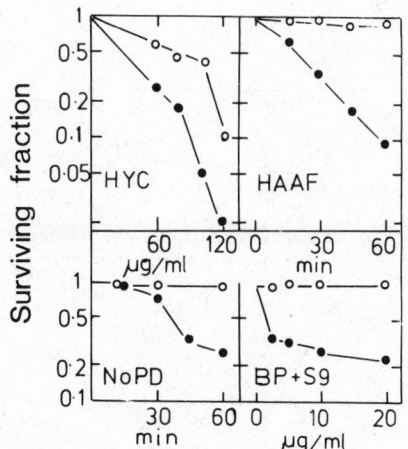

Fig. 1. Effect of hycanthone methanesulphonate (120 min treatment), hydroxyacetylaminofluorene (40 µg/mℓ), N-o-phenylenediamine (40 µg/mℓ), and benzo(a)pyrene plus S9 supernatant fraction on the viability of *E.coli* WP2 (●) and AB2480 (o).

37°.   AB2480 was more resistant than WP100 to all these treatments. In contrast to these frameshift mutagens, a number of direct-acting mutagens produced essentially the same cytotoxic effect in both strains (Figure 2).   These were AF2 (a nitrofuran), bleomycin, ethyl methylsulphonate and methyl methanesulphonate.   Previously Toogood (8) had shown that aniline mustard is equally cytotoxic to AB2480 and WP100.   Only with mitomycin C was a slightly greater resistance observed for AB2480.   This may reflect its need to penetrate the double helix in order to effect the cross-links which are, at least in part, responsible for its cytotoxic action.

Figure 3 illustrates the effect of three intercalating agents which do not bind covalently to DNA: acriflavine, acridine orange and ethidium bromide.   In all cases AB2480 showed greater resistance to the lethal effect of these agents in nutrient broth than did WP100.

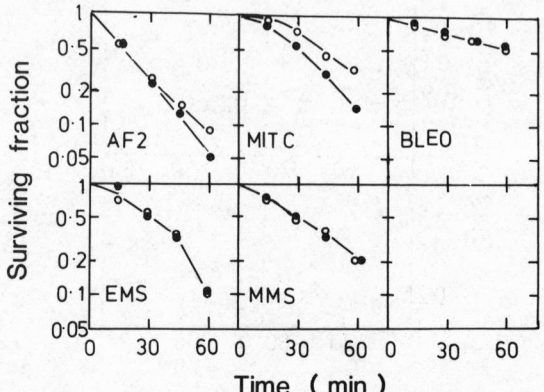

Fig. 2.   Effect of AF2 (0.02 µg/mℓ), mitomycin C (0.015 µg/mℓ),
          bleomycin (4 µg/mℓ), ethylmethanesulphonate (80 µg/mℓ)
          and methyl methanesulphonate (2.5 µg/mℓ) on the viability
          of *E.coli* WP2 (●) and AB2480 (o).

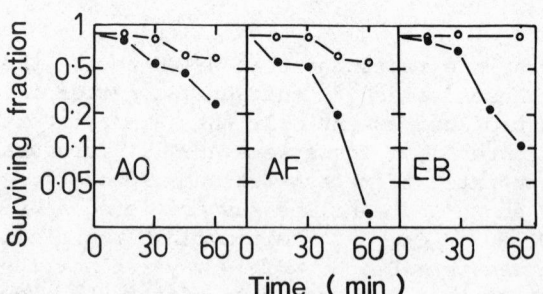

Fig. 3.   Effect of acridine orange (20 µg/mℓ), acriflavine (15 µg/mℓ)
          and ethidium bromide (300 µg/mℓ) on the viability of
          E.coli WP2 (●) and AB2480 (o).

    Taken together, the data suggest that the DNA of AB2480 is less
accessible than that of WP100 to agents which intercalate or are
known to be inducers of frameshift mutations in bacteria.   The
effects cannot be explained by differences in the intracellular
content of the chemicals at least in the cases of benz(a)pyrene
and acriflavine where fluorescence measurements have been made
(data not shown).   This is in agreement with previous results with
8-methoxypsoralen (1).   The tentative conclusion may be drawn that
the accessibility of DNA to a number of mutagens and carcinogens in
E.coli can be greatly modified by certain as yet unknown inter-
cellular factors.

INDUCIBLE REPAIR OF 8-MOP + UVA DAMAGE?

    Evidence was recently reported for a new, inducible, and
extremely rapid pathway for the repair of UVA damage (9).   We have
never seen anything to suggest that a similar system might operate
after exposure to 8-MOP plus UVA (which produces quite different
molecular damage from UVA alone), the possibility that K-12 strains
might utilize this pathway in contrast to B/r strains was nevertheless examined, particularly since the newly described UVA pathway
operates in *recA* strains of *E.coli*.

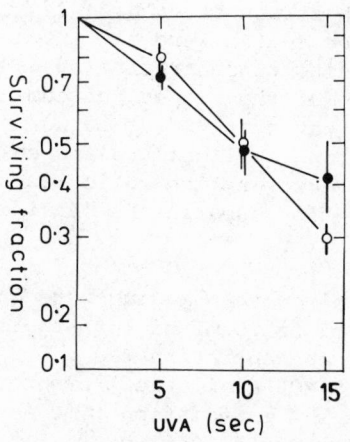

Fig. 4.   Effect of chloramphenicol on sensitivity of *E.coli* AB2480 to
to 8-MOP plus UVA.   Bacteria were incubated with glucose-minimal
casamino acids medium with 8-MOP (20 µg/mℓ) and with (●) or without
(o) chloramphenicol for 5 min before and 30 min after exposure to
UVA (utilizing shutter arrangement (11)).

When AB2480 bacteria were treated with 8-MOP plus UVA in the presence
of chloramphenicol during and for 30 min after UVA exposure, there was
no effect on cell killing (Figure 4).   The difference in sensitivity
between these two strains could not, therefore, be accounted for by a
rapid inducible repair pathway present only in the K-12 strain.
Experiments in which attempts were made to induce additional repair
by giving small inducing doses of 8-MOP plus UVA from 5 to 15 min
before larger inactivating doses also failed to alter the subsequent
sensitivity of either strain to 8-MOP plus UVA (unpublished data).

BINDING OF RADIOLABELLED 8-MOP TO DNA

We have attempted to measure the binding of ^{3}H-labelled 8-MOP
to the DNA of bacteria exposed to UVA.   The bacterial strains AB2480
uvrA recA (K-12) and WP100 *uvrA recA* (B/r) were grown to exponential
phase in Davis-Mingioli minimal medium (10) supplemented with casamino acids (Difco 4 g/l), L-tryptophan (25 mg/l) and thiamine(0.1
mg/l).   This was also used as plating medium when solidified with
15 g/l Davis New Zealand agar.

Since the concentration of 8-MOP present in the labelling experiments was unavoidably lower than that used in most of our other work,
some experiments were carried out to determine the killing of the
two strains by UVA given to bacteria suspended in buffer containing
different concentrations of 8-MOP.   Three full UVA survival curves
were obtained for each strain at each 8-MOP concentration and the
slopes of these curves (which were all exponential) were established.   With increasing concentration of 8-MOP, the slopes
increased.   The ratio of the slopes for the two strains in this
series of experiments was nearly 4 at 20 μg/ml compared with 6 to 7
obtained previously using a slightly different growth medium.   At
lower concentrations, however, the ratio was reduced to 2.8 at
1 μg/ml (the concentration used in the binding experiments) and 1.4
at 0.2 μg/ml.

The cells in 20 ml culture medium were harvested and resuspended
at about 2 x 10^8 per ml in 10 ml of buffer, to which ^{3}H-8-MOP was
added.   The ^{3}H-8-MOP concentration was 1.0 μg/ml, 3.8 μCi/ml.   The
limited amount of ^{3}H-8-MOP available prevented the use of higher
concentrations.   The UVA irradiation time was 20 min, after which
the cells were harvested by filtering.

DNA was isolated from filtered and harvested cells using the
standard methods of lysozyme spheroplasting and Sarcosyl NL-97
lysis.   RNA and proteins were eliminated by successive pancreatic
RNAase (Sigma) and proteinase K (Noehringer-Mannheim) treatments
before chloroform-isamylalcohol (24:1) extraction and ethanol
precipitation.   The precipitated DNA was redissolved in 2 ml of
1 mM EDTA in 0.01M Tris buffer, pH8.1 and dialysed against the same
buffer overnight at 0°C.   The DNA content was measured spectro-

photometrically, after which the solution was made 5% with trichloro-
acetic acid (TCA), and filtered through glass fibre (Whatman GF/C)
filters.   The filters were washed once with 5% TCA and once with
ethanol, dried, immersed into toluene based scintillation fluor and
counted in an Intertechnique SL-3000 liquid scintillation counter.

When the radioactivity recovered in isolated DNA in ^{3}H-8-MOP
experiments was compared with the activities in the original cell
lysates, it was found that only about 1-2% of the total ^{3}H-MOP
activity became bound to DNA following the UVA exposure.   The
isolated DNA was, however, reasonable free from either RNA or
protein contamination judging from experiments using ^{3}H-uracil
labelled cells and the DNA isolation method described above, or from
the A260/A280 ratios of final DNA solutions (data not shown).   The
activities recovered in DNA (listed in Table 1) therefore most
probably reflect the real ^{3}H-8-MOP-DNA adduct yields obtained in
these experimental conditions.

Table 1.   Binding of ^{3}H-8-MOP to DNA in *E.coli* WP100 and AB2480
during 20 min Exposure to UVA

Experiment	Strain	Irradiation temperature	Activity recovered/ 10 µg DNA (cpm)	WP100/AB2480 activity ratio
1	WP100	room	143	2.1
	AB2480	temperature	67	
2	WP100	room	375	4.1
	AB2480	temperature	91	
3	WP100	0°	204	2.7
	AB2480		76	
4	WP100	0°	596	5.0
	AB2480		120	
5[a]	WP100[b]	room	67	
	AB2480[b]	temperature	54	
	WP100	room	649	3.5[c]
	AB2480	temperature	218	

[a] = control experiment for non-covalently bound ^{3}H-8-MOP
[b] = no near UV irradiation
[c] = calculated after the subtraction of the background activity of
the unirradiated cells.

It can be seen from Table 1 that there was from two to five times as much 8-MOP bound to the DNA of WP100 as to that of AB2480 after exposure to UVA (Experiments 1 to 4).   In experiment 5 a correction was made for non-specific 8-MOP binding by using control bacteria not exposed to UVA.   In this experiment the amount of 8-MOP binding associated with UVA exposure was 3.5 times greater in WP100 than in AB2480.   These values are consistent with the ratio of 2.8 between these strains for sensitivity to the lethal effect of 8-MOP plus UVA at the same concentration.

The amount of 8-MOP bound when irradiation was at 0°C was if anything, greater than that found at room temperature and the binding ratios in the two strains were similar at both temperatures.   This tends to discount the hypothesis that a very rapid repair process might be responsible for the lower photoproduct yield in AB2480.

It is also possible from these data to make a rough estimate of the number of 8-MOP molecules bound to DNA per lethal hit.   The exponential, non-shouldered nature of the killing curve suggests that the whole DNA in each cell should be used for this calculation. On this basis one may calculate that the number of molecules of bound 8-MOP per lethal hit is about 2 (range ~ 1 - 4) in both strains. If, on the other hand, the unreplicated genome (3.2×10^{6} base pairs) is used as the target size, the above values must be approximately doubled.   Notwithstanding, these rough values are consistent with the previous estimate based on indirect evidence "that WP100 is probably inactivated by one or a very small number of adducts" (11).

DISCUSSION

The greater sensitivity of *E.coli* WP100 to 8-MOP plus UVA compared with *E.coli* AB2480 is associated with, and presumably results from, a greater amount of covalent binding of 8-MOP to DNA as measured shortly after treatment.   This is consistent with earlier results (1) that DNA synthesis in the first 20 min after treatment is depressed to a much greater extent in WP100 than AB2480.   It is also in agreement with the observation that the number of photoproducts capable of reaction during a further 40 min UVA exposure to form crosslinks is much greater in WP2 *uvrA* than in AB1886.

Although AB2480 does seem to have some small capacity for liquid holding repair of 8-MOP plus UVA damage that is not present in WP100 (1) there is no evidence that this is responsible for a significant proportion of the difference in sensitivity.   Indeed the present results impose severe limitations on any explanation that depends upon a difference in repair capacity.   Any repair process postulated to exist in K-12 but not B/r strains must

(a) be complete in 20 min or less, (b) need no protein synthesis, and (c) function apparently as well at 0° as at room temperature. While such a repair process may exist it would seem to be rather unlikely. Regrettably, the limited amount of labelled 8-MOP available to us has prevented further study of possible repair using labelling techniques.

Thus while the involvement of repair may still be regarded as an open question, it would seem sensible to explore the remaining possibility, namely that the photoproduct yield following 8-MOP plus UVA treatment is indeed greater in WP2 strains than K-12 strains. Since the necessary preliminary to covalent binding is non-covalent intercalation it would seem most likely that any difference in photoproduct yield would arise as a consequence of a different pattern of intercalation. This could come about as a result of differences in chromosomal structure.

The most obvious such difference to affect an intercalating agent would be a difference in superhelicity. Von Wright (unpublished data) has, however, found no gross difference in the degree of superhelicity of nucleoids from B/r and K-12 strains using the ethidium bromide sucrose gradient sedimentation method. The general shapes of the titration curves for WP2 and AB1157 were practically overlapping and were in good agreement with those reported by Drlica and Snyder (12) and Worcel and Burgi (13).

Nevertheless, differences between B/r and K-12 strains in the sedimentation pattern of nucleoids were observed. When the repair proficient strains B/r and AB1157 were used, the membrane-free nucleoids of WP2 sedimented near the middle of the gradient while those of AB1157 sedimented near the bottom. The membrane-bound nucleoids of both strains sedimented similarly near the bottom of the gradient. With lysis procedures producing a mixed population of nucleoids, membrane-free nucleoids predominated in WP2, while membrane-bound nucleoids predominated in AB1157.

With the *uvrA recA* strains WP100 and AB2480 the membrane-free and mixed nucleoid populations were similar to those with the corresponding repair-proficient strains. The membrane-bound nucleoids of AB2480, however, sedimented near the middle of the gradient and not near the bottom in contrast to WP100. What relevance, if any, these differences in sedimentation pattern have to the difference in response to 8-MOP plus UVA remains to be determined.

While the possibility of a fast repair process operating in K-12 but not B/r WP2 derivatives cannot be completely excluded, the data seem to point to a difference in chromosomal structure that influences the intercalation pattern of 8-MOP and perhaps other

intercalating molecules.   That such differences may exist within
the same bacterial species raises the possibility that similar (or
even greater) differences may exist between different species, between different cells in multicellular organisms, and between different chromosomal regions within individuals cells.

REFERENCES

1. Bridges, B. A. and Mottershead, R. P.: Inactivation of
   Escherichia coli by near ultraviolet light and 8-methoxy-
   psoralen: different responses of strains B/r and K-12.
   J. Bacteriol., 139:454 (1979)

2. Clarke, C. H. and Wade, M. T.: Evidence that caffeine, 8-
   methoxypsoralen and steroidal diamines are frameshift mutagens
   for *E.coli* K-12. Mutation Res., 28:123 (1975)

3. Bridges, B. A. and Mottershead, R. P.: Frameshift mutagenesis in
   bacteria by 8-methoxypsoralen (methoxsalen) in the dark.
   Mutation Res., 44: 305 (1977)

4. Howard-Flanders, P., Theriot, L. and Stedeford, J. B.: Some
   properties of excision-defective recombination-deficient
   mutants of *Escherichia coli* K-12. J. Bacteriol., 97: 1134 (1969)

5. Bridges, B. A. and Sedgwick, S. G.: Effect of photoreactivation
   on the filling of gaps in deoxyribonucleic acid synthesised
   after exposure of *Escherichia coli* to ultraviolet light.
   J. Bacteriol., 117: 1077 (1974)

6. Boyle, J. M. and Symonds, N.: Radiation-sensitive mutants of
   T4D.I.T4y; a new radiation-sensitive mutant: effect of the
   mutation on radiation survival, growth and recombination.
   Mutation Res., 8: 431 (1969)

7. Hubbard, S. A., Green, M. H. L., Bridges, B. A., Wain, A. J.
   and Bridges, J. W.: The fluctuation test with S9 and
   hepatocyte activation, in Short Term Tests for Carcinogens
   (Report of the International Collaborative Programme)
   Plenum Press, New York (1981).

8. Toogood, S. M.: The role of metabolism in some short term tests
   for chemical carcinogens and mutagens, M. Phil. Thesis,
   University of York (1979)

9. Peters, J. and Jagger, J.: Inducible repair of near-UV radiation
   lethal damage in *E.coli*. Nature, 289: 194 (1981)

10. Davies, B. D. and Mingioli, E. S.: Mutants of *Escherichia coli*
    requiring methionine or vitamin B_{12}. J. Bacteriol., 60: 17
    (1950)

11. Bridges, B. A., Mottershead, R. P. and Knowles, A.: Mutation
    induction and killing of *Escherichia coli* by DNA adducts and
    crosslinks: a photobiological study with 8-methoxypsoralen.
    Chem.-Biol. Interactions, 27: 221 (1979)

12. Drlica, K. and Snyder, M.: Superhelical *Escherichia coli* DNA;
    relaxation by coumermycin. J. Molec. Biol., 120: 145 (1978)

13. Worcel, A. and Burgi, E. : On the structure of the folded
    chromosome of *Escherichia coli*. J. Molec. Biol., 71: 127 (1972)

DISCUSSION

DR. SAMSON: When we first looked for removal of
O^6-methylguanine from adapted bacteria, the adapted bacteria
removed O^6MeG so quickly that we couldn't even catch them in the
act of removing it, even when we put the cells at zero degrees.
The only way that we really knew there was an active repair and
that it wasn't really just prevention of making the
O^6-methylguanine lesion was by looking at cell extracts.
Lindahl looked at extracts of adapted bacteria which could remove
the O^6MeG and now there's a mutant that can't remove
O^6-methylguanine and its adaptation deficient. Just a comment.

DR. BRIDGES: I agree. It's very difficult to prove a negative.

DR. SAMSON: Unless you use extracts.

DR. BRIDGES: You've got to be lucky to have something that's
going to work in a cell-free extract, and the trouble is that a
lot of these things actually don't.

DR. SAMSON: You can use DNA that's already damaged and give that
to the cells, and see exactly the same extent of damage between
the B and the K strains.

DR. BRIDGES: I did that with T3, and it was the same; so if it is
a repair process, it doesn't operate on T3.

DR. SHAPIRO: You said that bacteria with different degrees of
superhelicity still bound your reagent to the same extent?

DR. BRIDGES: They have the same sensitivity to psoralens and UVA.

DR. SHAPIRO: But you didn't measure binding?

DR. BRIDGES: No. Unfortunately, we had only a limited amount of
labeled psoralen, insufficient for repair-type experiments, and we
haven't been able to get another supply yet.

DR. SHAPIRO: If there are different degrees of this
superhelicity, as your own reasoning shows, they must bind
psoralen to different extents. If they're showing only the same
killing, that would imply some repair.

DR. BRIDGES: Well, you can't assume that they will actually bind
psoralen to different extents, because Psoralens are very weak
intercalators. It's not like ethidium bromide which forces its
way in and will actually remove the superhelicity. I don't think
psoralens would do that. But you have a point.

DR. GLICKMAN: There's a question of repair. It could be tolerance. If you had two pieces of DNA with different helicity, the identical repair process working, you might get different survivals on the basis of what one might call tolerance. And that's not really repair.

DR. LOEB: Can't you tell whether the Psoralen is intercalated with the DNA in the bacteria by using different wave lengths of light, and seeing if it quenches? Then you'd have a measure of the amount that's in the DNA. I need someone else who's more knowledgeable about it.

DR. BRIDGES: So do I. I have no comment on that. The trouble is, you see, you look at bacteria in suspension, which means that most of the psoralen you're looking at is free in suspension. As soon as you take them out of suspension, psoralen will just come out of the cells.

DR. LOEB: The other question is the gyrase. Are gyrase mutants sensitive to Psoralen?

DR. BRIDGES: The uvrA recA gyrase mutants are not sensitive to psoralen. Gyrase affects the repair pathways, the known repair pathways, as I said yesterday. But it doesn't seem to affect photoproduct yield.

DR. JACKYMCZYK: I wonder if you have ever compared the excision of the products of psoralen between these two strains by let's say sucrose gradient or something like that.

DR. BRIDGES: The only way to do it, I think at the moment, is to use labeled psoralen, which we don't have. We ran out.

DR. JACKYMCZYK: I'm asking because although I have never worked with bacteria, I had quite a similar situation in yeast. I was working with two strains of yeast which are also excision deficient. There are some differences between them in the speed of excision of crosslinks and in break formation in DNA. Maybe this is the same case.

DR. BRIDGES: What we'd really like is an endonuclease which will attack psoralen adducts. None of the T4 or M. luteus endonucleases will do that. The E. coli one will, but nobody's got some to give us, so at the moment we don't have that. But that would be a beatiful enzyme because then you could just use straightforward alkaline sucrose gradients, just as we use for dimer sites.

DR. WALKER: A comment on the use of chloramphenicol; this troubles me anytime you add a protein synthesis inhibitor when

you're trying to decide whether repair process is inducible or
not. If something happens to the damage when you add
chloramphenicol, then you can get a measure of whether the system
is inducible. If all that happens is that things just sit there,
the damage stays in the DNA and nothing's removed, and you're
going to measure survival as your ultimate endpoint, at some point
you have to take out the protein synthesis inhibitor. If the
damage is still in the DNA, a system that's inducible should now
be induced and repair, and you'd never really know. Such an
experiment will only work in a situation where there's a competing
system, one that didn't need protein syntheis, that could alter
the damage. So I'm never quite sure how to interpret those
experiments.

DR. BRIDGES: You can if you've got some measure of the existence
or fixation of damage, as you have with UV, where you can look at
loss of photoreversibility; this tells you whether or not
something's happened to the damage. We couldn't do that with
psoralen. In isolation you're comment is absolutely valid. But
what we were trying to do was to test specifically the Jagger
repair pathway which does work during the first few minutes,
during treatment in fact; it's very fast. We found no evidence
for it. Now, that doesn't tell you that there isn't an inducible
repair pathway. But I think that since we had chloramphenicol
there for 30 minutes and this pathway, whatever it is, has to be
very rapid, the evidence suggests that it isn't an inducible
pathway, if it's a pathway of repair at all, and I'm not convinced
of that.

THE USE OF DNA-REPAIR-DEFICIENT MUTANTS OF CHINESE HAMSTER OVARY
CELLS IN STUDYING MUTAGENESIS MECHANISMS AND TESTING FOR
ENVIRONMENTAL MUTAGENS

Larry H. Thompson

Biomedical Sciences Division, L-452
Lawrence Livermore National Laboratory
P.O. Box 5507, Livermore, CA 94550

SUMMARY

Our laboratory has taken a somatic-cell-genetics approach to
the study of mutagenesis by utilizing mutant strains of Chinese
hamster ovary (CHO) cells that are deficient in DNA repair
processes. From more than 150 UV-sensitive strains tested, five
complementation classes were identified, and representative mutants
were found to be defective at, or before, the incision step of
excision repair. A representative mutant, strain UV-5, was compared
with the parental strain in terms of cytotoxicity and dose-response
curves for mutation induction after treatment with UV and several
chemicals that are known to produce large adducts in DNA. Excision
repair in normal CHO cells protects against both cytotoxicity and
mutagenesis, but the degree of protection depends on both the agent
and the genetic marker used for detecting mutations. Upon treatment
with low doses (100% cell survival) of the polyaromatic hydrocarbon
7-bromomethylbenz(a)anthracene, repair-deficient UV-5 cells had
linear responses for mutation induction to thioguanine resistance or
azaadenine resistance, whereas the normal repair-proficient cells
showed curvilinear responses in which the slope increased with
dose. This behavior suggests that in the normal cells the repair
system acting on potentially mutagenic lesions becomes saturated at
doses that produce cytotoxicity. In no instance was a lower
mutation frequency induced in UV-5 cells than the parental cells, at
a given dose of mutagen, suggesting that the excision repair system
is error-free in normal CHO cells.

217

INTRODUCTION

In recent years an increased concern with the risks of exposure
to environmental pollutants has led to widespread development and
application of short-term tests for detecting chemicals that may be
mutagenic or carcinogenic in humans (1,2). In mammalian cells in
culture, several systems have been validated and standardized for
assaying specific locus mutations (3-5), and sister chromatid
exchange (SCE) has become a very popular chromosomal endpoint (6,7)
reflecting structural alteration at the level of the duplex DNA
molecules. SCE can be readily measured in humans in peripheral
blood lymphocytes (8,9), and efforts are underway to quantify
specific locus mutations in human tissues as well as in experimental
animals (10-12). However, the molecular mechanisms by which
mutations and SCEs arise in mammalian cells are still poorly
understood. Greater knowledge of these processes requires
studying: the metabolic conversion of promutagens to active
species, the chemical attack of reactive species at various
nucleophilic sites in the DNA molecule, the action of enzymatic
repair processes on the breaks and adducts present in the DNA, and
the interaction of the DNA replication machinery with those DNA
lesions. DNA repair processes play a particularly important role.
In many situations the great majority of the lesions are repaired
(13,14), and the minor fraction of unrepaired damage may be
responsible for the observed genetic changes. Repair is generally
considered to be a recovery process, but if it is inaccurate at the
level of individual DNA nucleotides, surviving cells may carry
elevated mutation frequencies (i.e., so-called error-prone repair).
Clearly, DNA repair is an integral aspect of mutagenesis.

Our present understanding of DNA repair processes in mammalian
cells derives mainly from studies of cells from individuals with the
genetic disease xeroderma pigmentosum (XP). The cells from these
people are characteristically defective in the excision repair
pathway that recognizes and replaces segments of DNA containing
pyrimidine dimers (induced by ultraviolet radiation (UV)) or certain
bulky chemical adducts (15). As a consequence of this defect, XP
fibroblasts in culture are hypersensitive to killing and to mutation
induction by a variety of mutagens (16). Because people with XP
experience a greatly elevated risk of skin cancer, the increased
rate of induced mutagenesis shown in culture establishes a direct
link between mutagenesis and carcinogenesis (17).

A series of experiments carefully performed by Maher,
McCormick, and coworkers using both normal and XP fibroblasts lead
to the conclusion that the excision-repair system acts on UV and
chemical damage in an error-free manner (13,14). However,
exceptional results were presented in one study from another
laboratory (18), and several other studies suggest that error-prone
repair may be operating in mammalian cells under some conditions.

With UV-treated host cells, the elevated mutation frequencies observed in irradiated virus suggested a mutagenic repair process (19,20). In rat hepatoma cells, X-ray-induced mutation frequencies were enhanced under conditions of uncoupling of oxidative phosphorylation; these conditions were associated with increased repair synthesis (21).

Although the studies using diploid human fibroblasts have been extremely valuable, these cells have the limitations of senescence, low plating efficiency, and the requirement for very low cell density to minimize contact feeding (22) during drug selection for mutant colonies. Thus, obtaining precise dose-response information is difficult with these cells. Because Chinese hamster ovary (CHO) cells can be manipulated more readily, they are widely used for mutation detection studies (3,5). In addition, CHO cells have proven to be a rich source of somatic-cell-mutant strains (23). Thus, we reasoned that the isolation of DNA-repair mutations in this cell line might be feasible as well as desirable. Initially, the extent of excision repair capability in normal CHO cells was in question (24). Although CHO cells had been shown to excise thymine dimers quite poorly (25), the resistance of the cells to UV in terms of cell survival is comparable to that of normal human cells (14,27). Also, other Chinese hamster cells (line V79) were known to excise chemical adducts at rates comparable to those of transformed human cells (26). Initial efforts to isolate UV-sensitive DNA-repair-deficiency mutations in CHO proved successful in our laboratory and in one other (27,28), and subsequent studies with representative mutant strains provided the results discussed here. More than 150 UV-sensitive mutants were analyzed and classified into five genetic complementation groups (29,30). This number compares with seven reported groups for XP (31,32). We also isolated several other mutant strains of CHO cells (27) that appear to differ from all of the reported human mutagen-sensitivity syndromes (33). Results of studies on these alkylation-sensitive cells will be discussed elsewhere (34).

MATERIALS AND METHODS

Cells and Culture Conditions

The parental repair-proficient strain, CHO clone AA8, was a spontaneous isolate shown to be functionally heterozygous at the adenine phosphoribosyltransferase (aprt) locus (35). The excision repair-deficient strains, UV-5 and UV-20, were isolated as described (27) and have been shown to belong to different genetic complementation classes (29). Cells were maintained in suspension culture as indicated using α-MEM medium supplemented with 10% fetal bovine serum, 100 μg/ml streptomycin, and 100 units/ml penicillin (27,35).

Exposure to Mutagens

For treatment with UV or chemical mutagens requiring metabolic activation, cells were exposed on 100 mm plastic petri dishes (Corning). Conditions for UV exposure were described (35). Exposure to Trp-P-2 (3-amino-1-methyl-5H-pyrido[4,3-b]-indole) (Nard Institute, Ltd., Osaka, Japan) was performed in serum-free medium for 4 h in the presence of a liver S9 homogenate fraction (prepared from Aroclor-1254-induced hamsters), supplemented with cofactors as described by Krahn and Heidelberger (36). After exposure, monolayer cells were rinsed, trypsinized, plated for survival, and allowed to express mutations in suspension culture as described (35). Treatment with 7-bromomethylbenz(a)anthracene (7-BrMeBA), was done in suspension with rapid mixing in a spinner flask containing 150 ml of medium (cell concentration, 1×10^5/ml). In this case the mutagen was not removed because of its very short half-life (37); survival was measured 1 h after addition of 7-BrMeBA.

Expression and Assay of Mutations

To minimize sampling error cultures for mutation expression contained a minimum of 6×10^6 cells. In experiments with UV or 7-BrMeBA, cultures were diluted every two days and plated for mutations to 6-thioguanine resistance (TG^r) or 8-azaadenine resistance (AA^r) on day 6 or later. This procedure was previously validated (35); dilution of the cultures at 3-day intervals has been shown to give comparable results. Thus, in experiments with Trp-P-2, cultures were diluted on day 3 and plated on day 6. In experiments in which ouabain resistance (Oua^r) was measured, a minimum of 6×10^7 cells was maintained during the expression; 500-ml cultures were diluted daily and plated on days 2 and 3 for mutations (38). Control cultures were run in duplicate and treated with solvent. All plating for colonies was done using 100 mm plastic dishes. Plating efficiency was measured using 300 cells per dish (4 replicates). TG^r (2 μg/ml) and AA^r (80 μg/ml) were assayed at 6×10^5 cells per dish (6 replicates), conditions that were previously validated (35). Oua^r was assayed at 3mM ouabain (Sigma Chemical Co.) using 1.5×10^6 cells per dish (20-30 replicates). All platings for drug resistance and plating efficiency were done in 10% dialyzed fetal bovine serum (K.C. Biological, Inc.) using 20 ml of medium per dish for mutations and 10 ml per dish for plating efficiency.

Alkaline Elution of DNA from Filters

The alkaline elution procedure introduced by Kohn and coworkers (39) was used as an assay for the incision step of excision repair (40,41) after treatment of cells with 7-BrMeBA. Elution was performed under the conditions previously described (39) with the following modifications: The lysis buffer consisted of 0.02 M EDTA, 2% sodium docecyl sulphate (SDS), and 0.1 M glycine. The elution buffer was as described (39) except for the addition of 0.1% SDS, and filters were polycarbonate with 2 μm pore diameter (Nucleopore, Pleasanton, CA). Before elution the extract was treated for 1 h with proteinase K at 0.5 mg/ml (EM Laboratories) in lysis buffer. For scintillation counting the eluted fractions were mixed with 6 volumes of Instagel (Packard) containing 0.15% acetic acid. DNA was labeled by adding to the cultures 0.005 μCi/ml 2-^{14}C-thymidine (59 mCi/mmol; 50 μCi/ml; Amersham) 48 h before mutagen treatment; starting 3 h before treatment, cells were incubated in fresh medium without labeled thymidine.

RESULTS

Mutations Affecting the Incision Step of Excision Repair

The two mutant strains to be discussed here (UV-5 and UV-20) were isolated through their hypersensitivity to killing by UV and were shown to be deficient in repair ·replication in response to UV (27). These mutants represent the first two complementation classes of UV-sensitive CHO mutants and are the two classes obtained most frequently (29). Among the more than 150 UV-sensitive mutants isolated using automated procedures (28), three additional complementation classes were identified (29,30). Although the five classes were initially identified using UV as the insult, the same classes obtained when 7-BrMeBA was used (29).

Using the technique of alkaline elution of DNA (39), which provides a measure of the relative size of DNA molecules, we obtained evidence that the defect in mutants from each complementation class is evident at the incision step of excision repair. An example of the analysis is shown in Fig. 1 for the mutant UV-20 treated with the mutagen 7-BrMeBA. In the parental strain, referred to as AA8, the rate of DNA elution was faster after treatment with 7-BrMeBA, and this effect was much more pronounced after treatment with hydroxyurea and cytosine arabinoside, agents known to retard the closing of nicks produced by the incision step of repair (40). In contrast, the mutant UV-20 treated under the same conditions showed only a small increase in the rate of elution, either with or without the DNA synthesis inhibitors. These results indicate that UV-20 has very little, if any, capacity to perform the incision step, which is an early step in excision repair. The mutant UV-5 behaved similarly in response to 7-BrMeBA treatment (results not shown), and mutants from each of the five complementation classes showed a similar response to UV exposure. Thus, we conclude that the defects in each of the UV-sensitive CHO mutants operate at or before the incision step. Similar defects have been described for different

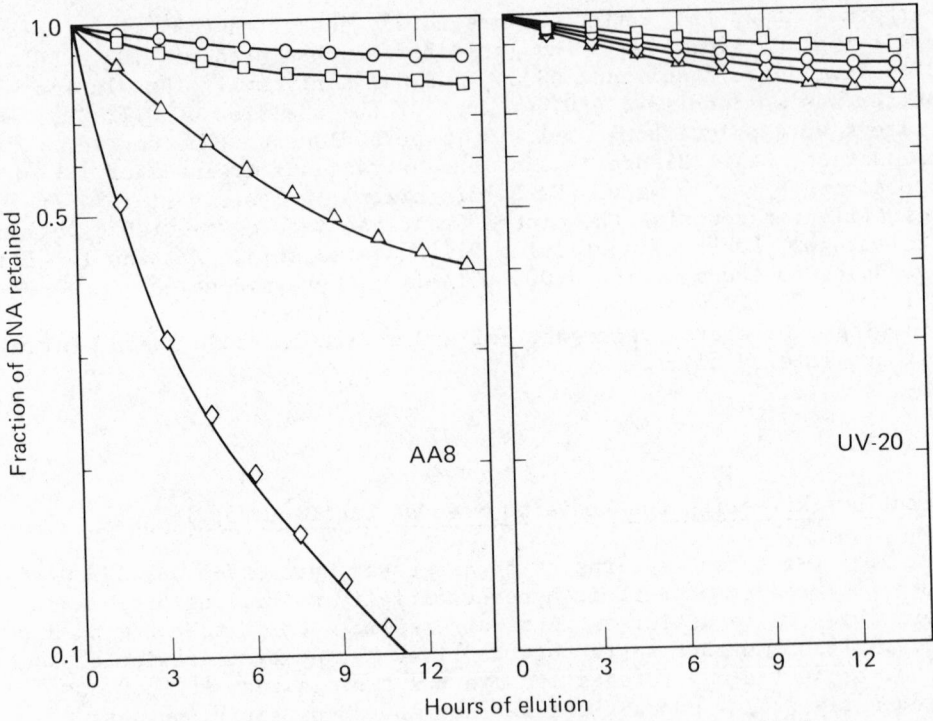

Fig. 1. Alkaline elution of DNA from UV-20 and parental AA8 cells
after treatment with 7-BrMeBA. The fraction of DNA remaining on the
filter is plotted vs elution time. Cells were treated for 5 min at
37°C with 7-BrMeBA at 8 x 10^{-7} M in suspension. They were then
diluted 10-fold into phosphate-buffered saline at 4°C, centrifuged,
and resuspended in fresh medium at 37°C. Cultures were next
incubated for 60 min with or without DNA systhesis inhibitors
present. (0---0), untreated control cells; (□---□), cells
treated only with hydroxyurea at 2 mM and ara-C at 10 µM;
(△---△), cells treated only with 7-BrMeBA; (◇---◇), cells
treated with both 7-BrMeBA and hydroxyurea + ara-C. Left, AA8;
Right, UV-20.

complementation groups of XP cells (41-43), which are also
characteristically hypersensitive to UV exposure. Thus, the CHO
mutants are phenotypically analogous to the XP mutations, but an
equivalence in terms of specific genotypes has not been established.

Hypersensitivity of Repair-Deficient Strains to Killing and Mutation Induction by UV

The survival response of the parental AA8 cells to UV is
compared in Fig. 2 with that of two DNA-repair-deficient mutants,

strains UV-5 and UV-20. AA8 shows a broad shoulder characteristic
of most mammalian cells. Both mutants were much more sensitive than
AA8. The D_{37} dose (dose required to produce 37% survival) was
approximately 1 J/m^2 and 7 J/m^2 for the mutants and for AA8,
respectively. Using $\underline{D}_q$ (see figure legend) as a measure of
resistance, the parental cells were about 10 times more resistant
than both mutants. These results indicate that the repair system,
which is defective in the mutants, is able to correct a high
percentage of the potentially lethal damage produced by UV in the
AA8 cells. The survival curves for the mutants also have small
shoulders, suggesting that in these cells killing is more than a
single-hit function of dose. Mutants from three other
complementation classes also exhibited UV sensitivity comparable to
that of UV-5 and UV-20 (29). These data provide good genetic
evidence that CHO cells effectively repair UV-damaged DNA. Thus,
the poor pyrimidine-dimer excision in CHO cells is somehow not
indicative of the extent of biologically relevant repair (25).

Mutations induced in AA8 and UV-5 were determined in order to
assess the role of the excision repair process in UV mutagenesis in
CHO cells (Fig. 3). Mutations were measured for two enzyme-deficiency
markers, which are thought to be sensitive to a wide variety of gene-
inactivating events: thioguanine resistance (TG^r) at the hprt
locus and azaadenine resistance (AA^r) at the autosomal aprt
locus. In repair-deficient UV-5 cells, mutations at both markers
were induced with linear dose-response kinetics over a range of
survival from 1.0 to 0.1. (In experiments performed on four other
UV-sensitive strains from other complementation classes, the results
were similar except the slopes of the mutation-induction curves were
slightly lower (manuscript in preparation)). The behavior of the
repair-proficient parental cells, determined over the same range of
survival, differed markedly in several respects: 1) The fluences of
UV required to produce mutagenesis in AA8 comparable to that
occurring in UV-5 were more than 10 times higher. 2) The shapes of
the dose-response curves were noticeably nonlinear for both markers;
mutation induction appeared to saturate at high doses. (This
behavior is unlikely to result from incomplete expression of
mutations; frequencies did not differ appreciably for expression
beyond day 6.) 3) The highest frequencies of mutations detected in
AA8 were 2 to 3 times lower than the maximal frequencies obtained
with UV-5.

Several conclusions can be drawn from these mutagenesis
results. First, the repair system in AA8 is very effective at
eliminating lesions that are potentially mutagenic. The protection
that the repair confers in terms of mutation induction is
significantly greater than the protection conferred against cell
killing. This latter point is emphasized in Fig. 4, where the data
have been plotted in terms of induced mutations vs cell survival.
UV-5 shows greater ratios of mutation per lethal hit than does AA8.

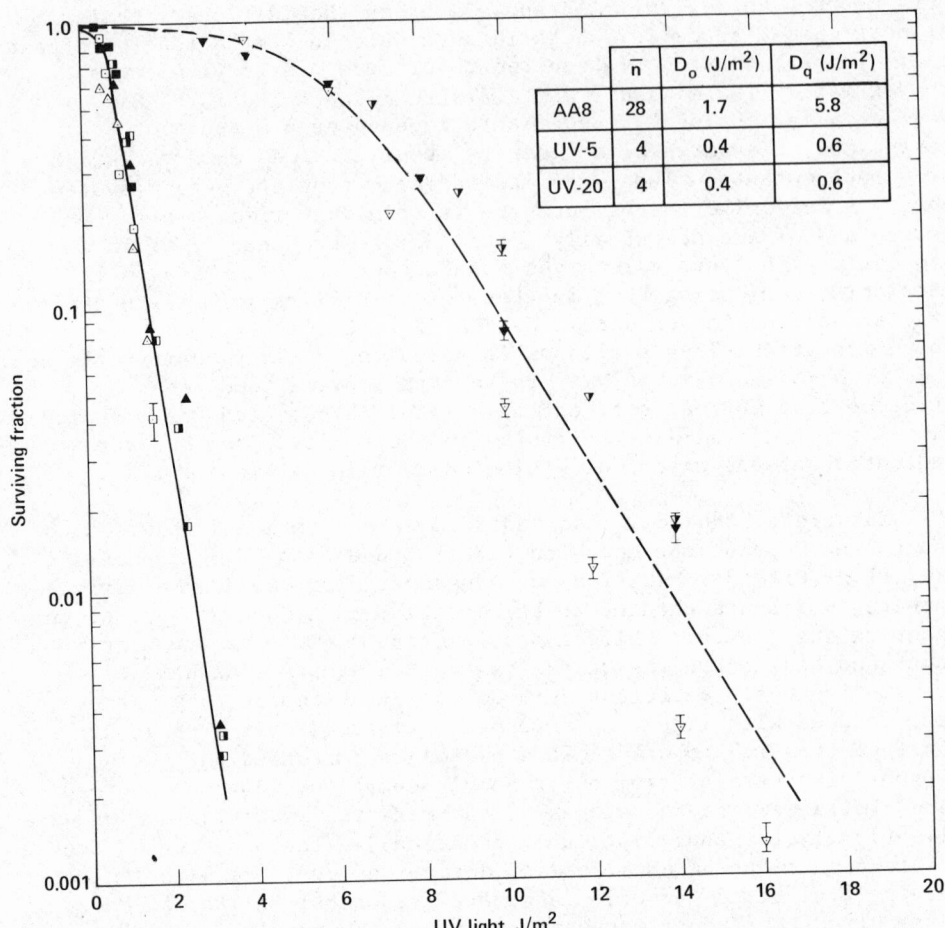

Fig. 2. Survival curves of parental CHO cells (clone AA8) and two repair-deficient strains, UV-5 and UV-20, in response to UV. (∇, $\blacktriangledown$, $\mathbb{V}$), parental AA8; ($\square$, $\blacksquare$, $\mathbb{L}$, $\mathbb{I}$), UV-5; ($\triangle$, $\blacktriangle$), UV-20. Error bars indicate standard errors of the means for colony counts on replicate dishes. Different symbols represent different experiments. ($\underline{D}_q$ is defined as the intercept on the abscissa obtained by back-extrapolating the exponential portion of the curve.) Data redrawn from reference 27.

Second, mutation induction in AA8 appears to saturate at higher fluences, at which the repair system may be saturated (fluences for which survival falls below 50%). This behavior suggests that as a consequence of the repair system the occurrence of detectable mutations becomes relatively less efficient, with increasing dose, than cell killing. One explanation for this phenomenon is that at high doses there is an effective transient block to DNA

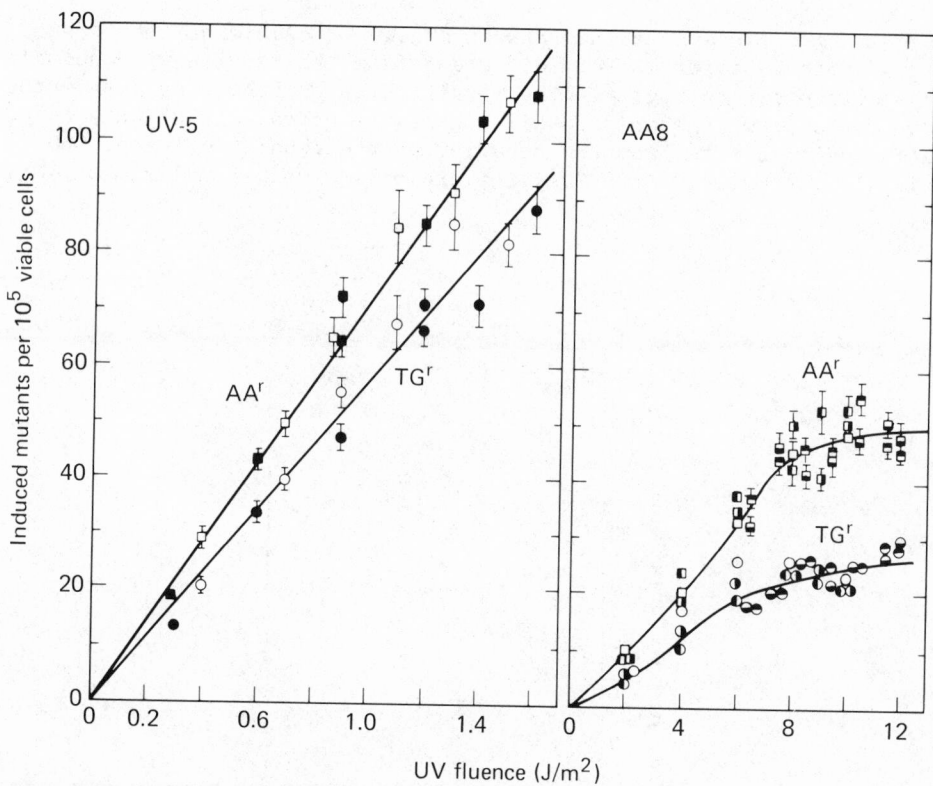

Fig. 3. Induced mutations to TG^r or AA^r as a function of UV fluence. <u>Left</u>, strain UV-5: squares, AA^r; circles, TG^r; <u>Right</u>, parental AA8: squares, AA^r; circles, TG^r. Half-shaded symbols at the same fluence represent measurements at different days of expression after UV treatment. Otherwise, different symbols represent separate experiments. Error bars are standard errors of the mean obtained by compounding the counting errors associated with plating efficiency and drug resistance. The background frequencies that were subtracted ranged from 0.3–3.1 x 10^{-5} for TG^r and 1.0–2.5 x 10^{-5} for AA^r.

replication. This may allow time for relatively more efficient repair of potentially mutagenic lesions than of potentially lethal damage since the latter may represent complex interactions of the adducts with the DNA replication and DNA repair machinery. An alternative explanation could be based on cell cycle phase differences. Perhaps the cells that are in the most sensitive phase in terms of killing are also the most sensitive in terms of mutation induction, resulting in preferential loss of mutant cells. Some evidence in support of this interpretation has been presented (44), and further experiments with synchronized populations might resolve this point.

suggesting a close coupling between mutagenic and lethal lesions
(46). This behavior is very different from the results with normal
CHO cells shown in Fig. 4. The results with XP fibroblasts show the
same ratio of mutagenesis to killing as the normal fibroblasts (45),
which again differs from the behavior of the repair-deficient CHO
cells. In summary, the CHO system exhibits a more complex behavior.

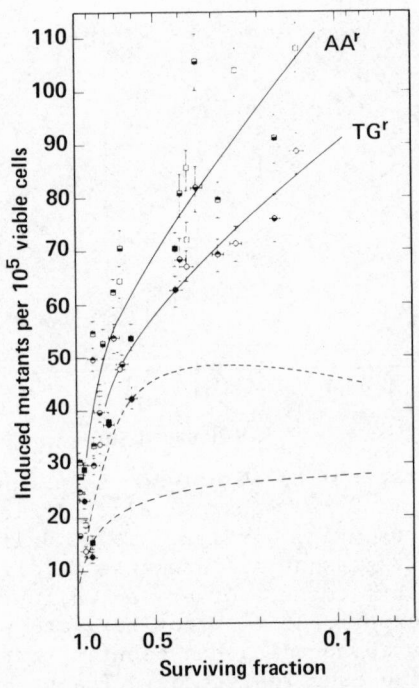

Fig. 4. Induced mutations vs surviving fraction after treatment
with UV. Squares, AA^r in UV-5; circles, TG^r in UV-5; upper
dashed line, AA^r in AA8; lower dashed line, TG^r in AA8. (See
Materials and Methods for exposure conditions.)

A third general point is that these results in CHO cells differ
markedly in some respects from the results presented for diploid
human fibroblasts. The normal human fibroblasts show a threshold
for UV-induced mutation at the hprt locus (14), which parallels the
threshold for cell killing. When induced mutations were plotted vs
log(surviving fraction), a linear relationship was seen (45),

Hypersensitivity of Repair-Deficient Strains to Cell Killing and Mutation Induction by 7-BrMeBA

7-BrMeBA was chosen as a prototype chemical mutagen because it is a direct-acting polyaromatic hydrocarbon, is known to produce a differential response in normal vs XP human fibroblasts (16,47), and is known to produce adducts that are repaired in Chinese hamster cells (26). Also the chemical identities of the DNA adducts are known; the reaction between 7-BrMeBA and DNA occurs at the exocyclic amino groups of adenine, guanine, and cytosine (48). As anticipated, repair-deficient CHO cells were hypersensitive to the cytotoxic effects of 7-BrMeBA. Fig. 5 shows the survival curves of AA8 cells in comparison with the mutants UV-5 and UV-20. AA8 exhibits a pronounced shoulder ($\underline{n}$ = 4) followed by an exponential decline, similar to the response of normal human fibroblasts (47). The mutant strains had approximately the same extrapolation number as AA8, but their $\underline{D}_0$ values were 4 to 5 times lower than that of AA8. UV-sensitive mutants from the other three complementation classes were also found to be similarly sensitive to killing by 7-BrMeBA (29). Thus, the UV-damage repair system in AA8 appears to be quite effective at repairing potentially lethal DNA damage produced by 7-BrMeBA, although the degree of protection afforded is slightly less than seen with UV.

Dose-response curves for mutation induction by 7-BrMeBA were determined in AA8 and UV-5 cells for three genetic markers. Fig. 6 compares induced TG^r mutations for doses of 7-BrMeBA that covered a range of survival down to 0.1. In AA8, TG^r mutations were induced at low doses according to a curvilinear response with slope increasing with dose. In the middle to high dose-range the slope was approximately constant, then decreased slightly. UV-5 differed from AA8 by having a steeper initial slope and by having a linear response. Induced AA^r mutations behaved somewhat like TG^r in AA8 (Fig. 7), but the nonlinearity of the dose-response curve was more obvious. At low doses (0 to 10 x 10^{-8} M 7-BrMeBA) the response was curvilinear with increasing slope, as in the case of TG^r. However, at doses above 20 x 10^{-8} M the frequency was constant. UV-5 was noticeably hypersensitive to induced AA^r and gave a response that deviated little, if significantly, from linearity.

The differential effects of the repair system on cytotoxicity and mutations caused by 7-BrMeBA can be more readily seen by plotting induced mutations vs survival, as shown in Fig. 8 for TG^r. In UV-5, induced TG^r mutations remained essentially constant as survival fell from 0.5 to 0.1, whereas in AA8 the frequencies increased significantly. Similar behavior was seen with Oua^r, as indicated in Fig. 9. At equal levels of survival, much higher Oua^r frequencies were obtained in AA8 than in UV-5, which reflects our finding that excision repair gives little or no protection against

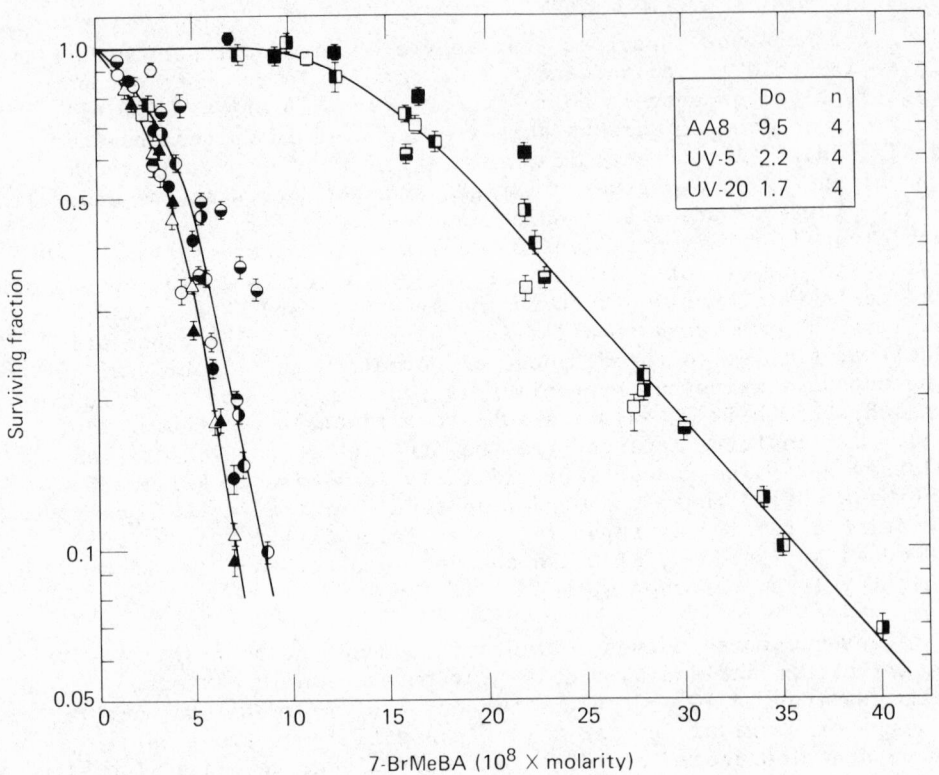

Fig. 5.   Survival curves of parental AA8 cells and the
repair-deficient strains, UV-5 and UV-20, in response to 7-BrMeBA.
Squares, AA8; circles, UV-5; triangles, UV-20.   Error bars are
defined in Fig. 2, and different symbols represent different
experiments.

Oua[r] mutagenesis in AA8.   (At low doses, the frequency of Oua[r]
mutations was similar in AA8 and UV-5; results not shown.)   Thus,
these results with 7-BrMeBA are in marked constrast to those with UV
shown in Fig. 4; with UV, the higher mutation frequencies occurred
in the repair-deficient cells.   The situation with 7-BrMeBA is
probably more complex than with UV because of the multiple adducts,
which may have differential effects on mutation and cytotoxicity and
which may be repaired at different rates.

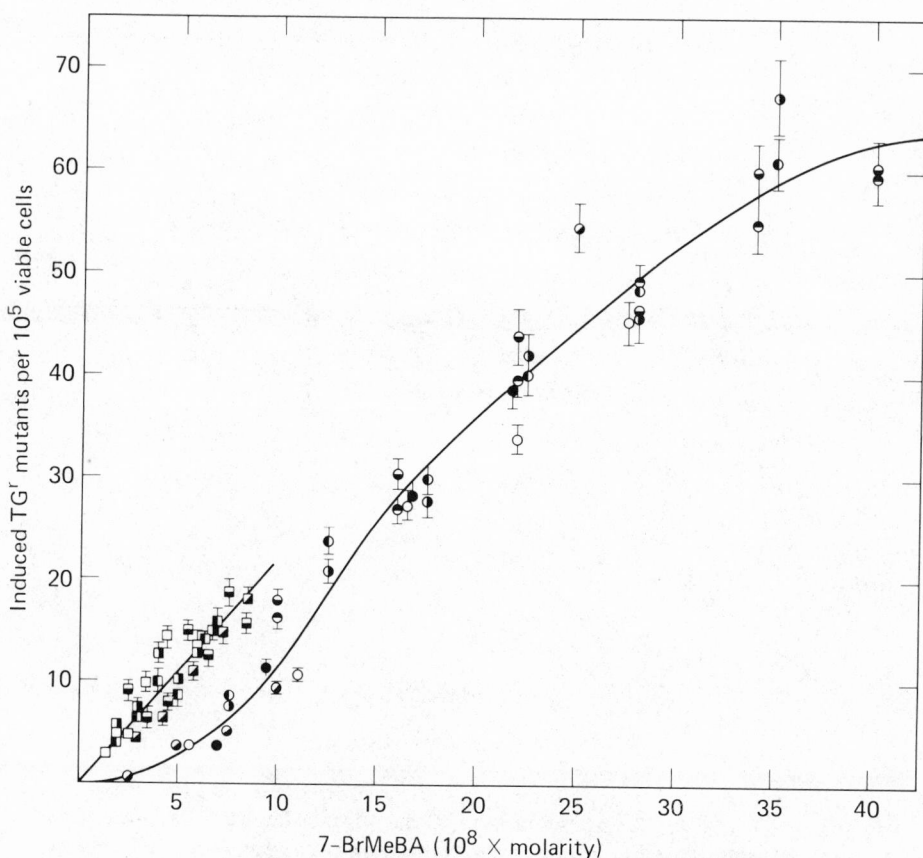

Fig. 6. Induced TG^r mutations as a function of exposure dose of
7-BrMeBA. Squares, UV-5; Circles, AA8. Symbols at the same dose and
shaded on opposite sides represent measurements on different days of
expression in the same experiment. The average values of
spontaneous mutations that were subtracted were 0.5×10^{-5} for AA8
and 1.5×10^{-5} for UV-5.

Hypersensitivity of Repair-Deficient Cells to Killing and Mutation
Induction by Trp-P-2 and Other Promutagens in the Presence of
Microsomal Activation

An important application of the repair-deficient strains lies in
testing chemicals that are environmental contaminants and may be
genotoxic. The repair mutants offer a sensitive system for
detecting mutagenicity and DNA damage caused by many of these
agents. The majority of these chemicals are not direct mutagens and
require conversion to active forms by enzymes that are present in
liver or certain other tissues. One compound of this type that we

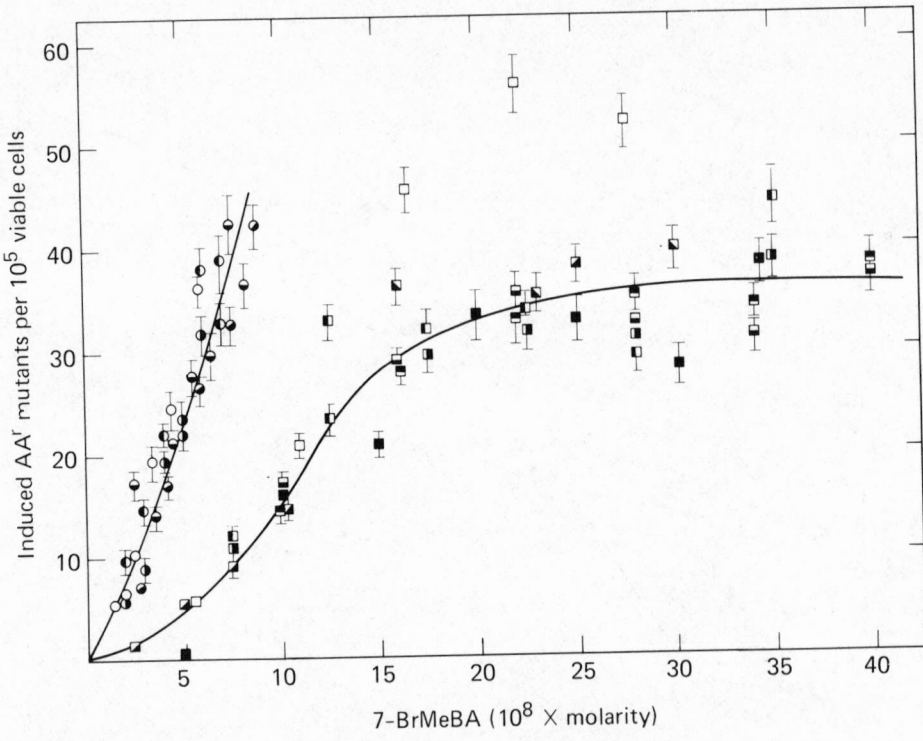

Fig. 7. Induced AA^r mutations as a function of exposure dose of
7-BrMeBA. Circles, UV-5; squares, AA8. Symbols at the same dose
and shaded on opposite sides represent measurements on different
days of expression in the same experiment. The average values of
spontaneous mutations that were subtracted were 2.5 x 10^{-5} for AA8
and 2.3 x 10^{-5} for UV-5.

used initially is known as Trp-P-2, a tyrptophan pyrolysate that is
known to be present in various cooked meats (49,50). Trp-P-2 was
also of interest because it is one of the most potent mutagens that
has been analyzed in the Ames <u>Salmonella</u> test (50).

   Strain UV-5 responded to lower concentrations of Trp-P-2 than
did AA8 in terms of both cell killing and mutation induction (Fig.
10). UV-5 was at least two-fold more sensitive to killing, and
these results were quite reproducible. The dose-response curves for
mutation induction were nonlinear in each cell line. UV-5 showed a
threshold response followed by a linear increase for each genetic
marker. The curve for AA8 cells was nonlinear over a broader dose
range although the precision of the data is lower in this case. (It
is not clear why mutation frequencies were lower in one of the three
experiments with AA8 because the survivals were similar in all

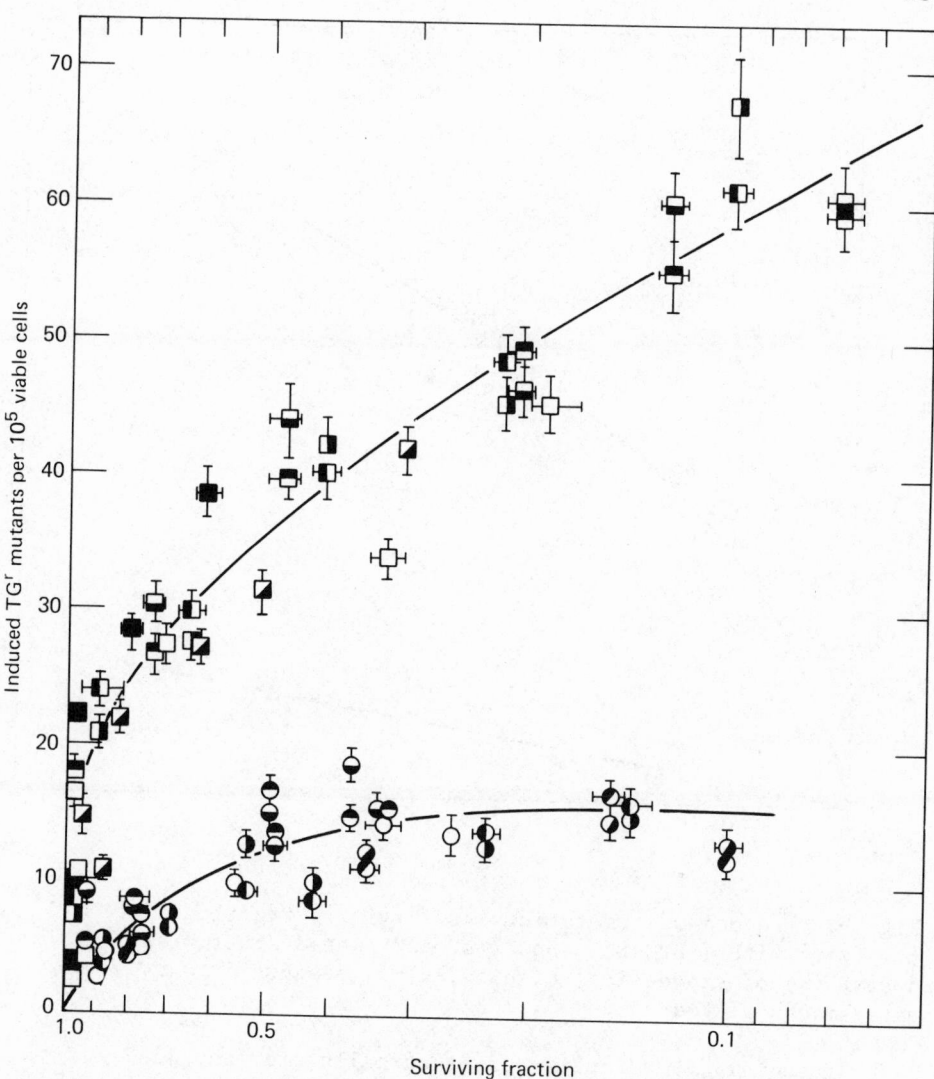

Fig. 8.   Induced TG^r mutations vs surviving fraction after
treatment with 7-BrMeBA.   Squares, AA8; circles, UV-5.

cases.)   However, the shapes of these curves should be interpreted
with caution because the actual dose of active metabolite(s) may not
be linearly related to the applied dose of Trp-P-2.   If more than
one active metabolite is produced, the situation becomes complex;
for example, cytotoxicity could result from a metabolite that
damages cellular structures other than DNA.   In any case, UV-5 is
more sensitive to mutation than AA8 by more than a factor of two for
each genetic marker.

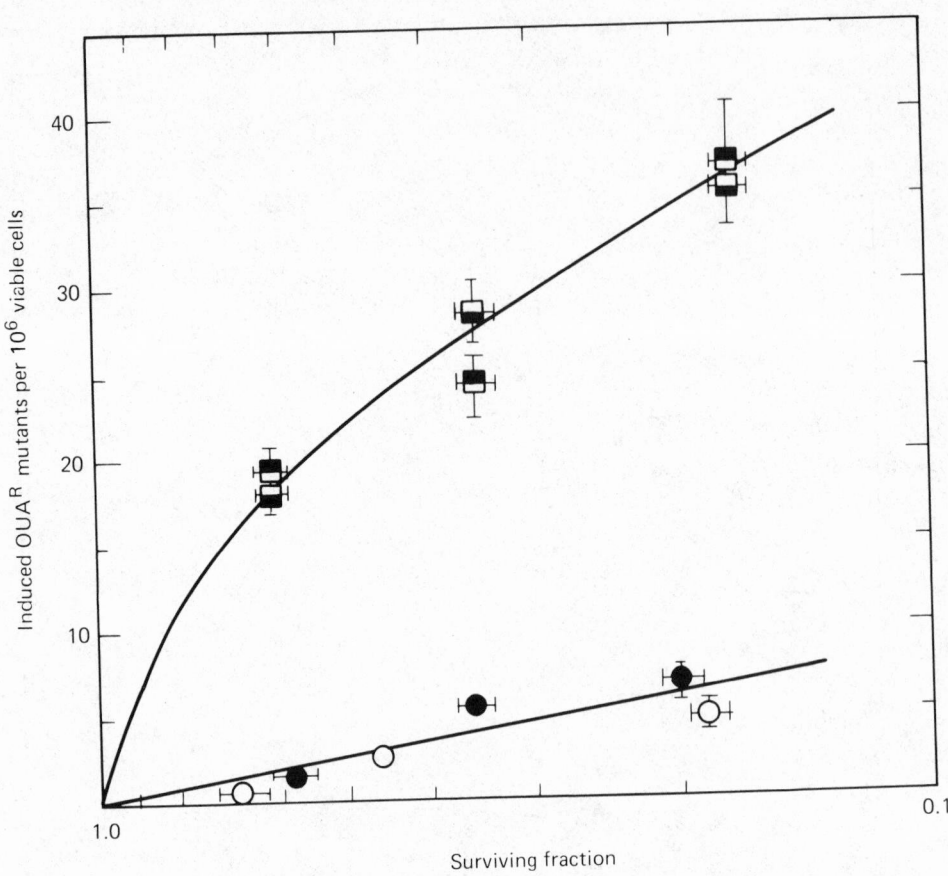

Fig. 9.   Induced Oua^r mutations vs surviving fraction after treatment with 7-BrMeBA.   Squares, AA8 plated after 2 days (■) and 3 days (■) of expression; circles, UV-5, results from two experiments plated on day 3.

In summary, the repair-deficient cells provide added sensitivity in detecting the mutagenic and cytotoxic action of Trp-P-2. In other experiments we observed hypersensitivity in repair-deficient cells with respect to mutation induction by other pure compounds and complex mixtures: These include benzo(a)pyrene, 2-acetylaminofluorene (2-AAF), and the basic fraction of a tar sample that was collected from the effluent stream of an underground coal gasification experiment. In the latter case the repair-mutant strain is especially advantageous because in the test sample toxic chemical species that are not genotoxic tend to mask the genotoxic activity. In the case of 2-AAF, the solubility limit of the compound prevented obtaining a clear-cut dose-response curve in AA8 cells whereas the results with UV-5 were not confounded by this problem. Thus, our experience leads us to conclude that the

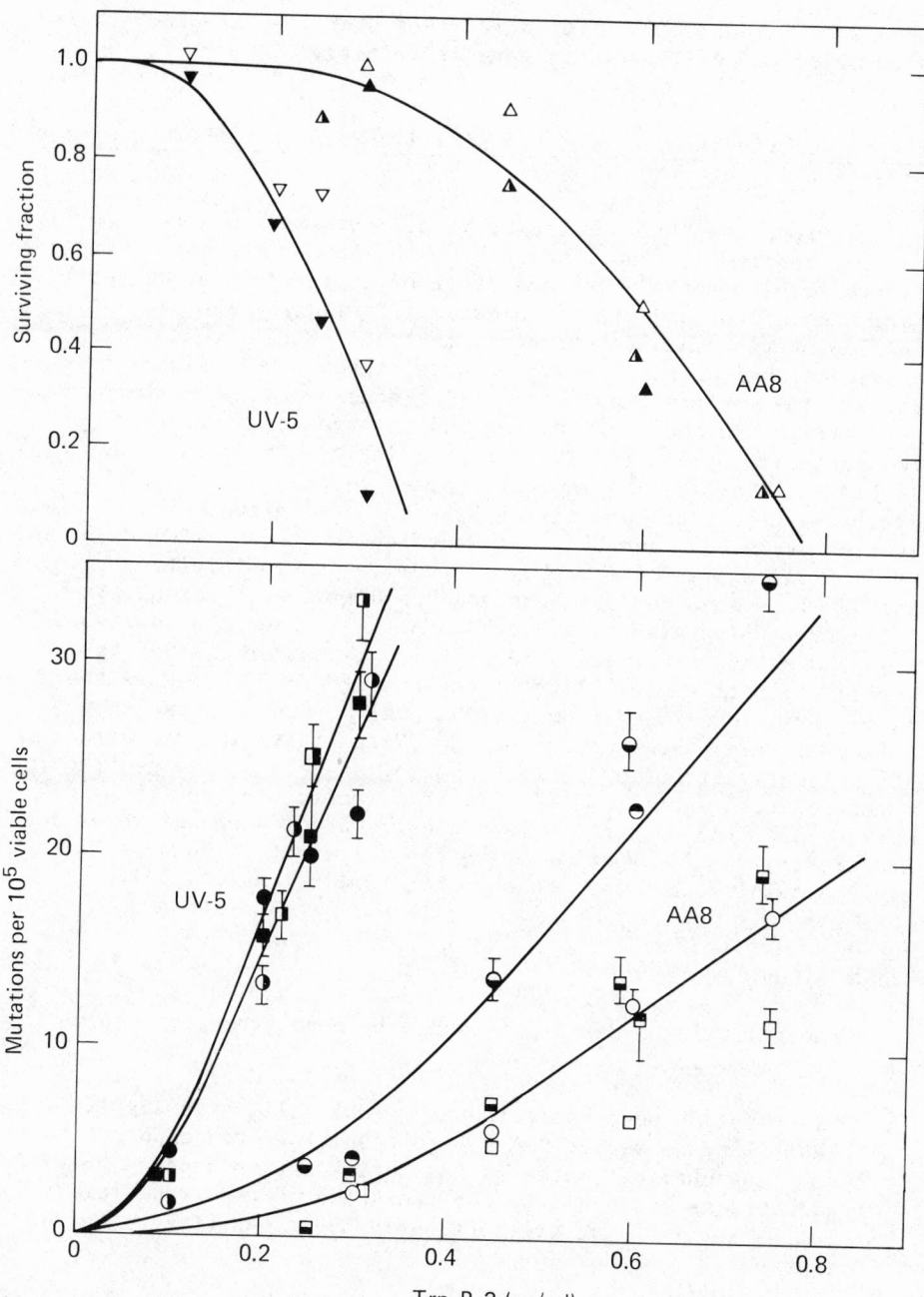

Fig. 10. Cell survival and mutation induction in AA8 and UV-5 cells treated with Trp-P-2. Upper: (∇, ▼) UV-5; (△, ▲, ▲) AA8. Lower: (●, ◐) TG^r and (■, ◨) AA^r in UV-5; (O, ◓, ◑) TG^r and (□, ◪, ◨) AA^r in AA8. (See Methods for exposure conditions.)

repair-deficient cells offer advantages over the parental strain
when using CHO cells for mutagenesis testing.

Phenotypic Diversity of the UV-Sensitive Mutants in Response to DNA Cross-linking Agents

In response to UV or the chemical 7-BrMeBA, mutants from five
genetic complementation classes had similar sensitivity as indicated
by the results shown above and other unpublished data. To date we
have seen no instance in which mutants from different
complementation classes responded much differently to a
monofunctional agent. However, in response to DNA cross-linking
agents, the mutants show a surprising heterogeneity of phenotype.
Initially, UV-5 and UV-20 were found to have 4- and 80-fold,
respectively, hypersensitivity to killing by mitomycin C (MMC)
(27). Mutants representing three other complementation classes also
exhibited levels of sensitivity like those of either UV-5 or UV-20.
Strains UV-24 (29) and UV-135 (unpublished) behaved like UV-5, and
UV-41 (29) behaved like UV-20. Extreme hypersensitivity of UV-20
and UV-41 is seen for the compound cis-diamminedichloroplatinum (II)
(cis-DDP), which also cross-links DNA (51). Recent studies have
also shown that the extreme sensitivity to MMC or cis-DDP is
associated with an inability of the mutant UV-20 to remove the
respective cross-links from its DNA (51). Thus, the excision
defects in at least some of the CHO UV-sensitive mutants appear to
affect the repair of cross-link damage, as well as damage from large
monofunctional agents and UV. In this regard, whether these CHO
mutants are like XP strains is unclear. Studies on the removal of
DNA cross-links in XP cells using two different techniques have
yielded results that lead to opposite conclusions (52,53).

CONCLUSIONS

The following general points can be made from the results that
have been discussed:

1. Wild-type Chinese hamster ovary (CHO) cells have significant
   capability to repair DNA damage resulting from exposure to UV
   and many chemical mutagens that are known to produce bulky DNA
   adducts, as evidenced by the isolation of repair-deficient
   mutant strains that are noticeably hypersensitive to such
   agents.

2. The UV-sensitive mutants of CHO resemble human XP mutant cells
   in terms of the defect being at or before the incision step of
   excision repair and also with respect to the types of
   DNA-damaging agents to which the cells are hypersensitive.

3. The relative amount of protection that excision repair confers
    against killing and mutation in CHO depends on the particular
    agent and also on the genetic marker used for detecting
    mutations.

4. At low doses (100% cell survival) of the polyaromatic
    hydrocarbon 7-BrMeBA, the shape of the dose response curve for
    mutation induction in repair-proficient CHO cells is distinctly
    nonlinear; the slope increases with increasing dose. This
    behavior may be attributable to saturation of the
    excision-repair system because in repair-deficient cells
    (strain UV-5) the dose responses were linear.

5. Excision repair appears to be an error-free process in CHO
    cells; in no instance was a higher induced mutation frequency
    observed in repair-proficient cells than in repair-deficient
    cells, at a given dose of mutagen.

6. In response to DNA cross-linking agents, the five
    complementation classes of UV-sensitive mutants show phenotypic
    diversity that is not seen with UV or monofunctional
    chemicals. Mutants in complementation classes 2 and 4 show
    extraordinary hypersensitivity (~100-fold) to killing by
    several crosslinking agents, thus providing a model system for
    studying the repair of this type of DNA damage.

7. The hypersensitivity of the repair-deficient strains to killing
    and to mutation induction confers distinct advantages in
    testing for environmental mutagens. The properties of the
    cells suggest devising a simple assay for genotoxic chemicals
    based on the differential killing of normal vs repair-deficient
    CHO cells.

ACKNOWLEDGMENTS

     This work was performed under the auspices of the U.S.
Department of Energy by the Lawrence Livermore National Laboratory
under contract number W-7405-ENG-48.

REFERENCES

1. Hollstein, M. and McCann J.: Short-term tests for carcinogens
    and mutagens. Mutat. Res., 65: 133-226 (1979).

2. Ames, B.N.: Identifying environmental chemicals causing
    mutations and cancer. Science, 204: 587-593 (1979).

3.   Hsie, A.W.: O'Neill, J.P., Couch, D.B., SanSebastian, J.R.,
     Brimer, P.A., Machanoff, R., Fuscoe, J.C., Riddle, J.C., Li,
     A.P., Forbes, N.L., and Hsie, M.H.: Quantitative analysis of
     radiation- and chemical-induced lethality and mutagenesis in
     Chinese hamster ovary cells. Radiat. Res., 76: 471-492 (1978).

4.   Clive, D., Johnson, K.O., Spector, J.F.S., Batson, A.G., and
     Brown, M.M.M.: Validation and characterization of the
     L5178Y/TK$^{+/-}$ mouse lymphoma mutagen assay system. Mutat.
     Res., 59: 61-108 (1979).

5.   Hsie, A.W., O'Neill, J.P., and McElheny, V.K., Eds.: Banbury
     Report 2, Mammalian Cell Mutagenesis: The Maturation of Test
     Systems. Cold Spring Harbor Laboratory (1979).

6.   Wolff, S. and Perry, P.: Differential Giemsa staining of sister
     chromatids and the study of sister chromatid exchanges without
     autoradiography. Chromosoma, 48: 341-353 (1974).

7.   Perry, P.E.: Chemical mutagens and sister chromatid exchange.
     Vol. 6 in Chemical Mutagens, Ed. by F.J. deSerres and A.
     Hollaender. Plenum Press, New York, pp. 1-39 (1980).

8.   Carrano, A.V., Minkler, J.L., Stetka, D.G., and Moore II,
     D.H.: Variation in the baseline sister chromatid exchange
     frequency in human lymphocytes. Environmental Mutagen., 2:
     325-337 (1980).

9.   Carrano, A.V. and Moore II, D.H.: The rationale and
     methodology for quantifying siter chromatid exchange in
     humans. In Mutagenicity: From Bacteria to Man. Ed. by J.A.
     Heddle. Academic Press, New York, in press.

10. Strauss, G.H. and Albertini, R.J.: Enumeration of
     6-thioguanine-resistant peripheral blood lymphocytes in man as
     a potential test for somatic cell mutations arising in vivo.
     Mutat. Res., 61: 353-379 (1979).

11. Ansari, A.A., Baig, M.A., and Malling, H.V.: In Vivo germinal
     mutation detection with "monospecific" antibody against lactate
     dehydrogenase-X. Proc. Natl. Acad. Sci. USA, 77: 7352-7356
     (1980).

12. Ansari, A.A., Baig, M.A., and Malling, H.V.: Development of in
     vivo somatic mutation system using antibody against
     hemoglobin. Preparation and use of an anti-hemoglobin antibody
     for identifying C57BL/6 red cells in artificial mixture of
     DBA/2 and C57BL/6 red cells, Mutat. Res., 81: 243-255 (1981).

13. Yang, L.L., Maher, V.M., and McCormick, J.J.: Error-free excision of the cytotoxic, mutagenic N^2-deoxyguanosine DNA adduct formed in human fibroblasts by (±)-7β, 8α-dihydroxy-9α, 10α-epoxy-7,8,9,10-tetrahydro-benzo(a)pyrene. Proc. Natl. Acad. Sci. USA, 77: 5933-5937 (1980).

14. Maher, V.M., Dorney, D.J., Mendrala, A.L., Konze-Thomas, B., and McCormick, J.J.: DNA excision-repair processes in human cells can eliminate the cytotoxic and mutagenic consequences of ultraviolet irradiation. Mutat. Res., 62: 311-323 (1979).

15. Cleaver, J.E.: Xeroderma pigmentosum. In Metabolic Basis of Inherited Disease, Ed. by J.B. Stanbury, J.B. Wyngaarden, and D.S. Fredrickson, 4th edition, McGraw-Hill, New York, pp. 1072-1095 (1978).

16. Maher, V.M., McCormick, J.J., Grover, P.L., and Sims, P.: Effect of DNA repair on the cytotoxicity and mutagenicity of polycyclic hydrocarbon derivatives in normal and xeroderma pigmentosum human fibroblasts. Mutat. Res., 43: 117-138 (1977).

17. Setlow, R.B.: Repair deficient human disorders and cancer. Nature, 271: 713-717 (1978).

18. Simons, J.W.I.M.: Development of a liquid-holding technique for the study of DNA-repair in human diploid fibroblasts. Mutat. Res., 59: 273-283 (1979).

19. Sarasin, A. and Benoit, A.: Induction of an error-prone mode of DNA repair in UV-irradiated monkey kidney cells. Mutat. Res., 70: 71-81 (1980).

20. DasGupta, U.B. and Summers, W.C.: Ultraviolet reactivation of herpes simplex virus is mutagenic and inducible in mammalian cells. Proc. Natl. Acad. Sci. USA, 75: 2378-2381 (1978).

21. Laval, F.: Effect of uncouplers on radiosensitivity and mutagenicity in x-irradiated mammalian cells, Proc. Natl. Acad. Sci. USA, 77: 2702-2725 (1980).

22. Corsaro, C.M. and Migeon, B.R.: Comparison of contact-mediated communication in normal and transformed human cells in culture. Proc. Natl. Acad. Sci. USA, 74: 4476-4480 (1977).

23. Siminovitch, L.: On the nature of heritable variation in cultured somatic cells. Cell, 7: 1-11 (1976).

24. Gautschi, J.R., Young, B.R., and Cleaver, J.E.: Repair of
     damaged DNA in the absence of protein synthesis in mammalian
     cells. Exptl. Cell Res., 76: 87-94 (1973).

25. Meyn, R.E., Vizard, D.L., Hewitt, R.R., and Humphrey, R.M.: The
     fate of pyrimidine dimers in the DNA of ultraviolet-irradiated
     Chinese hamster cells. Photochem. and Photobiol., 20: 221-226
     (1974).

26. Dipple, A. and Roberts, J.J.: Excision of
     7-bromomethylbenz(a)anthracene-DNA adducts in replicating
     mammalian cells. Biochem., 16: 1499-1503 (1977).

27. Thompson, L.H., Rubin, J.S., Cleaver, J.E., Whitmore, G.F., and
     Brookman, K.: A screening method for isolating DNA
     repair-deficient mutants of CHO cells. Somat. Cell Genet., 6:
     391-405 (1980).

28. Busch, D.B., Cleaver, J.E., and Glaser, D.A.: Large-scale
     isolation of UV-sensitive clones of CHO cells. Somat. Cell
     Genet., 6: 407-418 (1980).

29. Thompson, L.H., Busch, D.B., Brookman, K., Mooney, C.L., and
     Glaser, D.A.: Genetic diversity of ultraviolet-sensitive DNA
     repair mutants of Chinese hamster ovary cells. Proc. Natl.
     Acad. Sci. USA, in press (1981).

30. Brookman, K.W., Thompson, L.H., Salazar, E.P., Dillehay, L.E.,
     and Mooney, C.L.: Genetic complementation, DNA repair, and
     mutation induction in UV-sensitive Chinese hamster ovary cell
     mutants. 12th Ann. Mtg. Environ. Mutagen Soc., p. 161,
     abstract (1981).

31. Keijzer, W., Jaspers, N.G.J., Abrahams, P.J., Taylor, A.M.R.,
     Arlett, C.F., Zelle, B., Takebe, H., Kinmont, P.D.S., and
     Bootsma, D.: A seventh complementation group in
     excision-deficient xeroderma pigmentosum. Mutat. Res., 62:
     183-190 (1979).

32. Arase, S., Kozuka, T., Tanaka, K., Ikenaga, M., and Takebe, H.:
     A sixth complementation group in xeroderma pigmentosum. Mutat.
     Res., 59: 143-146 (1979).

33. Arlett, C.F. and Lehman, A.R.: Human disorders showing
     increased sensitivity to the induction of genetic damage. Ann.
     Rev. Genet., 12: 95-115 (1978).

34. Thompson, L.H., Brookman, K.W., Dillehay, L.E., Carrano, A.V.,
     Mooney, C.L., Mazrimas, J.A., and Minkler, J.A.: A CHO-cell
     strain having hypersensitivity to mutagens, a defect in DNA
     strand-break repair, and an extraordinary baseline frequency of
     sister chromatid exchange. Mutat. Res., submitted.

35. Thompson, L.H., Fong, S., and Brookman, K.: Validation of
     conditions for efficient detection of HPRT and APRT mutations
     in suspension-cultured Chinese hamster ovary cells. Mutat.
     Res., 74: 21-36 (1980).

36. Krahn, D.F. and Heidelberger, C.: Liver homogenate-mediated
     mutagenesis in Chinese hamster V79 cells by polycyclic aromatic
     hydrocarbons and aflatoxins. Mutat. Res., 46: 27-44 (1977).

37. Venitt, S. and Tarmy, E.M.: The selective excision of
     arylalkylated products from the DNA of Escherichia coli treated
     with the carcinogen 7-bromomethylbenz(a)anthracene. Biochim.
     Biophys. Acta, 287: 38-51 (1972).

38. Thompson, L.H., Brookman, K.W., and Carrano, A.V.: The role of
     DNA repair in mutagenesis of Chinese hamster ovary cells by
     7-bromomethylbenz(a)anthracene. Proc. Natl. Acad. Sci. USA,
     submitted.

39. Kohn, K.W., Erickson, L.C., Ewig, R.A.G., and Freidman, C.A.:
     Fractionation of DNA from mammalian cells by alkaline elution.
     Biochem., 15: 4629-4637 (1976).

40. Hiss, E.A. and Preston, R.J.: The effect of cytosine
     arabinoside on the frequency of single-strand breaks in DNA of
     mammalian cells following irradiation or chemical treatment.
     Biochim. Biophys. Acta., 478: 1-8 (1977).

41. Fornace, A.J., Kohn, K.W., and Kann Jr., H.E.: DNA
     single-strand breaks during repair of UV damage in human
     fibroblasts and abnormalities of repair in xeroderma
     pigmentosum. Proc. Natl. Acad. Sci. USA, 73: 39-43 (1976).

42. Tanaka, K., Hayakawa, H., Sekiguchi, M., and Okada, Y.:
     Specific action of T4 endonuclease V on damaged DNA in
     xeroderma pigmentosum cells in vivo. Proc. Natl. Acad. Sci.
     USA, 74: 2958-2962 (1977).

43. Bootsma, D.: Xeroderma pigmentosum. In DNA Repair Mechanisms,
     Ed. by P.C. Hanawalt, E.C. Friedberg, and C.F. Fox. (Proc.
     ICN-UCLA Symp. on DNA Repair Mechanisms, February 1978,
     Keystone, Colo.) Academic Press, New York, p. 589-601 (1978).

44.   Riddle, J.C. and Hsie, A.W.: An effect of cell-cycle position
      on ultraviolet-light-induced mutagenesis in Chinese hamster
      ovary cells. Mutat. Res., 52:'409-420 (1978).

45.   McCormick, J.J. and Maher, V.M.: Mammalian cell mutagenesis as
      a biological consequence of DNA damage. In DNA Repair
      Mechanisms, Ed. by P.C. Hanawalt, E.C. Friedberg, and C.F.
      Fox. (Proc. ICN-UCLA Symp. on DNA Repair Mechanisms, February
      1978, Keystone, Colo.) Academic Press, p. 739-749 (1978).

46.   Munson, R.J., and Goodhead, D.T.: The relation between induced
      mutation frequency and cell survival--a theoretical approach
      and an examination of experimental data for eukaryotes. Mutat.
      Res., 42: 145-160 (1977).

47.   McCaw, B.A., Dipple, A., Young, S., and Roberts, J.J.: Excision
      of hydrocarbon-DNA adducts and consequent survival in normal
      and repair defective human cells. Chem.-Biol. Interactions,
      22: 139-151 (1978).

48.   Dipple, A., Brookes, P., Mackintosh, D.S., and Rayman, M.P.:
      Reaction of 7-bromomethylbenz(a)anthracene with nucleic acids,
      polynucleotides, and nucleosides. Biochem., 10: 4323-4330
      (1971).

49.   Sugimura, T., Kawachi, T., Nagao, M., Yahagi, T., Seino, Y.,
      Okamoto, T., Shudo, K., Kosuge, T., Tsuji, K., Wakabayshi, K.,
      Iitaka, Y., and Itai, A.: Mutagenic principle(s) in tryptophan
      and phenylalanine pyrolysis products. Proc. Japan Acad., 53:
      58-61 (1977).

50.   Sugimura, T.: Naturally occurring genotoxic carcinogens. In
      Naturally Occurring Carcinogens-Mutagens and Modulators of
      Carcinogenesis, Ed. by E.C. Miller et al., Japan Sci. Press,
      Tokyo/Univ. Park Press, Baltimore, pp. 241-261 (1979).

51.   Meyn, R.E., Jenkins, S.L., and Thompson, L.H.: Defective
      removal of DNA-cross-links in a repair-deficient mutant of
      Chinese hamster cells. Cancer Res., submitted for publication.

52.   Fujiwara, Y., Tatsumi, M., and Sasaki, M.S.: Cross-link repair
      in human cells and its possible defect in Fanconi's anemia
      cells. J. Mol. Biol., 113: 635-649 (1977).

53.   Kaye, J. Smith, C.A., and Hanawalt, P.C.: DNA repair in human
      cells containing photoadducts of 8-methoxypsoralen or
      angelicin. Cancer Res., 40: 696-702 (1980).

DISCUSSION

DR. MAHER: We have been using intact human epithelial cells capable of metabolizing various chemical carcinogens (mutagens) as our source of metabolic activation of promutagens and combining these with human fibroblasts as target cells. The fibroblasts are incapable of carrying out such activation. Our results with normal repair-proficient cells compared to those with excision repair-deficient xeroderma pigmentosum fibroblasts may suggest why you see such a great differential between your repair minus cells and your parental line. When our data comparing survival of the two human cell strains exposed to benzo(a)pyrene which must be activated before it can bind to DNA are combined with binding data, they suggest that DNA adducts are continually formed and excised by the normal cells during the incubation period, whereas they are continually formed and remain bound in the XP cells. Therefore, at the end of the exposure period when the cells are assayed for the biological effect of the benzo(a)pyrene, the two strains which were exposed to the same concentration of chemical for the same length of time actually have very different amounts of "initial" DNA damage. This could explain why they show a much greater differential than if they had been exposed to the reactive form of this compound, viz., the 7,8-diol-9,10-epoxide. Your differential may well reflect this difference in the number of DNA adducts present at the end of your 4 hr incubation.

DR. THOMPSON: Yes, I think what you're saying is that we may be amplifying the differential response because of the significant amount of time allowed for repair. Actually, in terms of testing that may be advantageous because you can enhance the difference.

DR. GLICKMAN: Would that be an explanation for what at least I think I saw, that is your mutant is hypomutable for the chemicals which don't require activation. If you took those lines for survival and mutagenesis where S9 is required, it looks like they would be about equal to the wild type in terms of mutants induced related to survival.

DR. THOMPSON: In some situations I think they are and that you are right. There are reservations about using the cytotoxicity data when you have activation, at least in some situations where we're certain that one has more than one metabolite. Let's say one metabolite is toxic but not mutagenic, and other metabolites may be both mutagenic and toxic. Doing an analysis of mutagenesis versus survival is probably very messy when you've got activation going on. But still your point is correct.

DR. GLICKMAN: Have you considered or are you able to do experiments just hybridizing XP line to your line with

polyethylene glycol and measuring repair replication in each
nucleus?

DR. THOMPSON: Well, we decided not to try that experiment, but
someone in Toronto is trying to do it and it's technically
difficult. That's an important question.

DR. MAHER: Do I understand that saturation of mutagenesis, so
that you don't continue to get a linear increase, just a leveling
off, starts about at the point where the killing by UV radiation
is just getting off the shoulder or does it begin at doses where
killing is greater?

DR. THOMPSON: With UV it appears to be at about 50% survival.

DR. MAHER: Is it about the same with 7-bromomethylbenz(a)-
anthracene?

DR. THOMPSON: With 7-bromomethylbenz(a)anthracene it was at
higher survival; I think maybe 80% in the case of aza-adenine
resistance.

DR. MAHER: Earlier you said that before you launched out into the
deep and starting isolating excision repair strains in CHO cells,
the problem that held you back, or held others back, was that even
in your parental strain the pyrimidine dimers aren't excised.
Now, do the dimers come out or not?

DR. THOMPSON: I believe the data says that the dimers don't all
come out.

DR. MAHER: What fluence do they test that at?

DR. THOMPSON: The same range that we're working in here, 5 and 10
Jm^{-2}.

DR. MAHER: At 5 Jm^{-2} you're still on the shoulder?

DR. THOMPSON: Yes.

DR. MAHER: There's a question of what's causing that saturation
effect. At 50% survival, the frequency of mutants has alreeady
reached 1×10^{-3}, which is very high.

DR. THOMPSON: We're forced to conclude that the repair of the
dimers does not require their complete excision.

DR. THILLY: Can I throw in something with regard to the
"plateau"? Could I suggest we call it a "plateau" until there's
evidence that it represents "saturation"? Veronica has brought

out a reasonable alternative, that is we may be dealing with cell cycle heterogeneity. One could expect, for instance, that if some cells are relatively resistant to killing, we would observe a plateau effect in any situation, after the sensitive cells have been removed by killing.

DR. THOMPSON: Yes, that was actually a point that I forgot to make. We're going to test that by using synchronized cells and that's a very likely explanation.

DR. ARLETT: I'd like to make a comment about plateaus before getting to my questions. A lot of mutation data in a lot of system, places and times have a plateau. As far as I can see, most people tend to stop their experiment before they get to the plateau because they really don't know how to explain it.

But my questions relate to things you didn't say. First of all, do your mutants have the same chromosome constitution as wild type, or are there obvious changes?

DR. THOMPSON: Not in the ones we've looked at. The growth rates of the mutants are invariably a little bit slower, on the order of 10-20%, sometimes more.

DR. ARLETT: I wouldn't expect mutagen-induced sensitive strains to have simply one base change. You've probably got quite a lot of background alterations.

DR. THOMPSON: There are probably thousands of base changes in these mutant strains.

DR. ARLETT: That's quite an important point.

DR. THOMPSON: I should emphasize, they were all isolated after heavy chemical mutagenesis with a variety of agents.

DR. ARLETT: You didn't actually make that clear. Has anybody looked at endonuclease sensitive sites in these things? Do you know that you're putting the same number of lesions into these cells with UV? We're a little bit concerned about the possibility of protection versus repair. Again, it's something which we tend to overlook because we're all searching for repair. But you haven't any data on these?

DR. THOMPSON: No, we don't. Let's see, there is the work that Ray Meyn in Houston has done with cross-linking agents. He has shown the same amount of damage in the normal and the UV-20 mutant early after treatment, with either mitomycin-C or the cis-D platinum compound. So the amount of initial damage in that situation is the same. The kinetics for removal of cross-link lesions are very different. So that speaks to the question with that particular class of agents.

DR. ARLETT: My final point, and I don't know if you're going to talk about this at all this afternoon, Veronica, is differential killing. I think perhaps we ought to make it clear that the plot of mutation against survival gives a common line for Xeroderma and wild-type cells only if a particular experimental technique is used; in situ experiments, for example. I imagine you're doing respread experiments, in the jargon of the trade. XP's and normal cells quite clearly do not give a common line with respect to UV response in resprayed experiments. It's not very different but it is different. There is no discrepancy there for CHOs.

DR. THOMPSON: Does yours go in the same direction as mine?

DR. MAHER: Yes. Let me comment on the so-called saturation that is found in the wild-type, but not the repair minus cells. Each time they replicate, repair-deficient cells may have the same chance of forming mutants because the lesion persists. Excision-deficient cells may be continually replicating and making more mutations during the second round, the third round, and fourth round of DNA replication. That would explain why the repair minus strain appears more mutable than the normal strain for the same degree of survival, even at the very low doses where you haven't reached the plateau. There are two things operating to keep the frequency lower in the parental cells, initial excision repair before the first mutations occur and then removal of the rest of the lesions before the next round of DNA replication occurs.

DR. THILLY: I think that's an interesting suggestion but doesn't go all the way to an explanation. For the record let it be said that opinion was not unanimous on this point.

John De Luca examined the UV survival and UV mutation for thioguanine resistance and trifluorothymidine resistance in human lymphoblasts, experiments analogous to the ones that Larry reported here. The results from XP and wild-type cells looked very similar to Larry's data, with the single exception that in the XP complementation group C, which he used, he did see a mutant fraction plateauing effect in the cell population. At low fluences where you indicated that your data were not sufficient to show whether the wild-type cells gave a linear extrapolation to the origin, John repeated the experiments a sufficient number of times to demonstrate that indeed it is non-linear for lymphoblasts. That is, similar to your 7-bromomethylbenzathracene response, in normal fibroblasts. The relationship you see between 7-bromomethylbenzanthracene mutation in normal and XP fibroblasts is very similar to the relationship between UV mutation in normal and XP lymphoblasts seen by De Luca at low fluences.

We are left with wondering whether there are different kinds of cell responses in the same animal. Are lymphoblasts different from fibroblasts? The experiments now on the table give dramatically different responses, and I am at a loss to begin to be able to explain them, although naturally one tries.

DR. THOMPSON: How long is the G-1 period in your system?

DR. THILLY: The G-1 period in our cells is approximately 6 hours, with a 16 hr doubling time.

DR. THOMPSON: How different is that, would you say Veronica, from his? Your cells have a longer doubling time, don't they?

DR. MAHER: 24-30 hrs.

DR. THILLY: This could also be important, not to mention all the progression effects. I'm reminded of Plant and Roberts discussion of methylnitrosourea, the comparison of HeLa cells and V-79s. If I remember correctly, Roberts reported that the Hamster cells when treated with the alkylating agent, immediately showed an inhibition of their DNA synthesis. However, this did not occur in the HeLa cells until the second round of DNA replication, leaving all of us to wonder what is going on. I think we're all in the dark. Bob Painter says he has some human cells which show inhibition of DNA synthesis in the first round of replication. We looked at a series of alkylating agents in two lymphoblastoid lines. Neither of them showed any immediate inhibition of progression when treated with alkylating agent concentrations which were sufficient to reduce their survival by a factor of 3 powers of 10. So, this is something which is also puzzling.

Also, I want to ask one question about the oubain locus. How come when we're working with the human cells we're looking at induced frequencies or background frequencies anywhere from 10^{-8} to 10^{-6}, but with the V-79s and CHOs that have been studied, reports indicate much higher values?

DR. THOMPSON: Well, the background in my system in about 10^{-6} for the oubain markers.

DR. THILLY: In the lymphoblasts our background is about 5 x 10^{-8}.

DR. THOMPSON: My knowledge of the literature is that in V-79, different Chinese hamster cells, the background tends to be considerably higher, which is something I don't understand. When it's very low, then big differences aren't very significant, are they?

DR. ARLETT: It's obvious that human cells are very much more sensitive to oubain; about 1,000 fold difference in toxic response. The Chinese hamsters, however, vary very considerably. The first V-79 cells with which we did our original oubain resistance work, I think we saw, in the course of the whole of our study one spontaneous mutant. We even got to the point where we just threw in the odd control plate to see if the system was behaving, and we estimated that the spontaneous mutation frequency was something like 1 in 10^8. In other Chinese hamster cells with which we've worked with, one can approach 1 in 10^5, so there is a real mystery here.

SESSION IV

MAMMALIAN SYSTEMS

MODERATOR: BETSY OHLSSON-WILHELM

VARIATION IN RESPONSE TO MUTAGENS AMONGST NORMAL AND

REPAIR-DEFECTIVE HUMAN CELLS

Colin F. Arlett and Susan A. Harcourt

MRC Cell Mutation Unit
University of Sussex
Falmer, Brighton, England

INTRODUCTION

Over the past five years we have been examining the mutability of a variety of human fibroblast strains established from both normal individuals and patients suffering from cancer-prone genetic diseases. The objective of these experiments was to determine if cancer proneness at the level of the individual is correlated with either hypersensitivity to the lethal effects or hypermutability to the mutagenic effects of DNA damaging agents. The existence of hypersensitivity and particularly hypermutability implies the presence of defects in repair of DNA damage and underlines the importance of effective repair processes for human health (1).

The now classical example of a cancer-prone syndrome with correlated hypersensitivity and hypermutability, in this case to ultraviolet light (UV), is provided by patients suffering from the autosomal recessive sun-sensitive disease xeroderma pigmentosum (XP). The individuals exhibit an array of symptoms (2) both dermatological and neurological. Notable amongst the dermatological effects are:- the marked acute sun-sensitivity in infancy with pigmented macules and achromic spots and telangiectasia in exposed areas followed ultimately by basal cell carcinoma, squamous cell carcinoma and malignant melanoma. Similar skin changes may be apparent in later life in normal Caucasian adults who have occupations or live in regions in which extensive exposure to sunlight is experienced. The XP patients can, therefore, provide us with a model for the pathology in normal individuals. The neurological abnormalities which attain their most extreme form in the so-called De Sancis-Cacchione syndrome (3) include, microcephaly with progressive mental deterioration, low intelligence, areflexia and ataxia. All these

249

symptoms are rarely present but many XP patients have one or more of
the neurological features.   At the cellular level some, but not all,
XP cell strains show a marked hypersensitivity to the lethal effects
of UV light and some chemicals such as 4-nitroquinoline-1-oxide
whose action may be described as "u.v.-like" (4).   All XP cell
strains studied, so far, appear to be hypermutable by u.v. light
(5,6,7).   Most of these observations are concerned with detection
of mutants resistant to 8-azaguanine or 6-thioguanine and thus
exhibiting defects in hypoxanthine, guanine ribosyltransferase
(HGPRT).   Hypermutability has also been shown when XP cells are
assessex for u.v. induced resistance to diphtheria toxin (8).

Since XP cells are defective in either the excision of pyrimidine dimers (9) or in daughter strand repair (10) it is usual to
suppose that the increased cancer incidence in such individuals is a
consequence of an increased mutation frequency which is itself
brought about by the defective repair processes.

A second cancer prone genetic disease where individual sensitivity is correlated with a hypersensitivity for cell killing by
ionizing radiation is ataxia-telangiectasia (A-T)(11).   In this
multi-system disease defects in the immune system often lead to
early death as a consequence of pulmonary infections.   Lympho-
reticular tumours are a frequent cause of death and the morbidity
following palliative radiotherapy (12,13) is what drew attention to
this syndrome as a potential "DNA repair defective" condition.   In
this syndrome, however, and in direct contrast to XP, there is no
evidence for hypermutability following treatment of cells with
ionising radiation (6,14).   At the level of repair there is
evidence for a defect as measured by repair replication (15).

The existence of the UV-sensitive XP cells and gamma-ray
sensitive A-T cells prompted a survey of spontaneous and induced
mutation as a set of cell strains which included these and other
possible cancer-prone syndromes as well as from a number of
normal individuals.   Most of our observations, to date, have used
u.v.or ionising radiation as the mutagen.

THE DESIGN OF EXPERIMENTS

The method for detecting mutation followed that of the
Sterilin bulk culture vessel (BCV) technique described by Cox
et al (16) with some modifications.   An initial assessment of the
toxic response of cells to various concentrations of 6-thioguanine
(TG) showed that normal and A-T cells had essentially similar
responses (Figure 1).   We elected to use a concentration of TG of
2.5 µg/ml as the dose to select for resistant mutants.

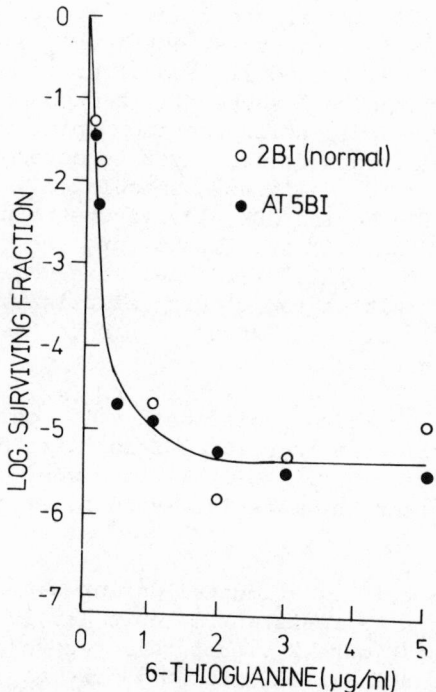

Figure 1.   The response of normal, 2BI and ataxia telangiectasia
cells to varying concentrations of 6-thioguanine.

        Cells were maintained in Eagles Minimal Essential Medium (MEM)
with 15% foetal calf serum (Flow Labs or Gibco-Biocult).   For
survival platings or for the selection of TG resistant mutants in
BCV's the serum was changed to 15% newborn calf serum.   Large
serum lots were used but obviously there have been changes in batch
during the course of the study.

        For a standard u.v. experiment a large number of 9 cm (Nunclon)
plastic petri dishes were each inoculated with 10^6 cells on day 1
and the next day the medium was removed and the cells irradiated with
graded doses of u.v. from an Hanovia type 12555 lamp (17).   The dose
rate was adjusted to accommodate the large differences in sensitivity
between XP and normal cells.   A parallel series of dishes were also
exposed to test for the lethal effects.   These cells were trypsinised
from the plates, counted and replated at the appropriate cell density
on a feeder layer immediately following the u.v. treatment (18).
With the plates for mutation the medium was replaced and the cells
incubated for a further day before being trypsinised and inoculated
into 800 ml flasks at ~ 1 x 10^6 cells per flask, the contents of one

dish being transferred to a single flask.   Two flasks were carried for
the control and three for each treatment level, usually three in
each experiment.   Five, 13 and 21 days after irradiation the flasks
were trypsinised and the cell suspension bulked and counted for each
treatment level.   The suspension was then split into three parts:
(A) plated on an homologous feeder layer to determine cloning
efficiency, (B) 6-7 x 10^6 cells were inoculated into a BCV which was
completely filled with Eagles MEM plus 15% newborn calf serum and
2.5 µg/ml 6-thioguanine, (C) two flasks for each treatment level
were reinoculated with 1.8-2.0 x 10^6 cells in Eagles MEM plus 15%
foetal calf serum.   Plates for cloning efficiency were stained after
16 days and the BCV's after 21 days.   In both the 50 cell criterion
was taken in scoring clones.

    In the case of ionising radiation 2 x 10^6 cells, either gamma-
irradiated or control were inoculated into pairs of 800 ml flasks
immediately after treatment.   Cell killing was monitored at the
same time.   Thereafter the material was handled as for a u.v.
experiment.

    The purpose of this experimental design was to permit full
expression of induced TG resistant mutants (19,20).   It is also
clear from our data (Figure 2), that full recovery from the lethal
effects of u.v. or gamma-irradiation may take at least 13 days.
Earlier experiments provided at least three estimates of mutation
for control and each treatment level.   In more recent studies we
attempted only to estimate mutation frequencies at 13 and 21 days
after treatment.   In all cases the estimated mutation frequencies
are given for populations where recovery is complete.

SPONTANEOUS MUTATION FREQUENCIES

    During the course of our investigations we have collected
data for spontaneous mutation on 27 cell strains.   In some of these
many repeat experiments and the use of multiple expression times,
as outlined above, have produced a large number of replicate
observations.   The results are summarised in Table I.   Five cell
strains from normal individuals gave a range of mutation frequencies
of 1.3-8.0 x 10^{-6} mutants per survivor, most of the observations
lying in the middle of this range because of the substantial contribution (30 observations) from cell strain 1BR(3.8 x 10^{-6}).   Amongst
the representatives from recessive cancer prone syndromes the six
XP cell strains, both excision and daughter strand repair types, gave
mutation frequencies within the normal range.   One, XP1BR from
complementation group D(21), gave a low value.   The heterozygotes

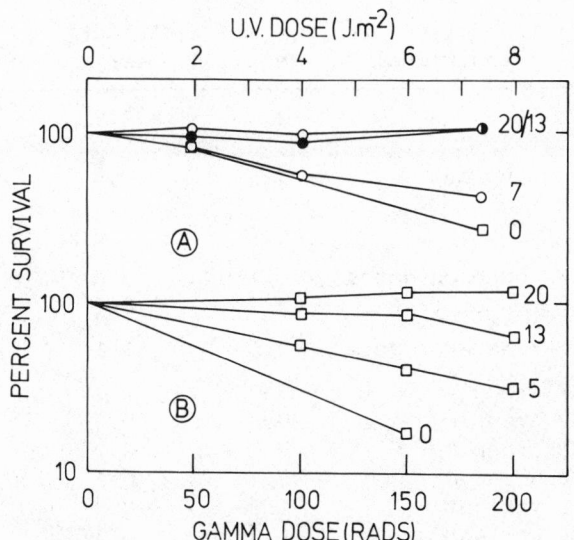

Figure 2.   The recovery from the lethal effects of (A) ultraviolet
light, (B) gamma-ray irradiaton.   The populations were
tested for survival immediately after treatment,= day 0;
sampled on days 7, 13 and 20 after treatment for u.v.
and days 5, 13 and 20 for gamma-irradiation.

XPHF4LO and XPHM4LO, both gave frequencies below the homozygote
XP4LO but, again, are consistent with the normal range. It seems
appropriate to include these heterozygotes within the set of normals.
The A-T cell strains which, in addition to being from a cancer-prone
condition, also exhibit enhanced chromosome breakage (12,22) gave
spontaneous mutation frequencies in the normal range as did a single

TABLE I. Spontaneous mutation frequencies in a variety of human
          fibroblast cell strains

Cell Strain	Phenotype of Individual	Mutation frequency x 10^{-6}		Number of observations
		Mean	± standard error	
1 1BR	)	3.8	0.9	30
2 48BR	)		8.0	2
3 54BR	) normal	1.3	0.9	4
4 2BI	)	1.5	0.4	18
5 GM730	)	7.4	2.3	6
6 XPHM4LO	XP heterozygotes	0.7	0.3	4
7 XPHF4LO	parents of XP4LO	3.2	0.7	7
8 ATH96TO	A-T heterozygote	2.5	1.1	3
9 XP4LO	XP group A	9.2	2.3	3
10 XP1BR	XP group D	0.4	0.2	4
11 XP3BR	XP group G	3.0	0.7	12
12 XP2BI	XP group G	3.4	0.6	6
13 XP30RO	XP variant	1.8	0.8	6
14 XP6DU	XP variant	1.2	0.3	4
15 CS697CTO	) Cockayne	3.4	1.3	6
16 CS698CTO	) syndrome	1.1	0.9	5
17 11961	sun-sensitive individual	1.5	0.4	18
18 D.L.	sun-induced keratosis		1.4	1
19 AT3BI	)	0.8	0.4	3
20 AT4BI	) A-T		4.0	2
21 AT5BI	)	2.6	0.8	16
22 62BR	) familial	4.0	0.7	3
23 A5570	) retinoblastoma	4.8	1.7	4
24 BCNS3BI	Gorlins syndrome	6.5	2.9	11
25 W.R.	) polyposis		3.1	2
26 G.P.	) coli		0.9	1
27 46BR	hypogamma-globulinaemia	0.5	0.2	8

XP = xeroderma pigmensotum   A-T = ataxia telangiectasia

heterozygote ATH96TO.   Retinoblastoma, Gorlins' syndrome and familial
polyposis coli represent dominant cancer prone syndromes (23).   The
spontaneous mutation frequencies were, again, within the normal range.
Claims have been made elsewhere (24) of enhanced spontaneous mutation
frequencies in some cell strains from Gorlins' syndrome.   Observations on cell strains from individuals not qualifying as cancer
prone, on two Cockayne syndrome (25) cell strains, a strain from a
sun-sensitive individual 11961 (26) and from 46BR established from a
patient with some novel features (27) could not be distinguished from
the normal range.

The overall range of spontaneous mutation frequencies observed
in this study was 0.4-9.2 x 10^{-6} mutants per survivor.   We wish to
regard the normal range, which includes heterozygotes from the
recessive autosomal cancer-prone syndromes XP and A-T as 0.7-8 x 10^{-6}
mutants per survivor.   It is clear that the spontaneous mutation
frequency is not exceeded by any of the other cell strains included
in the study.   Some, notably XP1BR and 46BR, might prove to have
reduced spontaneous mutation frequencies.   While it is obvious that
a proper estimate of spontaneous mutation would require the use of
fluctuation experiments our results lead us to conclude that cells
from cancer-prone individuals are intrinsically no more mutable than
cells from normal individuals.   The implication of such a conclusion
is that the increased incidence of cancer is a consequence of the
response of cells to DNA damage or the result of associated abnormalities (e.g. in the immune system) which reflect the progression
rather than the initiation of tumours.

ULTRAVIOLET-LIGHT INDUCED MUTAGENESIS

The response of four cell strains established from normal
individuals to u.v.-induced mutagenesis is illustrated in Figure
3A.   Cell strains 2BI and 1BR are our reference normal material
and thus replicate experiments are available for them and give some
guide to the experimental error.   Data for five other cell strains
are included in Figure 3B.   These include a strain from a bilateral
retinoblastoma patient (62BR) (28) and two A-T (AT3BI and AT5BI)
cell strains none of which show any sensitivity to the lethal effect
of u.v. light.   Results from two XP heterozygotes (XPHF4LO and
XPHM4LO), the parents of the patient who provided cell strain
XP4LO from complementation group A (29)are also displayed.   The
mother's cells (XPHM4LO) show no sensitivity to the lethal effect
of u.v., they may show a reduced frequency of spontaneous mutants
(Table I) but fall within the normal range for induced mutation.
The father's cells exhibit a slight hypersensitivity to the lethal
effect of UV, at low doses, the mutation induction data are within
the normal range.

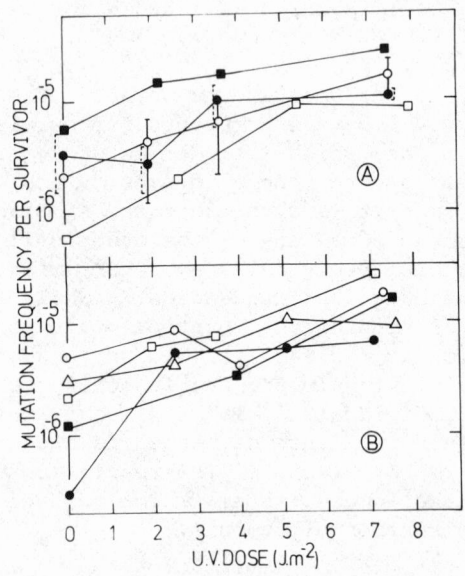

Figure 3.   The induction of 6-thioguanine-resistant mutants by u.v.
in (A) normal cell strains and (B) cell strains with
no sensitivity to the lethal effects of u.v.

(A)   O = 2BI, 3 experiments;   ● = 1BR, 2 experiments;
■ = GM730, 1 experiment;   ☐ = 54BR, 1 experiment.   The
vertical lines on 1BR and 2BI represent the limits of
observations in the replicate experiments.   For single
experiments each point is the data obtained from a single
BCV.

(B)   O = XPHF4LO, 1 experiment;   ● = XPHM4LO, 1 experiment;   △ = 62BR, 1 experiment;   ■ = AT3BI, 1 experiment;
☐ = AT5BI, 2 experiments.

These data give an overall picture of u.v.-induced mutagenesis
in nine human cell strains with no overt sensitivity for the lethal
action and allow us to assess the response of the u.v.-sensitive
cell strains.   Our observations on mutagenesis in the two XP heterozygotes do not confirm, at the cellular level the claim of Swift et
al (30) that in regions of high incident sunlight XP heterozygotes
show an increased cancer incidence when compared with homozygous
normal individuals.   Clearly more cell strains from heterozygotes
need to be examined to resolve this discrepancy.

The response of a set of sensitive cell strains to the mutagenic
effect of u.v. is summarized in Figure 4 together with the range from
our normal cell strains.   Six XP cell strains were tested, both
excision-defective and variant types.   XP4LO from complementation
group A, XP1BR from complementation group D and the two representatives from complementation group G, XP2BI and XP3BR, all showed
a dramatic increase in mutation frequency at low doses of u.v.
The two XP variants, XP30RO and XP6DU also gave enhanced mutation
frequencies.   Both types of XP gave enhanced mutation frequencies
compared with normals whether the data are displayed on a dose or
survival basis although the variants are substantially more mutable
than the excision-defective XPs when the comparison is made on the
basis of lethal events.   The simplest interpretation of these
results is that they suggest that daughter strand repair which is
defective in both classes of XP (10), is error prone.   Others (8,
31) have shown that excision repair may be predominantly or even
completely error-free.   Support for this comes from our observations on XP's from complementation group D (XP1BR) and group A
(XP4LO).   The group D cells have a larger residual excision
repair capacity, 30% (32), than the group A cells (less than 5%)
(33) and proved to be significantly less mutable.   This suggests
that the residual excision repair can remove lesions in an error-
free manner.

We have always found considerable difficulty in measuring
induced mutation in cells from Cockayne syndrome since such cells
appear to lose their potential for division at a greater rate than
others.   One full and one incomplete experiment on cell strain
CS96CTO produced an enhanced mutation frequency when compared with
normal cells.   Cockayne syndrome cells show an undoubted hypersensitivity to the lethal effects of u.v. light (34,35) but no
defects in excision or daughter-strand repair (36).   They exhibit
an impairment of the recovery of both DNA and RNA synthesis post-
irradiation suggesting the existence of some, as yet, undescribed
repair defect (37).   There is no evidence for increased tumour
incidence in this syndrome although early death from other causes
may conceal any increased tendency to form skin tumours.   While it
is appropriate to exercise some caution when interpreting results
on a single cell strain in this syndrome it is possible to suggest

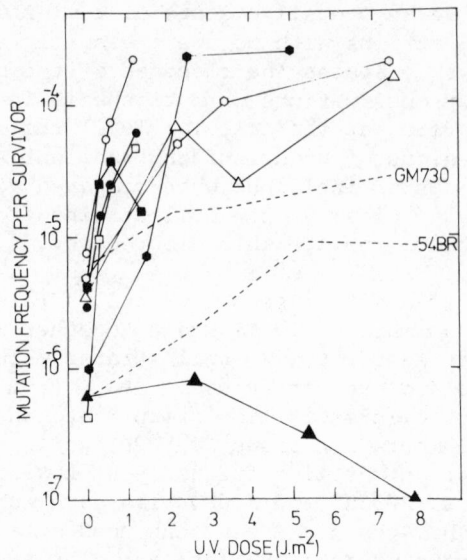

Figure 4. The induction of 6-thioguanine-resistant mutants with
u.v. in a set of u.v.-sensitive cell strains. The limits of the
response of normal cell strains, GM730 and 54BR is indicated by the
dotted lines.

○ = XP4LO, 2 experiments; ● = XP2BI, 1 experiment: □ = XP1BR, 1
experiment; ■ = XP3BR, 2 experiments; ⬢ = XP30RO, 1 experiment;
⬡ = XP6DU, 1 experiment;   Δ = CS697CTO, 2 experiments;
▲ = 46BR, 2 experiments.

that they may reveal an error-prone component of excision repair or
an enhanced error proneness of daughter-strand repair.

   Hypermutability is not an inevitable correlate of hypersensitivity to u.v. as shown by the response of cell strain 46BR. The
patient who gave this cell strain has a number of clinical features
in common with A-T (27) especially a deficiency in IgA and a lack
of stimulation into division of lymphocytes by mitogens. At the
cellular level there is hypersensitivity to a wide range of DNA
damaging agents including u.v., gamma-irradiation and methylating
agents but not ethylating agents. No defects have been observed
with respect to either excision or daughter-strand repair after u.v.

treatment (27).   The cells are clearly hypomutable implying the
absence of an error-prone repair process.

GAMMA-RAY INDUCED MUTAGENESIS

     The response of five cell strains established from normal
individuals is illustrated in Figure 5A.   Data for five other cell
strains with no overt sensitivity to the lethal effects of gamma-
irradiation are shown in Figure 5B.   These data show that the two
XP cell strains XP2BI and XP4LO, the bilateral retinoblastoma 62BR
and the XP heterozygotes, parents of XP4LO give responses very close
to normal.   These data can be used to assess the response of cell
strains with hypersensitivity to the lethal effects of gamma
irradiation.

     The response of a set of gamma-ray~ sensitive cell strains is
summarised in Figure 6.   The data are displayed as a plot of
mutation against survival in order that the large differences in
survival can be accommodated.   Information from three of the normal
cell strains from Figure 4A is included to scale the responses.
The xeroderma pigmentosum complementation group G strain XP3BR which
we had earlier believed to show a slightly enhanced induced mutation
frequency at low doses (38) appears to give an essentially normal
response now that more data on normal cells are available.   The
three A-T cell strains AT3BI, AT4BI and AT4BI appear hypomutable
as does the new cell strain 46BR.   The hypomutability of the A-T
cells may also be taken to indicate that these cells lack an error-
prone repair process which, in normal cells, confers resistance at
the price of mutation.   Other studies of mutability in A-T have
led to the conclusion that it is not possible to distinguish between
their response and that of normal cells (39).   They are clearly very
different in response to gamma irradiation than are XP cells to UV.
The results with 46BR present a further dilemma, the hypomutability
with both u.v. and gamma irradiation was unexpected.   Since spontaneous mutants arise, albeit at a reduced frequency, it seems
unlikely that the cells have a unique sensitivity to the selective
agent 6-thioguanine.   However, the possibility still exists that
the absence of induced mutants is a consequence of a selection
artifact.   Studies of induced mutagenesis using other selective
systems such as ouabain or diphtheria toxin resistance are necessary
to resolve this point.   In the absence of any evidence to that
effect we can only conclude that the cells are defective in either
two new error prone repair pathways or some step in repair which is
common to both u.v. or gamma-ray-induced mutagenesis.

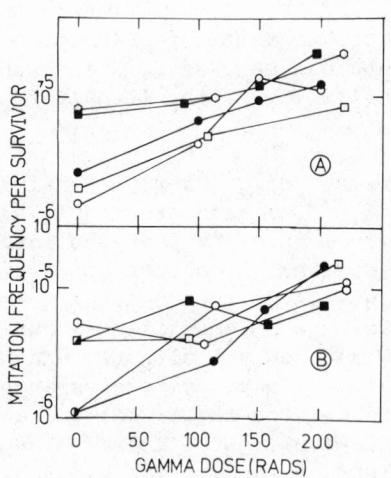

Figure 5.   The induction of 6-thioguanine resistant mutants by gamma
            irradiation in (A) normal cell strains and (B) cell
            strains with no sensitivity to the lethal effects of gamma
            irradiation.

            (A)   O = 2BI, 2 experiments;   ● = 1BR, 3 experiments;
            □ = 54BR, 1 experiment;   ■ = GM730, 1 experiment;
            ⬠ = 48BR, 1 experiment.

            (B)   O = XPHF4LO, 1 experiment;   ● = XPHM4LO, 1 experiment
            □ = XP2BI, 1 experiment;   ■ = XP4LO, 1 experiment;
            ⬠ = 62BR, 1 experiment.

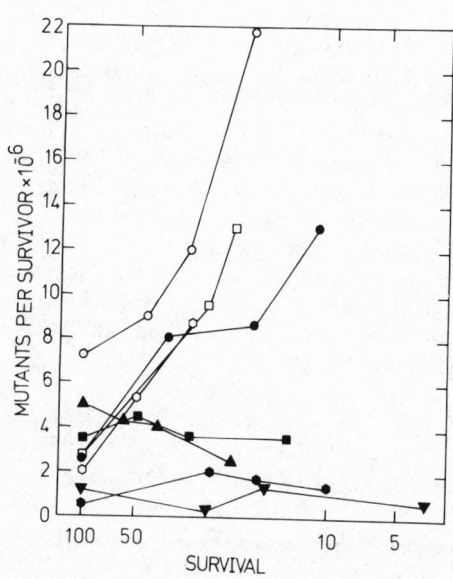

Figure 6.   The induction of 6-thioguanine-resistant mutants by gamma
            irradiation in a set of gamma-sensitive cell strains.
            Data for three normal cell strains, GM730 = O, 1BR = ☐
            and 54BR = O are included.   The mutation frequency is
            plotted against survival assayed in the treated population
            (day 0 from Figure 2).

            ●= XP3BR, 2 experiments; ▲ = AT4BI, 1 experiment;
            ■ = AT5BI, 2 experiments; ⬢ = 46BR, 2 experiments;
            ▼ = AT3BI, 1 experiment.

CONCLUSIONS

A large number of human variants whose cells show sensitivity to the lethal effects of either u.v. or gamma irradiation
are now known. Thus u.v. sensitivity is shown in a wide range of
genetically distinct XP's, Cockayne syndrome cells and cells (46BR)
from an individual with sensitivity to u.v., gamma and methylating
agents.   Cockayne syndrome cells and 46BR, unlike XP, have no defects
in either excision or daughter strand repair of u.v.-induced damage.
Cancer-proneness is characteristic of XP but not Cockayne syndrome
(46BR probably has a lymphoma (27)) but both show enhanced u.v.
mutability.   The response in XP can be explained on the assumption
that daughter strand repair is error prone.   The limited evidence
from Cockayne syndrome leads to the conclsuion that such cells are
defective in an error-free pathway, but it is equally possible to
invoke the existence of an error-prone component of excision repair
such as has been suggested for bacteria (40).   46BR cells are hypomutable, this leads to the conclusion that they lack an error-prone
repair pathway for u.v. damage.   The *recA* or *lexA* mutations
bacteria confer sensitivity to the lethal effects of u.v. but such
bacteria are not mutated by u.v. (42).

Hypersensitivity to the lethal effects of ionising radiation
have been demonstrated in A-T cells (11), XP3BR (38), and 46BR (27).
The A-T and 46BR cells are hypomutable and, by analogy with the
bacterial example above, this result suggests that they lack error-
prone repair processes which confer resistance at the price of
mutation in normal cells.   It remains to be seen whether these
putative error-prone repair processes are common or distinct in
A-T or 46BR.   XP3BR appears to give a normal response to mutation
induction by gamma irradiation.

Clearly the study of the mechanisms of mutation in cultured
human fibroblasts is still in its infancy.   The presence of
variation in response both with respect to hyper- and hypomutability suggests that we now have available tools to explore
mutation in human cells.

ACKNOWLEDGEMENTS

I am indebted to Professor B. A. Bridges for critical comment,
to Dr. R. Marshall for the use of results on Gorlins syndrome.
The study was supported in part by Euratom Contract No.166-76-1-BIO
UK.

REFERENCES

1. Arlett, C. F. and Lehmann, A. R.: Human disorders showing
      increased sensitivity to the induction of genetic damage.
      Ann. Rev. Genet., 12:95 (1978)

2. Kraemer, K. H.: Xeroderma pigmentosum. in Clinical Dermatology
   Ed. by D. J. Dennis, R. L. Dobson and J. McGuire (Unit 19.7:
   Vol. 4). Harper and Row, Hagerstown, p.1 (1978)

3. De Sanctis, C. and Cacchione, A.: L'idiozia xerodermica.
   Riv. Sper. Freniatr., 56:269 (1932)

4. Stich, H. F., San. R. H. C. and Kawazoe Y.   Increased sensitivity
   of xeroderma pigmentosum cells to some chemical carcinogens
   and mutagens. Mutation Res., 17:127 (1973)

5. Maher, V. M., Curren, R. D., Ouellette, L. M., McCormick, J. J.
   Effect of DNA repair on the frequency of mutations induced in
   human cells by ultraviolet irradiation and by chemical carcino-
   gens.   In Fundamentals in Cancer Prevention, Ed. by. P. N. Magee,
   S. Takayama, T. Sugimura, T. Matsushima.   Tokyo Univ.Press,
   Tokyo (1976)

6. Arlett, C. F.: Mutagenesis in repair-deficient human cell strains.
   In Progress in Environmental Mutagenesis, Ed. by M. Alacevic,
   Elsevier, Amsterdam, p.161 (1980)

7. Simons, J. W. I. M.   Effects of liquid holding on cell killing
   and mutation in normal and repair-deficient human cell strains.
   In DNA Repair Mechanisms, Ed. by P. C. Hanawalt, E. C. Friedberg
   and C. F. Fox. Academic Press, New York, p.581 (1978)

8. Glover, T. W., Chang, C. C., Trosko, J. E., and Li, S. S. L.
   Ultraviolet light induction of diphtheria toxin-resistant
   mutants in normal and xeroderma pigmentosum fibroblasts.
   Proc. Natl. Acad. Sci. USA, 76:3982 (1979)

9. Cleaver, J. E.   Defective repair replication of DNA in xeroderma
   pigmentosum.   Nature, 218:652 (1968)

10. Lehmann, A. R., Kirk-Bell, S., Arlett, C. F., Paterson, M. C.,
    Lohman, P. H. M., de Weerd-Kastelein, E. A., Bootsma, D.
    Xeroderma pigmentosum cells with normal levels of excision
    repair have a defect in DNA synthesis after u.v. irradiation.
    Proc. Natl. Acad. Sci. USA, 72:219 (1975)

11. Taylor, A. M. R., Harnden, D. G., Arlett, C. F., Harcourt, S. A.,
    Lehmann, A. R., Stevens, S., Bridges, B. A.   Ataxia telan-
    giectasia: A human mutant with abnormal radiation sensitivity.
    Nature, 258:427 (1975)

12. Bridges, B. A., and Harnden, D. G., Eds. Ataxia-telangiectasia: a
    Cellular and Molecular Link between Cancer, Neuropathology
    and Immune Deficiency, John Wiley and Sons, London (in press)

13. Cunliffe, P. N., Mann, J. R., Cameron, A. H., Roberts, K. D.,
    Ward, H. W. C. Radiosensitivity in ataxia telangiectasia.
    Br. J. Radiol.48:374 (1975)

14. Arlett, C. F. and Harcourt, S. A. Cell killing and mutagenesis
    in repair-defective human cells. In DNA Repair Mechanisms,
    Ed. by P. C. Hanawalt, E. C. Friedberg and C. F. Fox.
    Academic Press, New York, p.633 (1978)

15. Paterson, M. C. and Smith, P. J. Ataxia-telangiectasia: An
    inherited human disorder involving hypersensitivity to
    ionizing radiation and related DNA-damaging chemicals.   Ann.
    Rev. Genet. 13:291 (1979)

16. Cox, R. and Masson, W. K. X-ray-induced mutation to 6-thiogua-
    nine resistance in cultured human fibroblasts.   Mutation Res.,
    37:125 (1976)

17. Arlett, C. F., Harcourt, S. A. and Broughton, B. C. The
    influence of caffeine on cell survival in excision-proficient
    and excision-deficient xeroderma pigmentosum and normal human
    cell strains following ultraviolet irradiation.   Mutation
    Res., 33:341 (1975)

18. Cox, R. and Masson, W. K.   Changes in radiosensitivity during
    the in vitro growth of diploid human fibroblasts. Int.
    J. Radiat. Biol., 26:293 (1974)

19. Arlett, C. F. and Harcourt, S. A. Expression time and spon-
    taneous mutability in the estimation of induced mutation
    frequency following treatment of Chinese hamster cells by
    ultraviolet light. Mutation Res., 16:301 (1972)

20. Simons, J. W. I. M. Dose response relationships for mutants in
    mammalian cells in vitro.   Mutation Res., 25:219 (1974)

21. Lehmann, A. R. and Stevens, S. A rapid procedure for measure-
    ment of DNA repair in human fibroblasts and for complementation
    analysis of xeroderma pigmentosum cells.   Mutation Res., 69:
    177 (1980)

22. Taylor. A. M. R., Metcalfe, J. A., Oxford, J. M. and Harnden
    D. G. Is chromatid-type damage in ataxia telangiectasia
    after irradiation at G_0 a consequence of defective repair?
    Nature, 260:441 (1976)

23. McKusick, V. A. Mendelian Inheritance in Man: Catalogs of Auto-
    somal Dominent, Autosomal Recessive and X-linked Phenotypes.
    John Hopkins Univ. Press, Baltimore and London (5th ed), (1978)

24. Featherstone, T. Ph.D Thesis, University of Birmingham (1981)

25. MacDonald, W. B., Fitch, K. D., Lewis, I. C. Cockayne's syn-
    drome. An heredo-familial disorder of growth and development.
    Pediatrics, 25: 997 (1960)

26. Arlett, C. F., Lehmann, A. R., Giannelli, F. and Ramsay, C. A.
    A human subject with a new defect in repair of ultraviolet
    damage. J. Invest. Dermatol., 70:173 (1978)

27. Webster, D., Arlett, C. F., Harcourt, S. A., Teo, I. and
    Henderson, L. A new syndrome of immunodeficiency and
    increased cellular sensitivity to DNA damaging agents.
    In Ataxia-telangiectasia: a Cellular and Molecular Link
    between Cancer, Neuropathology and Immune Deficiency,
    Ed. by B. A. Bridges and D. G. Harnden. John Wiley and
    Sons, London (in press)

28. May, H. M. personal communication

29. Lehmann, A. R., Kirk-Bell, S., Arlett, C. F., Harcourt, S. A.,
    de Weerd-Kastelein, E. A., Keijzer, W., Hall-Smith, P.
    Repair of u.v. damage in a variety of human fibroblast cell
    strains. Cancer Res., 37:904 (1977)

30. Swift, M. and Chase, C. Cancer in xeroderma pigmentosum
    families. J. Natl. Cancer Inst., 62:1415 (1979)

31. Maher, V. M., Dorney, D. J., Mendrala, A. L., Konze-Thomas, B.
    and McCormick, J. J. DNA excision-repair processes in human
    cells can eliminate the cutotoxic and mutagenic consequences
    of ultraviolet irradiation. Mutation Res., 62:311 (1979)

32. Lehmann, A. R. personal communication

33. Lehmann, A. R. personal communication

34. Schmickel, R. D., Chu, E. H. Y., Trosko, J. E. and Chang, C. C.
    Cockayne syndrome - cellular sensitivity to ultraviolet light.
    Pediatrics, 60:135 (1977)

35. Marshall, R. R., Arlett, C. F., Harcourt, S. A., Broughton, B.
    C. Increased sensitivity of cell strains from Cockayne's
    syndrome to sister-chromatid exchange induction and cell
    killing by u.v. light. Mutation Res., 69:107 (1979)

36. Lehmann, A. R., Kirk-Bell, S., and Mayne, L. Abnormal kinetics
    of DNA synthesis in ultraviolet light-irradiated cells from
    patients with Cockayne's syndrome. Cancer Res. 39:4237.

37. Lehmann, A. R. and Mayne, L.   The response of Cockayne syndrome cells to u.v.-irradiation.   In Chromosome Damage and Repair. Ed. by E. Seeberg and K. Kleppe.   Plenum Press, New York (in press)

38. Arlett, C. F., Harcourt, S. A., Lehmann, A. R., Stevens, S., Ferguson-Smith, M. A. and Morley, W. N.   Studies on a new case of xeroderma pigmentosum (XP3BR from complementation group G with cellular sensitivity to ionising radiation.   Carcinogenesis 1:745 (1980)

39. Simons, J. W. I. M. Studies on survival and mutation in ataxia-telangiectasia cells after X-irradiation under oxic and anoxic conditions.   In Ataxia-telangiectasia: a Cellular and Molecular Link between Cancer, Neuropathology and Immune Deficiency.   Ed. by B. A. Bridges and D. G. Harnden.   John Wiley and Sons, London (in press)

40. Green, M. H. L., Bridges, B. A., Eyfjord, J. E. and Muriel, W. J.   Mutagenic DNA repair in Escherichia coli.   V. Mutation frequency decline and error-free post-replication repair in an excision-proficient strain.   Mutation Res., 42:33 (1977)

41. Witkin, E. M.   Mutatation-proof and mutation-prone modes of survival in derivatives of Escherichia coli B differing in sensitivity to ultraviolet light.   Brookhaven Symposium in Biology, 20:17 (1967)

DISCUSSION

DR. ROSENKRANZ: This may be far-fetched, but does patient 46-BR, the one with the immune deficiency, by any chance have, as many others have, a defect in purine metabolism such as a deficiency in adenosine deaminase which would show up when you use thioguanine-resistance as a marker and might not show up if you use another marker?

DR. ARLETT: I seem to recall that her purine levels were quite normal; she was not excreting purines, anyway. But that needs to be checked.

DR. ROSENKRANZ: Thioguanine toxicity was normal?

DR. ARLETT: Yes, I have the complete toxic response for those cells. I should add that we do get spontaneous mutants, but it is at the bottom of the scale. And certainly her cells and one of the XP's (XP1BR) were lower in spontaneous mutants than any of the other cell strains we have studied.

DR. GLICKMAN: You showed data in which survival from a dose of gamma rays, at 0, 5, 13 and 20 days went up to 100% survival. I really don't understand what 100% survival is after Day 20.

DR. ARLETT: This is actually a measure of the recovery from the lethal events. By 20 days, all the lethal damage has disappeared, and the cells have multiplied.

DR. GLICKMAN: I object to the use of the word survival. I'm not sure what it means in terms of looking at your plus independent isolate of mutations, and so on.

DR. ARLETT: That's a very good point. The proper way in which one constructs the experiment is the so called in situ method, where you will score mutation in a single event on a plate. We cannot do those experiments.

DR. GLICKMAN: I understand the technical difficulties. But because you chose to talk about it as survival I was a bit confused.

DR. ARLETT: If we score mutation on day 5, there are still a lot of lethal events. While we may put six million cells in the mutation assay vessel, only half of the cells still survive the lethal effects of the radiation. At 20 days, we've reduced the complication, which you're pointing out I think, of segregating mutational events.

DR. MAHER: When you put your cells in the pot on day 5, 13 and 20, and at the same time you put them in an assay for replating efficiency, is there a difference between day 5, 13 and 20 in the mutation frequency?

DR. ARLETT: You see a lot more mutants on day 5 than you do, per surviving cell, on days 13 and 20. Because I don't know how to handle that, I choose to use the one where we segregate out lethal and mutational events.

DR. THILLY: Can we address this by going way back to first principles, the assumption that the cloning efficiency at low cell density would be equal to the cloning efficiency at high cell density? Have we not learned that this is an assumption we make at great risk. Isn't the reason that you're carrying cells for a three week period the need to return all cells in the population to the same probability of forming a clone at high and low density? Don't we now that recently damaged cells are particularly sensitive to killing by plating at low density and this is an important factor in you experimental design?

DR. ARLETT: Well it's one of the many factors, yes.

DR. THILLY: And therefore you have not reported mutant fractions determined by plating at low cell density and at high cell density in the presence of 6-thioguanine at day 5.

DR. ARLETT: I have not reported it. I think I tried to make this point right in the beginning. I reported only observations on populations where recovery from the lethal events, was complete.

DR. THILLY: Should we not contemplate a general series of reconstruction experiments? Couldn't we create an artificially high mutant fraction at the beginning of the experiment, and pass this population through the entire experiment to see if there is any selective pressure for or against the phenotype which you select.

DR. GLICKMAN: Do you irradiate those cells as well?

DR. THILLY: Yes, irradiate or treat with a chemical.

DR. GLICKMAN: It's not really equivalent. You can't really do the reconstruction experiment having gotten the mutant, the thioguanine mutant but I guess it's the best you can do. The preexisting mutations that you add have already expressed and that doesn't affect the selectivity, whereas the induced mutants have to have an expression time. It's the best you can do, but it's not truly a reconstruction experiment because the mutants you're putting in have reached the expression potential.

DR. THILLY: This is in fact true of all phage, bacteria, and cell experiments. You're absolutely correct.

DR. ARLETT: In my manuscript I've said that we're very careful yet about the interpretation of 46 BR because there may be selection artifacts, some of which we've still got to wade our way through. One should always induce thioguanine-resistant mutants in 46 BR, by the mutagen you're actually testing against, and use those for the reconstruction experiment, but of course we can't because they are immutable. One tends to use Lesch-Nyhan as the prototype or model mutant cell, but I'm not sure actually that Lesch-Nyhan cells can necessarily give us the same story.

DR. GLICKMAN: I should say I understand the technical difficulties involved in working with mammalian system. My original reason for pointing out the problem was I still don't understand why you have to pretend it's like bacteria, for example, by talking about survival versus recovery or cloning efficiencies or the like.

DR. ARLETT: I apologize if I was pretending, I think it's implicit rather than explicit.

DR. MAHER: However, when you plotted your results on the basis of survival on the X-axis and your mutation frequencies on the Y-axis, that's initial survival data that you were using, isn't it?

DR. THILLY: As long as we've stopped on this point of reconstruction experiments, something's been bothering me. Barry, whenever you're looking at various genotypes and trying to compare the quantitative response from one locus to another, that should also apply, quite possibly, to various sites in the lacI locus. I was wondering if a set of reconstruction experiments have been performed at selected places through the lacI for their probability of expression so that we can understand the numerology.

DR. GLICKMAN: The answer to the question is that reconstruction experiments haven't been done properly. The assumption is that lacI is non-essential and is not expressed during the selection time. I mean amber and ochre mutants; one assumes that in the non-suppressed case the lacI fragment being produced would have no detrimental effect, etc. I don't think that that argument is particularly bad, but the reconstruction experiments haven't been done. Rather, I should say that they have been done only for two or three sites. However, spectra produced by Miller in the overnight growth situation, are in fact equivalent to the spectra produced by direct plating in top agar in which one can look at expression time by supplementing with glycerol or the like. And those spectra turn out to be the same. So there is no evidence for a differential selectivity of different sites. I guess for a reversion system one doesn't have the problem if one looks at only one site. But in forward systems, one is going to have to worry about that problem.

DR. THILLY: One last point about phenotypic expression for 6-thioguanine resistance, a thing that has bothered all of us in the field for many years. And that is that we have the non-mutant cell at the beginning of the experiment has several million copies of the protein for HGPRT and a cell at the end which is capable of growing in the presence of 6-thioguanine which, dependent upon the cell type, has anywhere from 0 to a few such copies. Since in reconstruction experiments both cell types behave equivalently, it has been assumed that anything intermediate with regard to the number of HGPRT copies would behave similarly. I hope this addresses your question of how we reason about this intermediate period during the expression of the phenotype.

DR. GLICKMAN: My worry in the case of the system you're describing is if there is some intermediate situation in which the purine metabolism is for some reason unbalanced, going from a non-expressed to an expressed mutation. I have no reason for thinking that such a thing is true, I'm just saying that you're not asking the identical question as in the reconstruction experiment. I think numerology would place some limits on the extent of such an error and undoubtedly the extent of such an error is likely to be below experimental significant error. I don't see a reconstruction problem, but I guess it depends on which point one choses to do the perfect experiment. In order to do the perfect experiment, however, you already have to know all of the answers!

RELATIONSHIP BETWEEN EXCISION REPAIR AND THE CYTOTOXIC

AND MUTAGENIC ACTION OF CHEMICALS AND UV RADIATION

Veronica M. Maher and J. Justin McCormick

Carcinogenesis Laboratory - Fee Hall
Department of Microbiology and Department of Biochemistry
Michigan State University, East Lansing, MI 48824

INTRODUCTION

We are investigating the mechanisms by which mutations are introduced into the DNA of diploid human fibroblasts following exposure to chemical carcinogens or ultraviolet (UV) radiation. In particular, we have asked, are the mutations induced by these agents introduced during the repair process by misrepair? or are they the result of misreplication during semi-conservative DNA synthesis? Our studies have been facilitated by the existence of excision repair-proficient diploid human cells derived from normal persons and repair-deficient fibroblasts derived from xeroderma pigmentosum (XP) patients. We have compared these fibroblasts for their response to the mutagenic and/or cytotoxic effect of UV radiation,[1-5] of reactive derivatives of a series of aromatic amines[6,7] or polycyclic aromatic hydrocarbons,[8-11] or of nitrosoureas.[12] Cytotoxicity is defined as the inability of a cell to form a colony, i.e., reproductive death. Mutagenicity is defined as an increase in 8-azaguanine or 6-thioguanine resistant cells in the population (resulting from the loss of active hypoxanthine(guanine)phosphorylbosyltransferase). The results of our studies indicate that mutations by these particular agents are not introduced during excision repair but result directly or indirectly from semi-conservative DNA synthesis on a template containing unexcised lesions, i.e., by misreplication or failure to replicate a portion of the DNA.

Our data further suggest that as soon as DNA lesions are introduced by these agents, there is an immediate response by excision repair processes and that the ultimate biological consequences, i.e., cell death and/or mutations, depend upon the rate of excision and the time available for excision before the onset of some critical cellular event responsible for death and/or responsible for induction of mutations. Thus, cells may be

271

protected from the potentially harmful effects of exposure to these agents by having a very rapid rate of excision or an extended length of time before these critical cellular events take place. Conversely, a cell population with a slow rate of excision or with a very short time before the onset of the critical event is predicted to be the one with the highest frequency of mutations for a given dose and the highest degree of sensitivity to the killing action of these agents. In analyzing the role of excision repair in preventing cell death or preventing mutation induction, we have taken two approaches. We have varied the rate of the excision of the cells by comparing these effects in a series of XP fibroblasts, each with a different rate of excision repair, and we have used synchronous populations and varied the length of time between exposure of the cells to the DNA damaging agent and the onset of semi-conservative DNA synthesis. The results are discussed below.

THE RATE OF EXCISION REPAIR DETERMINES THE CYTOTOXIC AND MUTAGENIC EFFECT OF THESE AGENTS

Three strains of diploid human fibroblasts, each with a cell cycle of approximately the same length but with a different rate of excision of UV-induced DNA damage,[13] were irradiated at the same time using the same lamp setting and the same size dishes in order to equalize the exposure. The surviving cells were allowed to form colonies or to undergo expression of mutations to 8-azaguanine resistance in situ. The results (Figure 1) indicate that the normal cells which have a rapid rate of excision are much more resistant to the potentially lethal and mutagenic effect of exposure to UV radiation than either XP strain. After exposure to $0.5J/m^2$, XP12BE cells, with little or no excision repair capacity,[14] exhibit ~5% survival; XP2BE cells, which excise damage slowly, exhibit a survival of ~25%; and normal cells exhibit 100% survival. Similarly, XP12BE cells show ~140 mutants induced per 10^5 survivors; XP2BE cells show ~75 mutants per 10^5 survivors; and normal cells with the most rapid excision repair exhibit no significant induction of mutations above background. This is the expected result if the slowly excising XP12BE cells were able to reduce their load of DNA damage to a level of about 40% of what they had initially received and the normal cells with a still more rapid rate had reduced their load to an insignificant number of photo lesions. Such a result predicts that there is a finite amount of time for excision repair between the initial radiation and the onset of the critical events responsible for cell killing and/or mutation induction in these cells.

Examination of Figure 1 will show that an approximately 2-fold larger dose of UV light must be given to the XP2BE cells to achieve the biological consequences observed in the very repair deficient XP12BE cells. Similarly, a dose ~16 times larger must be given to the normal cells in order to achieve the biological consequences seen in the XP12BE cells. However, at equicytotoxic doses the frequency of mutations induced in all three strains is approximately equal. Since an essential difference between these three strains is their respective rates of excision repair of UV induced

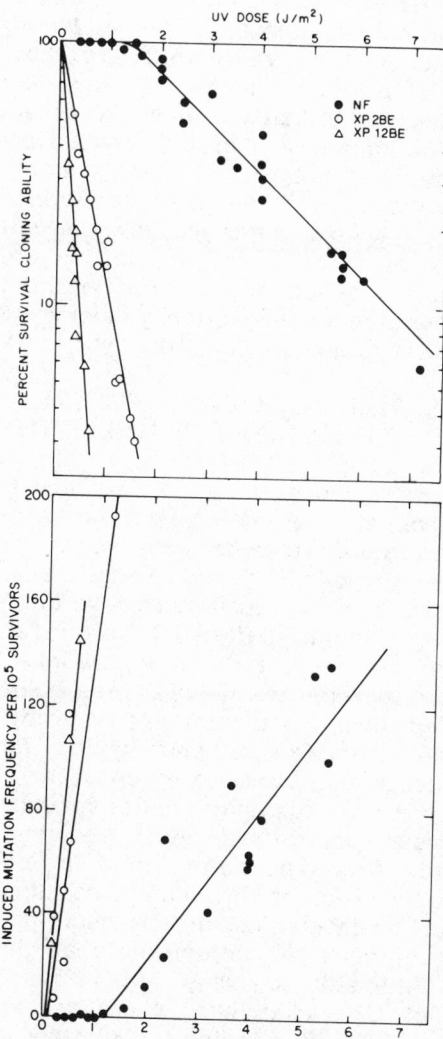

Figure 1. Comparison of the cytotoxic and mutagenic effect of increasing doses of UV radiation in normal fibroblasts (NF) and in XP cells with an excision rate ~20% that of normal (XP2BE) or with little or no detectable excision-repair capacity (XP12BE).Cells were plated into culture dishes at appropriate densities, allowed ~12 hr to attach, irradiated, and allowed to develop into colonies. Selection <u>in situ</u> with 20 uM 8-azaguanine was begun after 5-8 days of expression (>3 population doublings). Taken from Maher, et al.[1] with permission.

damage,[13] the data are consistent with the hypothesis that the loss of ability to form a clone and the frequency of mutations induced in these cells reflect the number of unexcised lesions remaining in the DNA at the time of some "critical cellular events". The results suggest that within a specified critical time the two strains which are capable of excision (NF and XP2BE) were able to remove many of the lesions initially introduced before these could result in lethality or mutation induction. Although one may measure the rate of excision of UV induced pyrimidine dimers in human cells with reasonable accuracy at doses greater than $10J/m^2$,[15] this is not true for the various low doses used in these biological experiments and so it was not practical to test this prediction directly by measuring the rate of disappearance of dimers in UV irradiated cultures. However, we have recently completed a similar comparative study using a reactive metabolite of benzo(a)pyrene radioactively labeled with a high specific activity.[10,11] The results obtained support the hypothesis (see below).

LENGTHENING THE TIME BETWEEN EXPOSURE AND ONSET OF S PHASE REDUCES THE BIOLOGICAL EFFECTS OF THESE AGENTS

If mutations are introduced not by excision repair but by semi-conservative DNA replication on a template containing unexcised lesions, then if one were to extend the time between the introduction of the lesions and the onset of DNA synthesis, it should be possible for a cell which has at least some capacity for excision repair to remove those lesions before they can cause mutations. Similarly, if cell killing refects replication on a damaged template then extending the time available for excision repair before allowing cells to replicate their DNA should result in increased survival. We tested this hypothesis by preventing human cells from replicating by growing them to confluence (density inhibition). After the cells reached confluence they were refed once and then held for 72 hrs without refeeding. (Autoradiography studies in cells maintained in the resting state under our conditions demonstrated that <0.5% of the cells incorporated tritiated thymidine during an 8 hr labeling period.)[1] We measured the rate of excision of thymidine-containing pyrimidine dimers[15] as well as of radioactive labeled chemical adducts from the DNA of these non-replicating human cells[6,10] and showed that they were capable of repair replication. We irradiated normal cells in the confluent state with a large enough dose of UV radiation to cause cell killing and to induce mutations in asynchronously growing populations, i.e., 5 to $10J/m^2$. However, we held them in this non-replicating state for 7 days before releasing them by trypsinization and plating them at lower densities to assay them for survival of colony formation or for mutation induction. The results are shown in Figure 2. The normal cells showed no cell killing and no mutation induction. In contrast, XP12BE cells held for 7 days in the confluent state exhibited the same degree of cell killing and frequencies of mutations as in cycling cells.

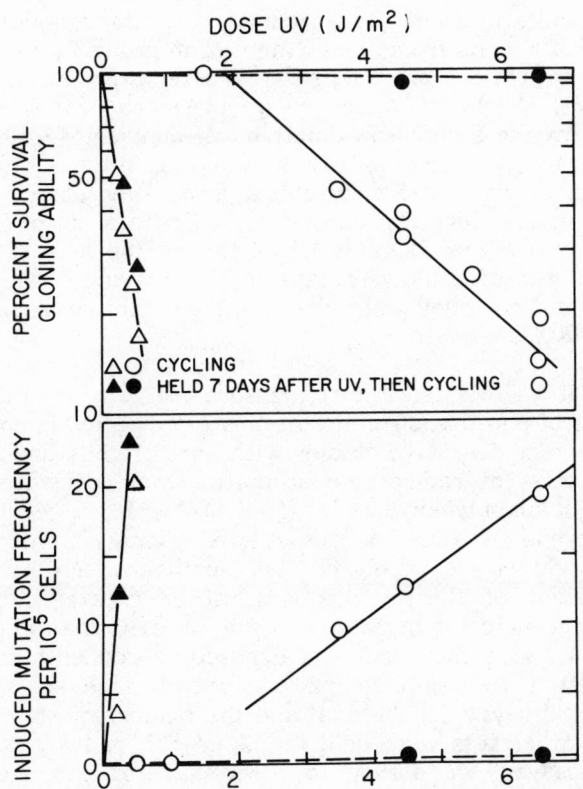

Figure 2. Loss of potentially cytotoxic and mutagenic lesions induced by UV irradiation in NF (circles) in comparison to XP12BE cells (triangles). Closed symbols represent cells grown in confluence, irradiated, and then held at confluence for 7 days before being assayed for survival or for frequency of induced mutations to TG resistance. Open symbols indicate cells irradiated as exponentially-growing cultures and then assayed for cytotoxicity and/or frequency of UV-induced mutations to 6-thioguanine (TG) resistance.

To determine how much time is required for the normal cells to remove potentially cytotoxic and/or mutagenic lesions we irradiated a series of cultures with 7 J/m^2 and released one immediately and the others at various times following radiation.[1] The results showed a gradual decrease in the lethal and mutagenic effect of the radiation. After ~15 hrs in the resting state the cells which had exhibited an initial survival of 20% and an induced mutation frequency of 50 x 10^6 exhibited ~100% survival and a frequency of mutations near background. A similar amount of time was required for XP5BE cells from complementation group D, with a slow rate of excision repair,[13] to exhibit 100% survival after exposure to 20% survival dose of UV (i.e., 1 J/m^2). The latter cells, exposed to a 7-fold lower dose of UV light had to excise a correspondingly lower number of lesions, but since they are reported to excise at a rate ~ 16% of normal cells, it is not surprising that they were able to exhibit a gradual increase in survival and that their rate of recovery was similar to the rate of recovery of normal cells given a much higher dose of UV.[1] During the 16 to 24 hrs held in confluence both sets of cells were capable of removing sufficient numbers of lesions so that when they were released from the non-replicating state they exhibited 100% survival.

Because it was not feasible to measure the cells' rate of removal of the low numbers of photoproducts during this 24 hr period in confluence, we conducted similar comparative studies with normal cells and XP12BE cells exposed to a series of radioactive aromatic amine derivatives, viz., N-acetoxy-4-acetylaminobiphenyl (N-AcO-AABP), N-acetoxy-2-acetylaminofluorene (N-AcO-AAF), N-acetoxy-2-acetylaminophenanthrene (N-AcO-AAP), and N-acetoxy-4-acetylaminostilbene (N-AcO-AAS).[6,7] A given concentration of these compounds caused greater cell killing and much higher frequencies of mutations in the XP cells than in normal cells. Cells were grown to confluence, and exposed to concentrations of these chemicals adjusted to result in ~20% survival. One set was released immediately and assayed for survival and the number of residues bound to the DNA. The other sets were held for 18, 36, 72, or 144 hrs before being released and assayed for ability to form colonies and the number of residues remaining in the DNA at the time of release. The results (Figure 3) showed that the rate of recovery was directly related to the rate of removal of the bound carcinogen residues from DNA. In the XP12BE cells there was little or no removal of the bound material and no evidence of recovery from the potentially lethal effects of the initial exposure, even after a period of 6 days.

We used this same approach to investigate the kinetics of removal of DNA adducts formed by the 7β,8α-dihydroxy-9α,10α-epoxy-7,8,9,10-tetra-hydrobenzo(a)pyrene (diol epoxide of BP) and the rate of recovery from its potentially cytotoxic and mutagenic effects.[10] Normal and XP12BE cells were grown to confluence, allowed 72 hrs without refeeding, and treated with radioactive diol epoxide of BP. One set of cells was harvested immediately and assayed for DNA adducts, suvival, and mutation induction. The rest were assayed after 2, 4, or 8 days in confluence. For mutation

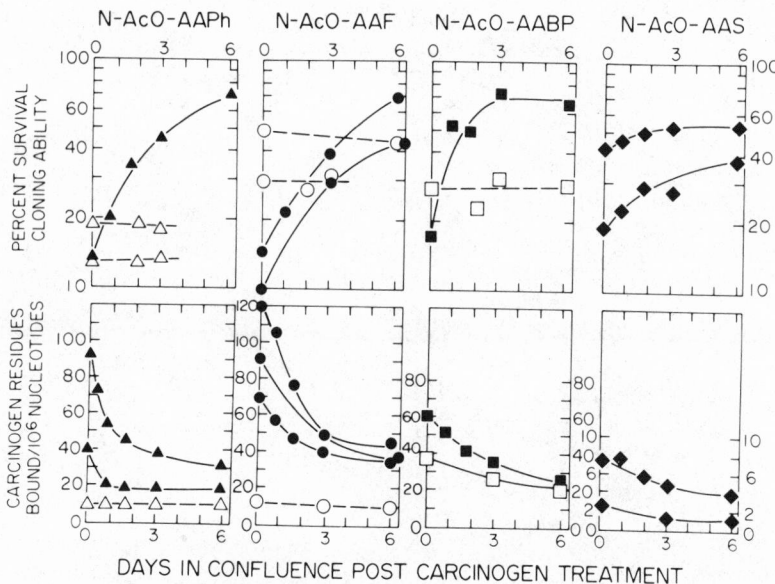

Figure 3. Comparison of the rates of recovery from the potentially lethal effects of four aromatic amine derivatives with the rate of removal of radioactively labeled residues from the DNA of NF (closed symbols) or XP12BE (open symbols). Cells were treated at confluence as described and then assayed after the designated period of time in the G_0 state. Some of this data was taken from Heflich, et al. [6]

studies the cells were released and assayed for the frequency of mutations only after an appropriate expression period (6 to 10 days) to allow introduction of mutations and elimination of preexisting HPRT enzyme. The data (Figure 4) show that residues were removed by the confluent normal cells over a period of 4 days but then excision repair slowed considerably. Similarly, the survival of the cells increased during that same period and the mutation frequency decreased to almost background. These data indicate that the potentially cytotoxic and mutagenic lesions were being removed with about the same kinetics as were the total number of DNA adducts. HPLC characterization of the DNA adducts indicated that the major adduct was the $\underline{N}^2$-guanyl derivative. The kinetics of decrease of tritium label in the specific HPLC peak corresponded to the kinetics of decrease of radioactivity in the total DNA with time and also with the kinetics of recovery from the mutations and the cytotoxicity. There may be unstable adducts whose existence could not be noted or a minor adduct(s) which escaped notice because the specific activity of the carcinogen did not permit their detection. Nevertheless, if such adducts

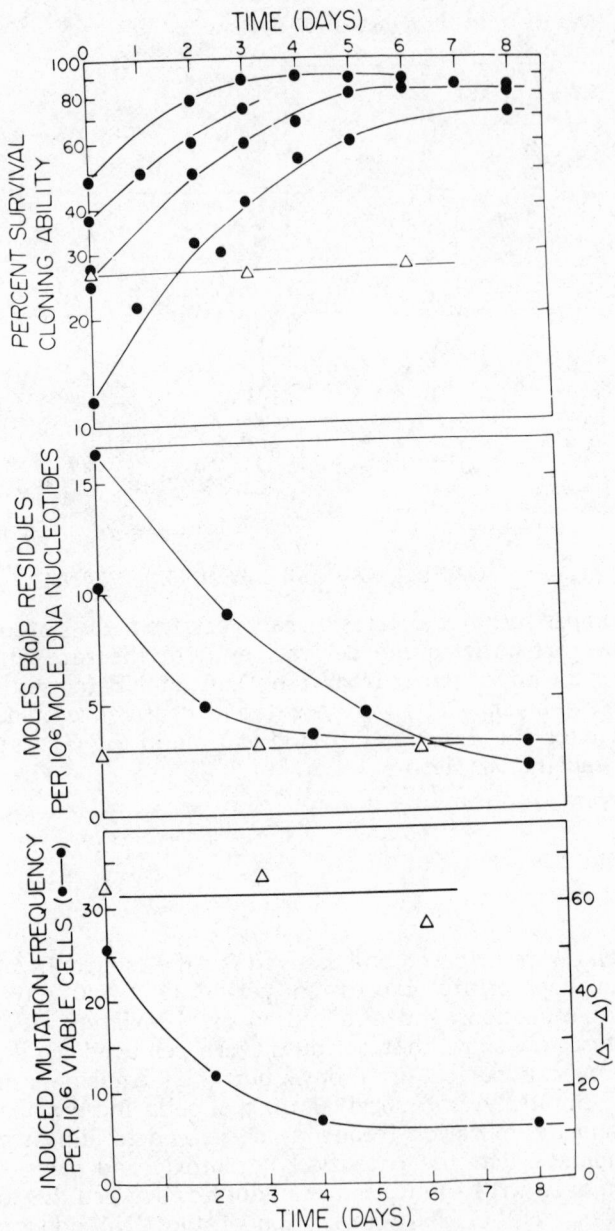

Figure 4. Kinetics of removal of covalently bound adducts (B) and recovery of NF (circles) or XP12BE cells (triangles) from the potentially cytotoxic (A) or mutagenic (C) effects of the diol epoxide of BP. The cells were treated in the G_0 state, released on the designated days, and assayed for survival of colony-forming ability, for the number of residues bound to DNA, and after a suitable expression period, for the frequency of induced mutations to TG resistance. Taken from Yang, et al.[10]

were responsible for the mutagenic and/or cytotoxic effects of the diol epoxide of BP, they would have to be lost with the same kinetics as the $\underline{N}^2$- guanyl derivative to account for these results.

The data in Figures 3 and 4 indicate that the frequency of mutations and the survival of cells exposed to chemical carcinogens are directly related to the number of lesions in the DNA at the time the cells are released from confluence and enter the cycling state. The fact that there was no loss of adducts from the XP12BE cells and no change in survival or mutation frequency indicates that the ability of the normal human cells to excise lesions from their DNA is responsible for reducing the potentially cytotoxic and potentially mutagenic effects of exposure to these chemical carcinogens as was true for UV radiation.

SHORTENING THE TIME BETWEEN EXPOSURE TO DNA DAMAGING AGENTS AND THE ONSET OF S PHASE INCREASES THE MUTAGENIC EFFECT

These data suggest that the ultimate biological consequences of exposure to DNA damaging agents depend upon a cell's rate of excision repair and the length of time available for excision before some critical cellular event which is responsible for the killing and mutation induction and suggest that the event is semi-conservative DNA replication. If so, it should be possible to shorten the time available for excision repair by synchronizing cells and irradiating them or treating them with chemical carcinogens at various times prior to the onset of S phase. The hypothesis that DNA replication is a critical event predicts that excision repair proficient normal cells exposed to these agents just prior to the onset of S will show a significantly greater cell killing and mutagenesis effect than cells exposed many hours prior to the onset of S. It further predicts that cells which are incapable of excision repair (XP12BE) will show no difference in sensitivity when irradiated or treated at various periods of time before the onset of S. We tested these predictions using UV radiation and the diol epoxide of BP as the DNA damaging agents.[5,11] The cells were synchronized by density-inhibition as described. Autoradiography studies indicated that cells released from the G_0 state after 72 hrs and replated at lower densities (3 to 9 x 10^3 cells/cm^2) in fresh medium containing 15% fetal bovine serum begin semi-conservative DNA synthesis (S phase) ~24 hrs after the replating.[5] To determine whether the time available for DNA excision repair between UV radiation and the onset of S was critical in determining the cytotoxic and mutagenic effect of UV we released cultures of normal and XP12BE cells from G_0, allowed them to reattach at lower densities, irradiated them in early G_1 (~18 hrs prior to the onset of S) or just prior to S phase, and assayed the frequency of mutations and survival. The results indicated that the slope of the dose response for mutations induced in normal cells irradiated or treated with diol epoxide of BP just prior to S was about 10-fold steeper than that of cells irradiated 18 hrs earlier. However, the two sets of normal cells irradiated at the two different times showed little difference in survival. The frequency of mutations induced in the XP cells was the same whether they were irradiated just before S or 18 hrs prior to S and the survival was the same. These results suggest that S phase is responsible for the induction of the mutations and that the frequency of mutations is directly related to the number of unexcised lesions remaining at the time the cell begins replicating its DNA. However, they suggest that something other than DNA replication is responsible for the cell killing by these agents.

We suggest that after DNA is damaged by these agents, the time available for repair of potentially lethal lesions is determined by the cell's need for critical cellular proteins and their respective mRNA's. If the DNA template for transcription of these mRNA's is still blocked by lesions at the time the cell has need of them, reproductive death, (i.e., inability to form a colony) is the result. This would explain why holding cells in a resting state following exposure to DNA damaging agents before releasing them into the cycling state results in a higher survival than does immediate release. Cells held in confluence have a lower metabolic state than do cells in exponential growth and, therefore, following irradiation fewer critical proteins are needed before the cell has time to remove the blocking DNA damage. We suggest that reproductive death from exposure to UV-radiation or these chemical agents results indirectly from faulty or blocked transcription from DNA containing photoproducts or adducts because of the resulting lack of required protein synthesis. This conclusion is consistent with the fact that the XP12BE cells which do not remove such lesions from their DNA show no dose modifying effect on being held in the G_0 state.[1,6,10]

STUDIES INVOLVING ETHYLNITROSOUREA AS THE DNA DAMAGING AGENT

It has been shown that following exposure to ethylnitrosourea (ENU) SV40-transformed cells derived from a xeroderma pigmentosum patient (XP12RO, complementation group A) remove O^6-ethylguanine from their DNA two to three times more slowly than do SV40-transformed cells from a normal individual (GM637).[16] However, both SV40-transformed cell lines removed several other DNA adducts formed by ENU at the same rate. If O^6-ethylguanine contributes significantly to the cytotoxic effect of ENU, then the population which can excise this lesion more rapidly than the other during the post treatment period should show the higher survival. We compared these two cell lines for survival following exposure to ENU and found the XP12RO cells 3.5-fold more sensitive than GM637.[12] We extended this study to non-transformed diploid human fibroblasts and compared the survival of normal cells with that of fibroblasts from three XP patients, XP12BE from group A, XP7BE from group D, and XP4BE (an XP variant). The slope of the survival curve for the XP12BE cells was 3-fold steeper than that of the exponential part of the curve for normal cells (see Figure 5). The slope of the curve for XP7BE was 2-fold steeper; that of the XP4BE 1.7-steeper.[12] This enhanced cytotoxicity was not observed in XP12BE cells treated with methylating agents such as MNU or MNNG.[12,17] We compared the normal cells and the XP cells for sensitivity to the

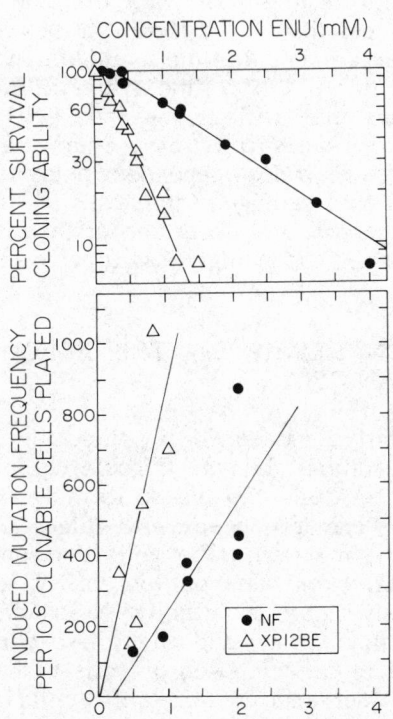

Figure 5. Comparison of the cytotoxic and mutagenic effect of ENU in NF (circles) and XP12BE cells (triangles). Large populations of cells in exponential growth were exposed to ENU for 3 hr and then replated at cloning densities to determine percent survival and at expression densities to allow expression to resistance to TG before selection in 40 uM TG. (Most of the data taken from Simon, et al.) [12]

mutagenic action of ENU and found 3.5-fold higher frequencies of mutants in the XP cells (Figure 5). These results are consistent with the hypothesis that O^6-ethylguanine contributes to both the lethal and the mutagenic effect of ENU. Although studies on the kinetics of loss of the various ethylated derivatives and of the kinetics of recovery from cytotoxicity or mutagenicity in these cells are required to test this prediction, the results are consistent with the hypothesis that following exposure to ENU there is a critical amount of time available for excision of the potentially cytotoxic and potentially mutagenic lesions. The fact that the differential between the mutation curves is less than the differential between the cytotoxicity curves suggests that the amount of time available for the removal of the cytotoxic lesions is greater than the time available for the removal of the mutagenic lesions. This may reflect the fact ENU, in contrast to UV radiation or the reactive derivatives of aromatic amines or polycyclic hydrocarbons does not cause a dose dependent delay in cell replication. If DNA replication occurs on a damaged template with little or no delay, the critical event for mutagenesis will occur sooner than it would for the other agents and there will be correspondingly less time for excision.

EFFECT OF EXCISION REPAIR ON THE FREQUENCY OF TRANSFORMATION OF HUMAN CELLS

We have recently succeeded in inducing loss of anchorage dependence in normal diploid human fibroblasts following exposure to chemical carcinogens.[18,19] Cells capable of forming colonies in semi-solid medium (soft agar) were produced with high frequency and in a concentration dependent manner. The cells derived from such colonies were isolated, propagated and assayed for ability to form tumors upon injection subcutaneously into sublethally X-irradiated athymic mice. Tumors formed in less than 10 days and when these were excised and placed in culture they proved to be composed of cells with a human karyotype. Upon reinjection into X-irradiated mice, the cells gave rise to tumors which were independently diagnosed by several pathologists as fibrosarcomas. Thus, we have demonstrated that in diploid human cells induction of anchorage independence acts like a genetic event which is induced early in direct response to exposure to a mutagen/carcinogen. Similar results in diploid mouse cells have been reported by Bellett and Younghusband.[20] The latter investigators carried out fluctuation tests with the mouse cells which indicated that the trait was induced by the mutagen. Similar results have been recently obtained in this laboratory (McCormick, et al., unpublished data).

We recently investigated the role of excision repair in preventing induction of anchorage of anchorage independence by exposing normal human cells and XP cells from complementation group D to UV radiation.[21] The XP cells irradiated with a 10-fold lower dose of UV than the normal cells exhibited approximately the same degree of cell killing and, more importantly, exhibited approximately the same frequency of UV-induced anchorage independent cells in the population as did the normal cells. Both sets of anchorage independent cells derived from soft agar colonies were propagated and injected into athymic mice. Both gave rise to fibrosarcomas in the animals. These results indicate that excision repair processes in normal human cells act to decrease the frequency of transformation of cells from normal to tumor - producing cells. Furthermore, they support the hypothesis that induction of anchorage independence is, indeed, the result of damage to DNA since the repair-deficient cells were abnormally susceptible to the event.[21]

CONCLUSION

In summary, these studies indicate that excision repair in diploid human fibroblasts is essentially an error-free process and that the ability to carry out excision repair of potentially cytotoxic and potentially mutagenic or transforming lesions induced in DNA by UV radiation or by these classes of chemical carcinogens determines their ultimate biological consequences. The data suggest that there is a certain amount of time available between the initial exposure and the onset of the cellular events responsible for mutation induction, for cell transformation, and for cell killing and that the critical event for the mutations is DNA synthesis on a damaged template. In contrast, the cytotoxicity studies indicate that although a population's survival is determined by the extent of excision repair of potentially lethal damage from DNA before some critical cellular event takes place, the critical event is not DNA synthesis on a damaged template.

ACKNOWLEDGEMENTS

We thank our colleagues Drs. R.H. Heflich, B. Konze-Thomas, K. C. Silinskas, L. Simon, and L. L. Yang for their invaluable contributions to the research summarized here. The excellent technical assistance of M. Antczak, R. Corner, D. J. Dorney, R. M. Hazard, S. A. Kateley, T. E. Kinney, A. Mendrala, T. G. O'Callaghan, L. Rowan, J. E. Tower, and T. VanNoord is gratefully acknowledged. Sincere thanks are extended to Dr. John Scribner of the Pacific Northwest Research Foundation for providing the labeled aromatic amine derivatives of high specific activity. The labeled diol epoxide of BP was generously provided by the Cancer Research Program of the National Cancer Institute. The research summarized in this report was supported in part by Contract EV-78-4659 from the Department of Energy, by Grants CA 21247, CA 21253, and CA 21289 from the Department of Health and Human Services, National Cancer Institute, N.I.H.

REFERENCES

1. V. M. Maher, D. J. Dorney, A. L. Mendrala, B. Konze-Thomas, and J. J. McCormick, DNA excision repair processes in human cells can eliminate the cytotoxic and mutagenic consequences of ultraviolet irradiation, Mutation Res. 62:311-323 (1979).

2. V. M. Maher and J. J. McCormick, Effect of DNA repair on the cytotoxicity and mutagenicity of UV irradiation and of chemical carcinogens in normal and xeroderma pigmentosum cells, in: "Biology of Radiation Carcinogenesis," J. M. Yuhas, R. W. Tennant, and J. B. Regan, eds., Raven Press, New York (1976).

3. V. M. Maher, N. Birch, J. R. Otto, and J. J. McCormick, Cytotoxicity of carcinogenic aromatic amines in normal and xeroderma pigmentosum fibroblasts with different DNA repair capabilities, J. Natl. Cancer Inst. 54:1287-1294 (1975).

4. V. M. Maher, L. M. Ouellette, R. D. Curren, and J. J. McCormick, Frequency of ultraviolet light-induced mutations is higher in xeroderma pigmentosum variant cells than in normal human cells, Nature, 261:593-595 (1976).

5. B. Konze-Thomas, R. M. Hazard, V. M. Maher, and J. J. McCormick, Extent of excision repair before DNA synthesis determines the mutagenic but not the lethal effect of UV radiation, Mutation Res. (submitted).

6. R. H. Heflich, R. M. Hazard, L. Lommel, J. D. Scribner, V. M. Maher, and J. J. McCormick, A comparison of the DNA binding, cytotoxicity and repair synthesis induced in human fibroblasts by reactive derivatives of aromatic amide carcinogens, Chem-Biol. Interactions, 29:43-56 (1980).

7. V. M. Maher, R. H. Heflich, and J. J. McCormick, Repair of DNA damage induced in human fibroblasts by N-substituted aryl compounds, in: "Carcinogenic Mutagenic N-Substituted Aryl Compounds," S. Thorgeisson, ed., National Cancer Institute (In press, 1981).

8. V. M. Maher, R. D. Curren, L. M. Ouellette, and J. J. McCormick, Role of DNA repair in the cytotoxic and mutagenic action of physical and chemical carcinogens, in: "In Vitro Metabolic Activation in Mutagenesis Testing," F. J. deSerres, J. R. Fouts, J. R. Bend, R. M. Philpot, eds., Elsevier/North-Holland Biomedical Press, Amsterdam (1976).

9. V. M. Maher, J. J. McCormick, P. L. Grover, and P. Sims, Effect of DNA repair on the cytotoxicity and mutagenicity of polycyclic hydrocarbon derivatives in normal and xeroderma pigmentosum human fibroblasts, Mutation Res. 43:117-138 (1977).

10. L. L. Yang, V. M. Maher, and J. J. McCormick, Error-free excision of the cytotoxic and mutagenic N^2-deoxyguanosine DNA adduct formed in human fibroblasts by ($\pm$)-7β,8α-dihydroxy-9α,10α-epoxy-7,8,9,10-tetrahydrobenzo(a)pyrene, Proc. Natl. Acad. Sci. U.S.A., 77:5933-5937 (1980).

11.   L. L. Yang, V. M. Maher, and J. J. McCormick, Relationship between excision repair and the cytotoxic and mutagenic effect of the "anti" 7,8-diol-9,10-epoxide of benzo(a)pyrene in human cells, Mutation Res. (submitted, 1981).

12.   L. Simon, R. M. Hazard, V. M. Maher, and J. J. McCormick, Enhanced cell killing and mutagenesis by ethylnitrosourea in xeroderma pigmentosum cells, Carcinogenesis, 2: 567-570 (1981).

13.   J. H. Robbins, K. H. Kraemer, M. A. Lutzner, B. W. Festoff, and H. G. Coon, Xeroderma pigmentosum -- An inherited disease with sun sensitivity, multiple cutaneous neoplasms, and abnormal DNA repair, Ann. Intern. Med., 80:221-248 (1974).

14.   R. A. Petinga, A. D. Andrews, R. E. Tarone, and J. H. Robbins, Typical xeroderma pigmentosum complementation group A fibroblasts have detectable ultraviolet light-induced unscheduled DNA synthesis, Biochim. Biophys. Acta, 479:400-410 (1977).

15.   B. Konze-Thomas, J. W. Levinson, V. M. Maher, and J. J. McCormick, Correlation among the rates of dimer excision, DNA repair replication, and recovery of human cells from potentially lethal damage induced by ultraviolet radiation, Biophys. J., 28:315-326 (1979).

16.   W. J. Bodell, B. Singer, G. H. Thomas, J. E. Cleaver, Evidence for removal at different rates of O-ethyl pyrimidines and ethylphosphotriesters in two human fibroblast cell lines, Nucleic Acids Res., 6:2819-2829 (1979).

17.   R. H. Heflich, D. J. Dorney, V. M. Maher, and J. J. McCormick. Reactive derivatives of benzo(a)pyrene and 7,12-dimethylbenz(a)anthracene cause S_1 nuclease sensitive sites in DNA and "UV-like" repair, Biochem. Biophys. Res. Commun., 77:634-641 (1977).

18.   J. J. McCormick, K. C. Silinskas, and V. M. Maher, Transformation of diploid human fibroblasts by chemical carcinogens, in: "Carcinogenesis, Fundamental Mechanisms and Environmental Effects," B. Pullman, P.O.P. Ts'o, and H. Gelboin, eds. D. Reidel Publ. Co., Dordrecht (1980).

19.   K. C. Silinskas, S. A. Kateley, J. E. Tower, V. M. Maher, and J. J. McCormick, Induction of anchorage independent growth in human fibroblasts by propane sultone, Cancer Res., 41:1620-1627 (1981).

20.   A. J. D. Bellett and H. B. Younghusband, Spontaneous, mutagen-induced and adenovirus-induced anchorage independent tumorigenic variants of mouse cells, J. Cell Physiol., 101:33-48 (1979).

21.   J. J. McCormick, L. Rowan, S. A. Kateley, K. C. Silinskas, and V. M. Maher, Xeroderma pigmentosum cells are more sensitive than normal diploid human fibroblasts to neoplastic transformation by UV radiation (manuscript submitted for publication).

DISCUSSION

DR. SHAPIRO: Just to return to the chemistry for benzopyrene diolepoxide for a moment, did you at any point detect any other adduct apart from the N^2 Guanine adduct?

DR. MAHER: No we didn't, nor with the N-acetoxy-acetylamino-fluorene (N-AcO-AAF). We found only one adduct, even though we had radioactive labeled material of high specific activity and could have seen more. We detected only one adduct. In the case of N-AcO-AAF, it was de-acetylated C8 guanine adduct, i.e. no acetyl left on the molecule.

DR. SHAPIRO: And what percent of the total are you picking up? Going back to benzopyrene diolepoxide, how much smaller a peak could you have detected?

DR. MAHER: We would have been able to see something that was around 5% of the main peak.

DR. SHAPIRO: In the early days of alkylation research, the N-7 of guanine was used as an indicator.

DR. MAHER: That is a good point. It is true that we could only study the kinetics of loss of the detectable adducts and were not able to study hypothetical unstable adducts or minor adducts present at undetectable levels. These may exist in the cell and may, in fact, be responsible for the mutagenic events we are measuring. However, if these other adducts are responsible for the mutations, they must be excised or must fall off the DNA with the very same kinetics as did the adduct we were able to detect. This is because the kinetics of loss of the potentially cytotixic and potentially mutagenic lesions correlated closely with the kinetics of loss of the adduct we could detect.

DR. JACHYMCZYK: I would like to ask you, it appeared from your data that if cells were held confluent, almost 2 days were required to remove 50% of the lesions. However, the decrease in mutant frequency observed in the cells treated 18 hr prior to S compared with that in cells treated just prior to S contradicts that.

DR. MAHER: You are correct. To determine this more directly we measured the rate of excision of the radioactive labeled adducts induced by 7,8-diol-0,10-epoxide of benzo(a)pyrene in cells under the two conditions. These studies by Yang, et al. will soon appear in Mutation Research. We had already shown that normally excising human fibroblasts held in the non-replicating state by density inhibition excise these adducts from their DNA and that the kinetics of this excision correlates with the kinetics of the cells recovery from the potentially cytotoxic lesions and and the potentially mutagenic lesions induced by this carcinogen, (L.L. Yang, et al., Proc. Natl. Acad. Sci., 1980). However, we noted that cells held in that confluent state slowed their repair after 3 or 4 days and yet, if the cells were replated into fresh medium at lower densities, they clearly exhibited a much higher survival (and much lower mutant frequency) than would excision repair-deficient XP cells containing the same number of DNA adducts at the time of release from confluence. Therefore, we measured the rate of excision of normal cells synchronized by release from confluence, and moving from early G_1 through S phase. We use BrdUrd to distinguish newly-replicated DNA from parental DNA for the DNA for increase in density so as to ascertain where the cells were in the cycle. The results indicated that cells released from confluence remove lesions with at least a 2-3 fold faster rate than cells held in confluence.

Interestingly, we saw that newly replicated DNA which had shifted to a higher density because of the BrdUrd in one strand still possessed unexcised diol-epoxide adducts. We also showed that excision repair minus XP cells can replicate their DNA without removing such adducts.

DR. GLICKMAN: You clearly have a correlation between the excision and something; survival, mutagenesis. But have you tried blocking excision in another way than using the mutants. I, of course, mean caffeine. I realize it's complicated but have you tried such kinds of things?

DR. MAHER: Caffeine doesn't block excision or "post-replication repair" in normal diploid human fibroblasts. We did that study in '75. I don't think there is a specific inhibitor for excision repair. If there is, I certainly would like to know. We have tried to see if we could, in effect, turn a repair-proficient normal cell into the equivalent of an XP cell by giving the DNA damage just before S so that there was not much time to excise the lesions. In theory, this could cause the response of the normal cell to be equal to that of the xeroderma cell. We found that we could achieve this if we used the diol epoxide of benzo(a)pyrene which is excised very slowly, but not if we used UV radiation as our agent. Therefore, our guess is that the normal human fibroblasts are able to excise UV induced lesions too rapidly and continue to do so even during the S period.

DR. GLICKMAN: What surprises me most from the data you've shown is the contrast to bacteria. Repair deficient bacteria have the ability to tolerate a great deal of damage. In mammalian repair mutants like XP, however, there appears to be little tolerance. How many lesions are required per lethal event?

DR. MAHER: The advantage of working with the particular strain of XP cells, XP12BE, is that we showed, using radioactive carcinogens, that these cells do not remove any of the bound adducts induced by a number of carcinogenic agents, such as a series of reactive hydrocarbon derivatives and aromatic amine derivatives. Therefore, with this strain one is able to discern how many initial DNA adducts are required to induce a certain number of mutations or to cause a mean lethal event. We, therefore, used it to compare a series of compounds and also UV-radiation. We found that the average number of adducts or pyrimidine dimers per 10^6 nucleotides required to cause one mean lethal event (37% survival) was rather uniform for a whole series of agents. The number averaged 1 to 4 adducts or photoproducts per 10^6 nucleotides. As might have been expected, this was not true for ENU. As B. Singer explained earlier, that agent causes a whole series of lesions and only one or two of the minor adducts appear to be correlated with cell killing. If we knew which ones we might find that their numbers added up to 1-4 per 10^6 nucleotides also.

A similar pattern was observed when we compared these various agents for the frequency of mutants induced in XP12BE cells by doses yielding 37% survival. Most of these mutation studies were carried out with asynchronously-growing populations of cells but in a few instances we treated cells in confluence and then released them and allowed them to replicate as synchronous populations. Because these particular XP12BE cells do not remove any adducts, the frequency of mutations was found to be quite similar using either protocol. (As I showed in my talk, this was not true for excision-proficient cells.) When we compared the initial number of lesions required to lower the survival of the repair-proficient cells to 37%, we found it varied from 4 to 60 per 10^6 nucleotides, and the initial number was larger for those agents, such as UV, whose lesions are excised more rapidly. We consider this as evidence that the repair-proficient cells remove many of these lesions before they can exert their potential to kill.

DR. GLICKMAN: It also has an interesting implication, which has to do with the model that as A.J. Clark proposed in E. coli, where you have differential repair in DNA which is replicating, non-replicating and so on.

DR. MAHER: Ray Waters says there is no difference in rate of repair of UV-induced damage between replicating and non-replicating DNA in human cells.

DR. GLICKMAN: What I'm saying is this kind of data suggests there is no difference, as compared to E. coli where there is one.

DR. BRIDGES: Do your cells actually go straight into S when you've UV'd them, when you assume they're going into S straight away?

DR. MAHER: It appears that if we irradiate just before S, they begin DNA synthesis on time but do not proceed through S as rapidly as unirradiated cells. Instead, S-phase is lengthened significantly.

DR. BRIDGES: Although you conclude that there must be a lag, have you ever demonstrated it? Bacterial cells clearly can monitor the level of photoproduct in DNA and they don't go into S. They wait until they've excised before they go in.

DR. MAHER: Because our method of synchronizing cells by release from confluence results is an abnormally lengthened S phase even in unirradiated cells, it is difficult to know if there is a short lag as the result of the irradiation. We certainly see an elongated S. A similar result is obtained with the 7,8-diol-9,10-epoxide of benzo(a)pyrene as the DNA damaging agent. But we still see that they actually replicate DNA containing those lesions. It takes them a longer time and they have a lag and they have an extended S phase that goes on for twice as long as it should, but they do replicate. I would have thought they would stop, wait until they got the lesions out of the way and then go on, but they actually replicate using a damaged template. This is, perhaps, why mutations are introduced.

DR. GLICKMAN: Do you have an estimate of mutation frequency compared to the number of lesions that are still in the DNA?

DR. MAHER: I can cite some figures based on UV experiments. Using XP12BE cells which do not excise dimers, we figure this to be about 10% efficiency for UV-induced thymine-containing pyrimidine dimers. This is probably an overestimate because we are assuming a target size of about a thousand base pairs for the gene coding for the basic subunit of HPRT. This, of course, is a minimum estimate. There may be many intervening sequences which cause the target size to be much larger. But from our in situ experiments, 0.5 J/m^2 of UV introduces 5 pyrimidine dimers per 10^6 nucletodies and induces 0.5 mutants per 10^3 surviving cells.

DR. ARLETT: If you look at recovery from potential lethal damage, using a DNA-damaging agent to which the XP is not sensitive, they behave impeccably.

DR. SARASIN: You explained the difference in survival curve when you treat the cells with UV or carcinogens by saying that the carcinogen lesions were removed more slowly than UV-induced dimers. However, I have published with P.C. Hanawalt results showing that DNA repair replication was exactly the same in aflatoxin- and UV-treated cells. So, could one deduce from your results that once you treat the cells with carcinogens you induce lesions on various targets which are not the DNA molecule which which could be important for cell survival.?

DR. MAHER: Our data with excision repair-proficient human cells held in confluence indicate that pyrimidine dimer removal is half completed in 8 hours, but it takes 48 hours to remove half of the diol epoxide adducts. Excision of the diol epoxide adduct is like excision of N-AcO AAF-induced lesions. Cells take 36 hours to remove half of the adducts formed by N-AcO AAF. It takes them longer to remove chemical adducts than dimers. In comparing rates of repair using thymidine incorporation, rather than adduct removal, patch size becomes an important problem as well as the length of time which is allowed for incorporation. Few investigators have measured repair replication over a period of 36 hrs or more. We have done so with a series of agents and have shown indirectly that the rate of repair replication correlates with rate of adduct removal (Heflich, et al., Chem.-Biol. Interactions, 1980).

THE INDUCTION OF RESISTANCE TO

ALKYLATION DAMAGE IN MAMMALIAN CELLS

Leona Samson   and   Jeffrey L. Schwartz

Biochemistry Department   Lab of Radiobiology
University of California   University of California
Berkeley, California   San Francisco, California

INTRODUCTION

We are continuously exposed to non-toxic levels of carcinogens both from our environment and from our diet. It would therefore seem important to determine how cells respond to this particular kind of treatment. Some years ago, an attempt to simulate this chronic environmental exposure to mutagens revealed the existence of an inducible antimutagenic DNA repair pathway in E. coli — the adaptive DNA repair pathway (1). The obvious progression from finding a new type of DNA repair in E. coli was to ask whether an equivalent pathway exists in mammalian cells, and ultimately whether it exists in human cells. We now have good biological evidence for the existence of an adaptive DNA repair pathway in cultured rodent and human cell lines (2). In order to set the scene for our recent findings in mammalian cells, we should first consider what is presently known about E. coli adaptation.

THE ADAPTIVE RESPONSE IN E. COLI

The chronic exposure of E. coli to very low non-toxic levels of the simple alkylating agent N-methyl-N'-nitro-nitrosoguanidine (MNNG) results in the de novo synthesis of the adaptive DNA repair pathway (1). Once induced, this pathway affords the cells considerable protection against both the killing and the mutagenic effects of alkylating agents (Fig. 1).

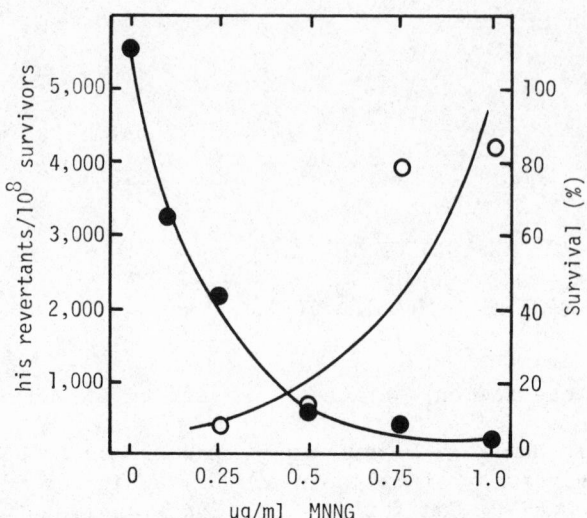

Fig. 1.   The effect of chronic MNNG pretreatment on the sensitivity of E. coli to the killing and mutagenic effects of
MNNG.   Six parallel cultures of E. coli F.26
(B/r thyA his) were first grown in minimal media for
75 min in the presence of 0-1.0 µg/ml MNNG, and then
challenged with 100 µg/ml MNNG for 5 min.   The resulting
mutation (•) and survival (o) were measured.

Adaptation is specific for alkylation damage; it can be induced
by and provide protection against all the simple alkylating agents
that have been tested so far (methyl-nitrosourea, MNU; methyl-
methanesulphonate, MMS; ethylmethane sulphonate, EMS; N-ethyl-N'-
nitro-nitrosoguanidine, ENNG); adaptation cannot be induced by nor
protect against UV light or 4-nitroquinoline-1-oxide (4-NQO) damage
(3,4).   However, it is noteworthy that the methylating agents are
far more efficient inducers of adaptation than the ethylating
agents (4).   This is our only clue, so far, as to the nature of
the induction signal for the adaptive response.

Although the adaptive resistance to the killing and mutagenic
effects of alkylating agents is known to occur via genetically
separable pathways (3,4,5,6), only the antimutagenic pathway has
been thoroughly investigated at the biochemical level.   (It is of
interest to note here that the events leading to alkylation-
induced mutation and cell death are also separable in human cells
(7)).   When DNA is exposed to simple monofunctional methylating
agents, several adducts are formed; in decreasing order of frequency, these are 7-methylguanine (7-MeG), 3-methyladenine (3-MeA)

and O^6-methylguanine (O^6-MeG), (8). Of these purine lesions, only O^6-MeG had been implicated as premutagenic (9,10) and precarcino-genic (11). Indeed, if left unrepaired, O^6-alkylguanine is recognized as adenine during DNA replication and so causes GC→AT transitions. It was therefore no surprise to find that adapted E. coli are exceedingly efficient at removing O^6-MeG from their DNA (12,13). This removal is achieved by an extraordinarily rapid (14) and hitherto undescribed DNA repair mechanism. The O^6-MeG lesion is not eliminated via nucleotide or base excision repair but rather by a direct transfer of the offending methyl group from the O^6 position of guanine to a cysteine residue on an acceptor protein, which, although this has not been definitively proven, is probably the adaptive enzyme itself (15,16,17). This is the first example of a DNA methyltransferase enzyme. However, it appears that the adaptive enzyme can only act once, and is therefore con-sumed during the DNA methyltransferase reaction (13,18). The mutagenic protection provided by the adaptive response therefore shows an unusual and quite characteristic saturation at very high doses of alkylating agent (19,12).

Initial studies indicated that adaptation does not enhance the repair of the other alkylated purines, 7-MeG and 3-MeA (12). However, it has now been shown that the 7-MeG- and 3-MeA-DNA glycosylase activities are increased markedly in adapted cells (20). These lesions are not currently thought to be involved in alkylation mutagenesis, and their enhanced removal may be respon-sible, at least in part, for the adaptive protection against the killing effects of alkylating agents. Indeed, it was noticed very soon after the discovery of adaptation that the kinetics of induction to mutational and killing resistance were quite different, and that mutational resistance was induced at much lower MNNG adapting doses than killing resistance (4,3, and see Fig. 1). Thus the discrepancy between these two reports on the repair of 7-MeG and 3-MeA in adapted bacteria can be accounted for by the earlier experiments (12) being carried out at rather low adapting doses. Given the differences between the kinetics of adaptation for killing and mutation, plus the existence of an E. coli mutant that cannot adapt for mutation but can adapt for killing (6), it should now be possible to directly show whether 7-MeG and 3-MeA are involved in cell killing.

THE ADAPTIVE RESPONSE IN MAMMALIAN CELLS

The existence of inducible mammalian DNA repair pathways, like adaptation, could significantly affect our approach to assessing the mutagenic and carcinogenic risk of chemicals in our environ-ment. Although it had already been reported that the chronic treatment of rats with alkylating agents induces a more efficient removal of O^6-MeG from liver DNA, the interpretation of these

results as the induction of an antimutagenic DNA repair pathway
was confused by a concomitant increase in hepatocarcinogenesis
(21,22,23). However, intense liver cell proliferation and thus
vastly increased DNA synthesis is also induced by chronic alkylation treatment. It seems imaginable that whether or not a
particular lesion leads to mutation (and, perhaps, to killing)
might depend upon whether this lesion is first encountered by a
repair enzyme or by a replication fork. Thus, a vast increase in
DNA synthesis could obscure any advantage gained from a moderate
increase in DNA repair. It therefore seemed judicious to search
for an adaptive response in established mammalian cell lines
which are not complicated by whole organ responses, such as these,
and which are amenable to measuring biological endpoints such
as cell killing and mutation as well as biochemical endpoints
such as adduct removal and enzyme activities.

The induction of resistance to cell killing and sister chromatid exchange

We anticipated that, as with E. coli, both the low dose used
to induce the adaptive response and the higher challenge doses
used to test for resistance would have to fall within critical
boundaries (4,3,19). For this reason we chose two rapid biological assays, cell killing and the induction of sister chromatid
exchange (SCE). In this way, it was feasible to test a large
range and several combinations of doses. SCE induction has been
correlated with mutation induction by the alkylating agents EMS
and ENU (24) and with a reduced capacity to repair O^6-MeG (25).

Many dose regimes were tried but only the most successful
ones shall be dealt with here. (This is not to say that there
is not some other more effective dose regime.) Chinese hamster
ovary (CHO) cells received a totally non-toxic dose of 10 ng/ml
MNNG once every 6 h for 48 h; this dose was repeatedly administered
because MNNG has a half life of about 1 h in tissue culture medium
(8,26). Pretreated and control cells were subsequently challenged
with various doses of MNNG up to 400 times the pretreatment concentration, and the resulting cell killing is shown in Figure 2.
Pretreated cells were clearly resistant to the killing effects of
MNNG, with about a twofold difference in the initial slopes of
the survival curves. (Pretreated CHO cells also became resistant
to the killing effects of MNU (2).)

CHO cells, and SV40-transformed normal human skin fibroblast
(GM637) cells that were chronically pretreated with non-toxic
levels of MNNG (see legend to Fig. 3) also became quite resistant
to the SCE-inducing effects of MNNG (Fig. 3), MNU and ENU (2),

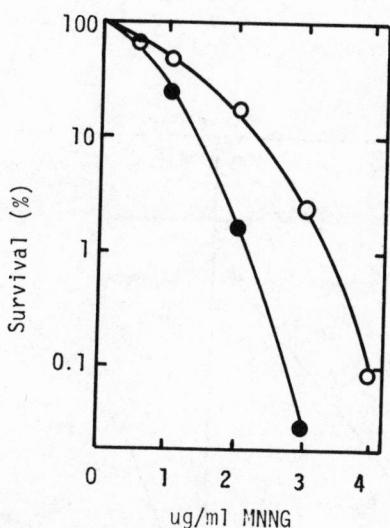

Fig. 2. The effect of MNNG pretreatment on the sensitivity of
CHO cells to the killing effects of MNNG. Cells were
grown in Ham's F12 medium supplemented with 5% fetal calf
serum, penicillin (100 U/ml), streptomycin (100 μg/ml)
and L-glutamine (0.03%). Three hours after the final
pretreatment dose, pretreated (o) and control (•) cells
were trypsinized, plated in 100 mm Petri dishes and
challenged with MNNG. After 10 d at 37°C, the duplicate
dishes of colonies were fixed, stained and counted.

but did not become resistant to SCE induction by UV (Fig. 4).
Moreover, it appears that the induced SCE resistance is saturated
at high challenge doses of MNNG (Figs. 3a and 3b) or, as shown for
CHO cells, when the pretreatment dose is increased (Fig. 3a).
In addition, resistance to both killing (for CHO) and SCE induction (for CHO and GM637) decays within 2-3 generations after
pretreatment is stopped. At this point we had clearly shown that
chronic exposures to non-toxic levels of alkylation rendered
these two types of mammalian cells refractory to normally very
toxic levels of alkylation, and that this response bore several
of the earmarks of the E. coli adaptive response. Our next task
was to determine just how, in a rather general sense, these mammalian cells achieved resistance.

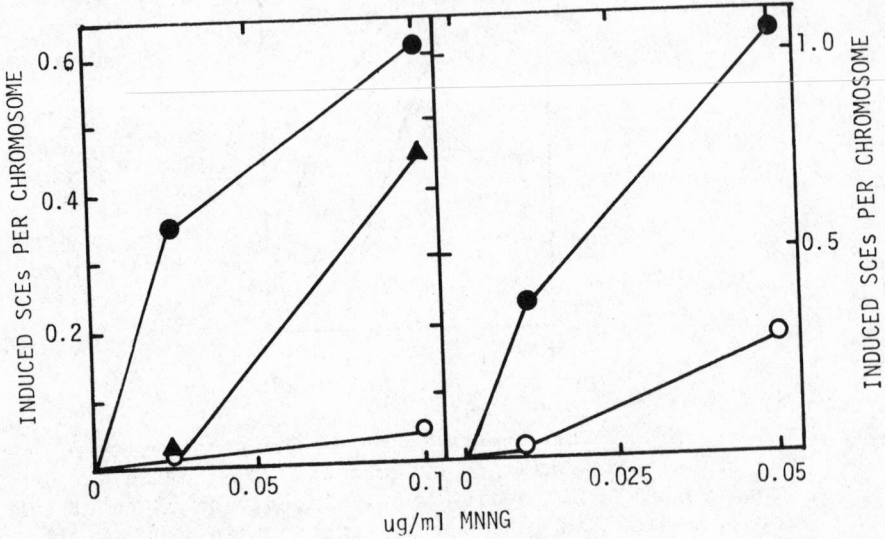

Fig. 3.   The effect of MNNG pretreatment on the sensitivity of
CHO and GM637 cells to the SCE-inducing effects of MNNG.
(a) CHO cells were grown as described in Fig. 2, and
treated with 0 (●), 5 (o) and 7.5 (▲) ng/ml MNNG once
every hour for 24 h.   30 min after the last pretreatment
dose cells were trypsinized, challenged with MNNG and
the induction of SCES measured as described in ref. 2.
(b) GM637 cells were grown in MEM medium supplemented
with 15% fetal calf serum, penicillin (100 U/ml), streptomycin (100 µg/ml) and L-glutamine (0.03%), and treated
with 10 ng/ml MNNG once every 6 h for 72 h.   3 h after
the last pretreatment dose cells were trypsinized,
challenged with MNNG, and the induction of SCEs measured
as described in ref. 2.

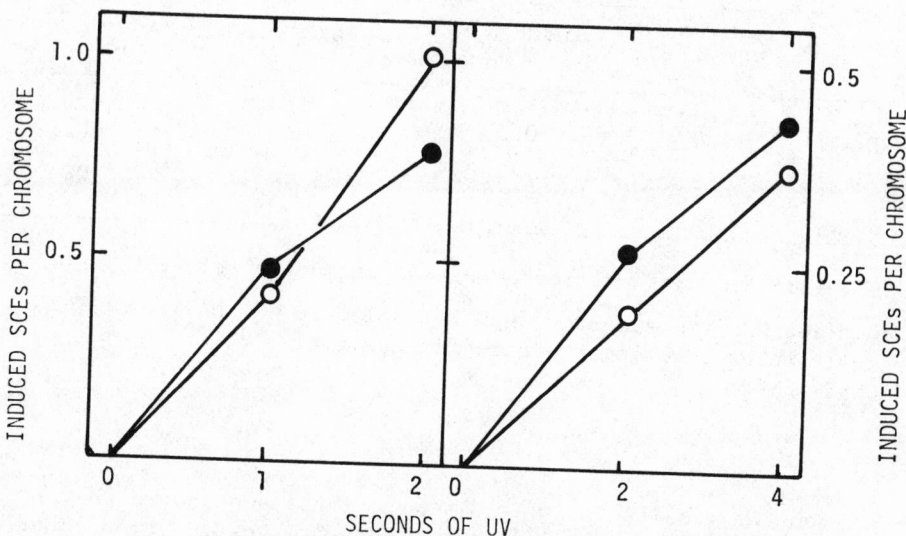

Fig. 4. The effect of MNNG pretreatment on the sensitivity of CHO and GM637 cells to the SCE-inducing effects of UV. (a) CHO cells were pretreated and then trypsinized as in Fig. 2, challenged with UV (12 erg/mm/s) and SCE induction measured as in ref. 2. (b) GM637 cells were pretreated and then challenged as in Fig. 3, challenged with UV (12 erg/mm/s) and SCE induction measured as in ref. 2.

There seemed to us four alternatives: (i) cells were becoming impermeable to, or somehow detoxifying, alkylating agents and were therefore not actually incurring damage; (ii) pretreatment might cause a redistribution of cells in the cell cycle, generating a population containing a larger than normal proportion of cells in a resistant phase of the cell cycle; (iii) pretreated cells somehow contrived to remain in G1 phase relatively longer than control cells and so gained extra time to repair damage before the onset of DNA replication in S phase; (iv) the cells were repairing DNA alkylation damage more efficiently. The first alternative was discounted when we found that the DNA purified from control and pretreated cultures that had been exposed to 2.5 µg/ml ^{3}H-MNU (120 Ci/mol) had similar specific activities (see Table I).

Table I. Alkylation and SCE induction in control and pretreated cell challenged with ^{3}H-MNU.

Cell type	Control		Pretreated	
	cpm/ug DNA[a]	Induced SCEs per chromosome	cpm/ug DNA	Induced SCEs per chromosome
GM637	13.4	0.59	9.03	0
CHO	0.57[b]	0.24	0.58[b]	0.04

[a] DNA isolation was carried out as described in ref. 27.

[b] Carrier DNA was added during the isolation of DNA from CHO cells, thus lowering the specific activities relative to GM637 DNA.

From the results in Table I we see that even though they receive the same gross insult to their DNA, pretreated cells incur fewer SCEs than do control cells. The second and third alternatives (cell cycle redistribution and/or an extended period in G1) were discounted by showing that pretreated and control homogeneous populations of mitotic CHO cells also show differential sensitivities to MNNG killing and SCE induction and that adapted and control cells (± a MNNG challenge) progress through the cell cycle at identical rates (2). We therefore deduce, by process of elimination, that chronic non-toxic MNNG exposure enables both CHO and GM637 cells to handle alkylation damage somewhat more efficiently and so become resistant to alkylating agents. The next task will be to determine the specific biochemical mechanism(s) by which these mammalian cells achieve resistance.

The existence of an adaptive response in mammalian cells, or any cells for that matter, poses questions not only about the mechanism of resistance but also about the mechanism of induction. Unfortunately our understanding of the induction of the adaptive response remains at a rather phenomenological level for both E. coli and mammalian cells. It is difficult to glean any information from the kinetics of induction as measured by cell killing and mutation, since these endpoints represent average values from what might be very mixed populations. For instance, a culture might appear partially adapted because all the cells are partially induced, or because some fraction of the cells are fully induced. In this regard, the use of SCE to monitor adaptation turned out to be extremely informative because here, in order to generate an average value for the whole population, one measures chromatid damage to individual cells. Thus Fig. 5 gives us a graphic representation and more meaningful kinetics of the induction of the adaptive response in a culture of GM637 cells. We measured the level of SCE's induced by the adapting dose itself during the induction of the adaptive response (GM637 cells were treated with 10 ng/ml MNNG once very 6 h for 48 h). When the cells experienced

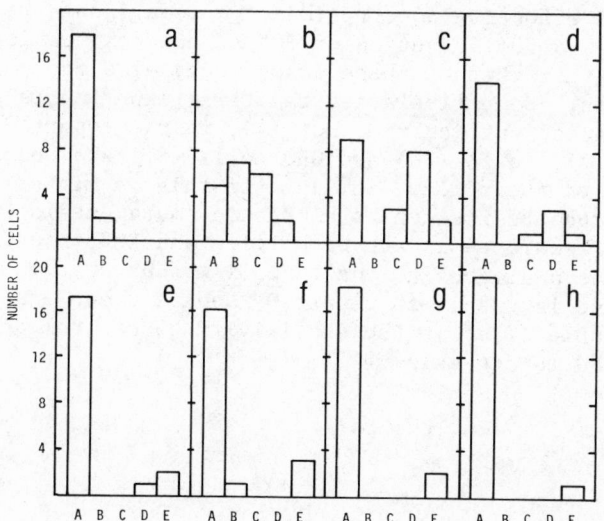

Fig. 5. The induction of SCE's during the adaptation of GM637 cells. Cells were treated with MNNG as described in the text, and SCE's were measured as described in ref. 2. On the abscissa, A, B, C, D and E represent 0-0.5, 0.5-1.0, 1.0-1.5, 1.5-2.0 and 2.0-2.5 SCE's per chromosome, respectively. Panel a shows the distribution of SCE's in control cells and panels b, c, d, e, f, g and h show the distribution of SCE's at 0, 6, 18, 24, 30, 42 and 48 h, respectively. These data are to be published elsewhere, manuscript in preparation.

their first adapting dose of MNNG (Fig. 1b) the amount of chromatid damage per cell showed the expected Poisson distribution (28). However, after 6 h and a second adapting dose (Fig. 5c) two quite separate populations of cells appeared; those that could now resist SCE induction, and those that could not. As adaptation continued (Fig. 5d-h) the proportion of fully resistant cells steadily increased with a concomitant decrease in sensitive cells. It would therefore seem that the switching on of adaptation within an individual cell is quite rapid since we do not observe cells of intermediate sensitivity. Furthermore, the fact that only a certain subpopulation of cells became resistant after the first adapting dose suggests that, in order to induce resistance, cells must experience the adapting dose while in a particular part of the cell cycle. Indeed, this is supported by the fact that in

order to become fully adapted, a population of cells — mammalian
and bacterial — must be chronically exposed to alkylating agent
for at least one generation time.

The induction of resistance to mutation

Having found effective dose regimes that can induce resistance
to cell killing and the induction of SCE's, the next obvious progression was to test whether these doses would also induce resistance to mutation. Some time after the first experiments on
mammalian adaptation were carried out (4), it was reported that
Chinese hamster V79 cells (30) and CHO cells (31) are unable to
remove O^6-MeG from their DNA. In light of this we might not expect
CHO cells to induce resistance to alkylation mutagenesis even
though they can become resistant to killing and SCE induction; this
would be somewhat analogous to the E. coli mutant that cannot adapt
for mutation (and is unable to repair O^6-MeG) but can adapt for
killing (6). Table II shows the sensitivity of control and adapted
CHO cells to MNNG mutagenesis.

Table II. MNNG mutagenesis of control and adapted CHO cells.[a,b]

Challenge dose of MNNG µg/ml	Thioguanine resistant colonies/10^5 cells[c,d]	
	Control	Adapted
0.1	38.7 ± 9.0	54.5 ± 12.2
0.2	95.7 ± 14.2	100.8 ± 7.6
0.5	271.2 ± 19.3	205.3 ± 29.6
1.0	512.9 ± 26.3	84.0 ± 11.5

[a] Average of four experiments.

[b] These results will be published elsewhere; manuscript in
preparation.

[c] Mutagenesis assays were carried out as described in ref. 29,
except that mutation expression was for 10d.

[d] Qualitatively similar results have been found for oubaine
resistance.

These results were quite surprising. Control and adapted cells showed no difference in sensitivity to MNNG mutagenesis at the low challenge doses, but there was a substantial difference (up to 5X) at the higher challenge doses. It occurred to us that the observed difference in mutation induction at high challenge doses might be generated by adapted cells taking relatively longer to express thioguanine resistance. However, as shown in Fig. 6, mutation expression reached a plateau after 10 d for both the control and the adapted CHO cultures, indicating that we are observing a real difference in mutation frequency.

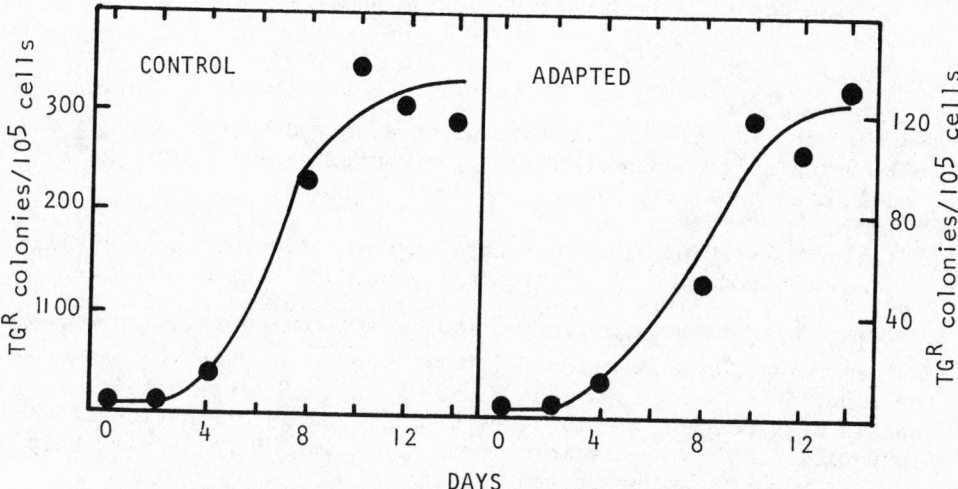

Fig. 6. Kinetics of the expression of thioguanine resistance in adapted and control CHO cells. CHO cells received 10 μg/ml MNNG and the resulting mutagenesis assayed as described in ref. 27. These results are to be published elsewhere, manuscript in preparation.

It also seemed possible that changes in mutagen permeability and detoxification might be responsible for the observed resistance at these high challenge doses of MNNG. However, using ^{14}C-MNNG, we found that control and adapted cells are equally susceptible to alkylation damage (see Table III). We therefore

Table III. Alkylation of control and adapted CHO cells
            with ^{14}C-MNNG.[a]

Challenge dose of ^{14}C-MNNG μg/ml	cpm/ug DNA[b]	
	Control	Adapted
0.75	2.17	2.41
1.0	3.73	3.82

[a] ^{14}C-MNNG was 14.8 Ci/mol.

[b] DNA isolation was carried out as described in ref. 27.

deduce, once again, that resistance is achieved by the adapted CHO
cells being able to handle alkylation damage somewhat more efficiently than control cells.

In summary, CHO cells that have become resistant to cell killing and SCE induction by alkylating agents also become resistant
to mutation induction. However, the kinetics of adaptation for
mutation are very different from both E. coli mutational adaptation
and mammalian SCE adaptation. (It will be interesting to measure
the kinetics of mutational adaptation in a cell line that is
capable of repairing O^6-MeG.) Hopefully, the determination of the
biochemical basis of mutational resistance in CHO cells will help
us understand these unexpected results.

DISCUSSION

Given that we are continuously exposed to alkylating agents, it
seems obvious to conclude that it must be beneficial to induce
resistance to these agents. However, the unexpected kinetics of
mutational adaptation in CHO cells may have uncovered other implications of the mammalian adaptive response. The mutation frequency
among cells that have survived the lower MNNG challenges is roughly
the same for adapted and control CHO cells; there is, however, a
considerable difference in their survival. Thus if we calculate a
mutation yield, e.g. the number of mutants per 10^5 challenged cells
(TG^R mutants/10^5 surviving cells x surviving fraction), we observe
another aspect of the adaptive response (Fig. 7).

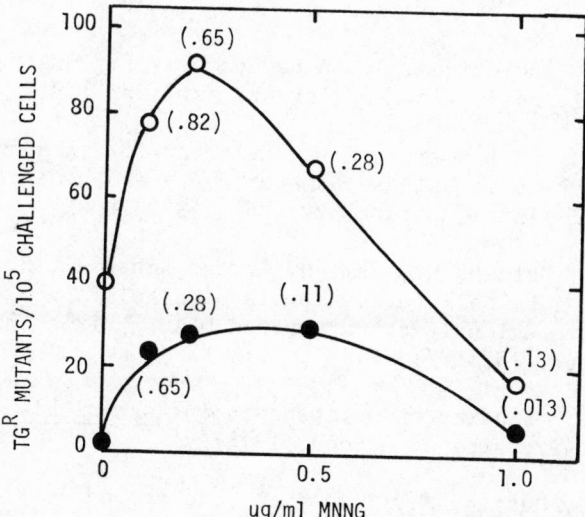

Fig. 7.   MNNG-induced mutation yield in adapted and control CHO
cells.   Mutation yield was calculated as described in the
text, from the mutation frequencies in Fig. 6 and the
surviving fractions shown in parentheses.   These data
will be published elsewhere, manuscript in preparation.

The adapted population acquired relatively more mutant cells
at each challenge dose, with very large differences in the low
dose range.   One might argue that the repair deficiency in CHO
cells reflects some peculiarity of cells that have been maintained
in tissue culture for countless generations.   However, it is clear
that, in rats at least, the very cells which are susceptible to
alkylation induced carcinogenesis are those that are unable to
repair O^6-alkylG (11,32).   Thus the induction of resistance to the
killing effects of alkylating agents in these particular cells
would enhance the rate of accumulation of mutations and would
therefore not be beneficial.   In other words, the consequences of
the absence of mutational adaptation are amplified in the presence of killing adaptation.

L. S. was supported by a SRC postdoctoral fellowship and the work
was supported by the U. S. Department of Energy.

REFERENCES

1. Samson, L. and Cairns, J: A new pathway for DNA repair in
   Escherichia coli. Nature (London), 267:281 (1977).

2. Samson, L. and Schwartz, J.L: Evidence for an adaptive DNA
   repair pathway in Chinese hamster ovary and human skin fibroblast cell lines. Nature, 287:861 (1980).

3. Jeggo, P., Defais, M., Samson, L. and Schendel, P: An adaptive
   response of E. coli to low levels of alkylating agent: Comparison with previously characterized DNA repair pathways. Molec.
   gen. Genet., 157:1 (1977).

4. Samson, L: Studies on mutagenesis in Escherichia coli. Ph.D.
   Thesis, University of London (1978).

5. Jeggo, P., Defais, M., Samson, L. and Schendel, P: The adaptive
   response of E. coli to low levels of alkylating agent: The
   role of pol A in killing adaptation. Molec. gen. Genet., 162:
   299 (1978).

6. Jeggo, P: Isolation and characterization of E. coli K-12
   mutants unable to induce the adaptive response to simple alkylating agents. J. Bact., 139:783 (1979).

7. Baker, R.M., Van Voorhis, W.C. and Spencer, L.A: HeLa cell
   variants that differ in sensitivity to monofunctional alkylating agents, with independence of cytotoxic and mutagenic
   responses. Proc. Natl. Acad. Sci. USA, 76:5249 (1979).

8. Lawley, P.D. and Thatcher, C-J: Methylation of DNA in cultured
   mammalian cells by N-methyl-N'-nitro-nitroseguanidine. Biochem.
   J., 116:693 (1970).

9. Loveless, A: Possible relevance of O-6 alkylation of deoxyguanosine to the mutagenicity and carcinogenicity of nitrosamines and nitrosamides. Nature, 233:206 (1969).

10. Gerchman, L. and Ludlum, D.B: The properties of O^6-methyl-
    guanine in templates for DNA polymerase. Biochim. biophys.
    Acta 308:310 (1973).

11. Goth, R. and Rajewsky, M.F: Persistence of O^6-ethylguanine
    in rat-brain DNA: correlation with nervous system-specific
    carcinogenesis by ethylnitrosourea. Proc. Natl. Acad. Sci.
    USA, 71:639 (1974).

12. Schendel, P-F and Robins, P.E: Repair of O[6]-methylguanine in adapted E. coli. Proc. Natl. Acad. Sci. USA, 75:6017 (1978).

13. Robins, P. and Cairns, J: Quantitation of the adaptive response to alkylating agents. Nature, 280:74 (1979).

14. Cairns, J: Efficiency of the adaptive response of E. coli to alkylating agents. Nature, 286:176 (1980).

15. Karran, P., Lindahl, T. and Griffin, B: Adaptive response to alkylating agents involves alteration in situ of O[6]-methyl-guanine residues in DNA. Nature, 280:76 (1979).

16. Foote, R.S., Mitra, S. and Pal, B.C: Demethylation of O[6]-methylguanine in a synthetic DNA polymer by an inducible activity in E. coli. Submitted to Biochem. biophys. Acta.

17. Olsson, M. and Lindahl, T: Repair of alkylated DNA in E. coli. J. Biol. Chem., 255:10569 (1980).

18. Lindahl, T: DNA methyltransferase acting on O[6]-methylguanine residues in adapted E. coli in Chromosome Damage and Repair, Ed. by E. Seeberg. Plenum Press, in press.

19. Schendel, P.F., Defais, M., Jeggo, P., Samson, L. and Cairns, J: Pathways of mutagenesis and repair in Escherichia coli exposed to low levels of simple alkylating agents. J. Bact., 135:466 (1978).

20. Lindahl, T: Personal communication.

21. Montesano, R., Bresil, H. and Margison, G.P: Increased exci-sion of O[6]-methylguanine from rat liver DNA after chronic administration of dimethylnitrosamine. Cancer Res., 39:1798 (1979).

22. Montesano, R., Bresil, H., Planch-Martel, G.P. and Pegg, A.E: Effect of chronic treatment of rats with dimethylnitrosamine on the removal of O[6]-methylguanine from DNA. Cancer Res., 40:452 (1980).

23. IARC Monographs on the Evaluation of the Carcinogenic Risk of Chemicals to Humans. 1:95 (1972).

24. Carrano, A.V., Thompson, L.H., Lindl, P.A. and Minkler, J.L: Sister chromatid exchange as an indicator of mutagenesis. Nature, 271:551 (1978).

25. Wolf, S., Rodin, B. and Cleaver, J.E: Sister chromatid ex-
     changes induced by mutagenic carcinogens in normal and xero-
     derma pigmentosum cells. Nature, 265:347 (1977).

26. Jensen, E.M., LaPolla, R.J., Kirby, P.E. and Haworth, S.R:
     In vitro studies of chemical mutagens and carcinogens.
     1. Stability studies in cell culture medium. J. Natl. Cancer
     Inst., 59:941 (1977).

27. Bodell, W.J. and Banerjee, M.R: Reduced DNA repair in mouse
     satellite DNA after treatment with methylmethansulphonate
     and methylnitrosourea. Nucl. Acids Res., 3:1689 ·(1976).

28. Wolf, S. and Perry, P: Differential giemsa staining of
     sister chromatids and the study of sister chromatid exchanges
     without autoradiography. Chromosoma, 48:341 (1974).

29. Jostes, R., Samson, L. and Schwartz, J.L: Kinetics of muta-
     tion and sister-chromatid exchange induction by ethylmethane-
     sulphonate in Chinese hamster ovary cells. Mutation Res.,
     91:255 (1981).

30. Warren, W., Carthorn, A.R. and Shooter, K.V: The stability of
     methylated purines and of methylphosphotriesters in the DNA
     of V79 cells after treatment with N-methyl-N-nitrosourea.
     Biochim. biophys. Acta, 563:82 (1979).

31. Goth-Goldstein, R: Inability of Chinese hamster ovary cells
     to excise O^6-alkylguanine. Cancer Res., 40:2623 (1980).

32. Lewis, J. G. and Swenberg, J. A.: Differential repair of
     O^6-methylguanine in DNA of rat hepatocytes and nonparenchymal
     cells. Nature, 286:185 (1980).

DISCUSSION

DR. RIPLEY: Why is the adapted yield of mutants lower than the initial number of mutants at zero dose (fig. 7)?

DR. SAMSON: They're all mutants that arose in the population during the course of adaptation, especially, as you might expect, for CHO cells because they can't repair O^6-methylguanine. So as you keep giving the pretreatment doses, mutants are just going to keep on accumulating.

DR. GLICKMAN: Most of us have gone through life thinking that O^6 products were very important, but I think there's a great deal of evidence accumulating that O^6-G is not really that important, or may not be the most important lesion. For example, uvrB$^-$ strains of E. coli don't excise O^6-G but mutate more than wild type. They give a 10-fold higher mutation at the same dose. But when one actually goes to look at the specificity of mutagenesis in these strains, it's not exclusively GC to AT any more. Some other lesion may also not be excised.

DR. SAMSON: People have not looked at the pyrimidine lesions. Bea Singer has done some studies on that but you know, in a system where you know you've got a differential in mutation like the adaptive system, one should really go and look at the pyrimidine lesions too.

DR. GLICKMAN: A second kind of evidence is that bacteriophage, when treated with alkylating agents, don't always mutate, even though you believe O^6-G to be in the DNA.

DR. SAMSON: The corollary to that is the fact that the cells in animals which don't remove O^6 methylguanine are the cells that give rise to tumors.

DR. GLICKMAN: That may be true but they also don't remove O^4 too or the correlation may exist with a large number of lesions.

DR. SAMSON: Absolutely.

DR. L. PRAKASH: You mentioned that both the challenge dose and the adaptive dose were important. Does it also matter what the site is that you're measuring mutation in? Do you find this with many different loci?

DR. SAMSON: With several different loci in bacteria, yes. By the way, this result in the CHO cells is the same, if you look at oubain resistance as well.

DR. SHAPIRO: Is there any idea what the killing lesion might be in bacteria?

DR. SAMSON: No. The perfect tool to identify the killing lesion is the ada⁻ mutant that will adapt for killing but won't adapt for mutation. So now you've eliminated the complication of the mutagenic lesions being removed you can look at the removal of those lesions responsible for that increased resistance to killing.

DR. GLICKMAN: That's probably explainable, at least if the latest information is correct, by there being an adaptive response to methyladenine as well as O^6-G.

DR. SAMSON: When you adapt cells, you also increase the activity of the N-7 methylguanine and the N-3 methyladenine glycosylases, and these may be responsible for giving resistance to killing. That would be easy to test with these mutants.

DR. GLICKMAN: The alk mutant is reputedly the inducible tag mutant. This 3-methyladenine glycosylase deficient mutant is very sensitive to killing by alkylating agents, suggesting that the 3-methyladenine lesion is an important lethal lesion.

DR. SAMSON: The tag gene is supposed to be the structural gene for the constituitive 3-methyladenine glycosylase, and the indicible one is presumably the one that is turned on in this adaptive response.

DR. THILLY: Benedick's measurements, indicate to me that as many as 7 or 8 particular alkylation products occur at sufficient frequencies to be considered potential nominees for pre-mutagenic lesions. Until we have more data, such as has been presented here today, I think it would be premature to eliminate any of them.

    Furthermore, going back to the very first part of your talk, I was just wondering if you've reflected on the actual level or the frequency of alkylation in live people undergoing continuous exposure?

DR. SAMSON: I can't give you a specific answer on that but we are exposed to a low level of nitrosamines in our food. Also what has recently been shown, again by Lindhal, is that S-adenosyl methionine will methylate DNA. This is such an efficient methyl donor that it actually doesn't need an enzyme. But it donates a methyl to the N-7 position of guanine and the N-3 position of adenine, the non-biological positions. So we're continually exposed to the dangers of methylation just from our own natural metabolites.

DR. ARLETT: How firm is the evidence for cross resistance? In the CHO particularly.

DR. SAMSON: Very firm. Adapted CHO cells are resistant to killing by MNU, and to SCE induction by MNU. The GM637 ells are resistant to SCE induction by MNU and ENU.

DR. ARLETT: I ask this because we tried to do these experiments following your protocol. We had greater MNU toxicity, so we used your preferred dose to adapt with MNMG, every 6 hours. We've done it twice now, and I can't convince myself that there is any effect. It looked to me as though the effect that you're see by adapting with MNMG to MNU was quite small, a factor of 2 at most.

DR. SAMSON: There's about a two-fold difference in the slopes of killing curves.

DR. ARLETT: Is it greater with MNMG than with MNU?

DR. SAMSON: It was about the same, about a two-fold difference in the initial slope of the curve. You have to try a lot of different doses, and I'm sure it will be different for each.

PARTICIPANT: I was wondering how the SCE distribution changes with time. Also, what would happen if you stopped administering the dose?

DR. SAMSON: We haven't done detailed kinetics on it but the cells do return to control SCE distribution, within the boundaries you would see for cells that had never seen mutagen, and this happens certainly within a couple of days after stopping the pretreatment.

DR. BRIDGES: I noticed that in the mutagenicity experiment, your surviving fractions at the dose where you get the difference are down by two log cycles in the unadapted cells. Now, that's a very big kill. It's way outside the range that most of us ever use with mammalian cells, for scoring mutation.

DR. SAMSON: Where would that put the bias?

DR. BRIDGES: I wouldn't like to predict that, but you have to be very careful, and reconstruction experiments are necessary.

DR. SAMSON: I agree with you. We really have to find the basis for that difference, but I don't think it changes anything in the mutation yield. That's a point well taken.

INDUCED MUTAGENESIS OF SIMIAN VIRUS 40

IN CARCINOGEN-TREATED MONKEY CELLS

Alain Sarasin, Claire Gaillard[*] and Jean Feunteun

Institut de Recherches Scientifiques sur le Cancer
BP 8, 94802 Villejuif Cédex, France and
[*]Institut de Recherches en Biologie Moléculaire
2 place Jussieu, 75221 Paris Cédex 05, France

INTRODUCTION

The mechanism of induced mutagenesis has been particularly well analyzed in E. coli after UV-irradiation. UV-mutagenesis appears to be an inducible phenomenon which requires the specific cleavage of the lexA gene product by the protease activity of RecA protein and the presence of the umuC gene product. Similar genetic requirements are necessary to detect mutagenesis of UV-irradiated bacteriophages (1, 2, see 3, 4, 5 for review). UV-irradiated λ phage has a better survival in UV-irradiated host cell than in unirradiated one. This UV-reactivation is only associated with mutagenesis of the damaged DNA phage (6). The most likely explanation for UV-mutagenesis mechanism is the induction of some type of DNA replication past pyrimidine dimers which will produce mutations on newly synthesized strand opposite to the lesions on the template strand (7). However, mutations in phage DNA appear to be produced not only in phage containing DNA damages but also in undamaged phage providing that the host cell has been UV-irradiated before infection. This result suggests that inducible host functions might decrease the fidelity of replication even on undamaged template. The exact mechanism of this untargeted mutagenesis is not yet understood. As in bacteria, UV-irradiation of mammalian cells gives rise to mutations. Studies with normal and classical xeroderma pigmentosum human cell lines indicate that excision repair synthesis is an accurate repair process (8). In consequence, mutations seem to be due to the presence of unrepaired UV-damages on DNA, probably pyrimidine dimers. On the other hand, xeroderma pigmentosum variant cells are more mutable by UV-light

311

than normal cells and seem to be deficient in an error-free post-replication repair pathway (9). Experiments carried out in mammalian cells in order to detect any inducible error-prone pathway have been negative so far. For example, no mutagenesis increase has been observed in split dose experiments (10). In contrast, experiments using animal viruses as a molecular probe, in the same way as bacteriophages, have shown that induced mutagenesis could in fact be detected in mammalian cells.

It is now well established that treatment of mammalian cells with various physical or chemical agents that damage DNA prior to infection with UV-irradiated virus enhances virus survival. This phenomenon which has been called induced virus reactivation or enhanced virus reactivation, by analogy with bacteriophages, was first demonstrated with UV-irradiated Herpes virus infecting UV-irradiated monkey kidney cells (11). This result has now been confirmed using a variety of UV-irradiated viruses which replicate their DNA in the cell nucleus such as simian adenovirus (12), simian virus 40 (13, 14) and several DNA parvoviruses such as Kilham rat virus (15), Lu III (16) or minute-virus-of mice (17). X-ray or γ-irradiation of human or monkey kidney cells also enhances the survival of UV-irradiated Herpes virus, simian adenovirus 7, SV40 (12) or human adenovirus 2 (18). Furthermore, Jeeves and Rainbow (19) have demonstrated that γ-irradiation of normal human fibroblasts leads to enhanced survival of either UV or γ-irradiated human adenovirus 2 although these two agents produce a different spectrum of DNA lesions. Chemical carcinogen treatment of monkey kidney cells has been shown to be very efficient in enhancing the survival of UV-irradiated SV40 (13). This effect was obtained both with compounds producing "UV-like" damages (such as metabolized aflatoxin B_1 or N-acetoxy-acetyl-aminofluorene) or those producing "X-ray" like damages (such as methyl-methane-sulfonate or ethyl-methane-sulfonate). These results have been confirmed using UV-irradiated Herpes virus (20) and Lu III parvovirus (16). SV40 reactivation was also promoted in monkey cells which were treated by drugs inhibiting DNA replication such as hydroxyurea or cycloheximide (13). We have hypothesized that the inhibition of scheduled DNA synthesis is a direct or indirect signal inducing a new repair process able to replicate UV-irradiated DNA (13). This induced pathway might be related to the phenomenon in which enhanced post-replication repair is observed in carcinogen-treated Chinese hamster or human fibroblast cells (21, 22).

If this virus reactivation phenomenon resembles bacteriophage reactivation, it should be accompanied by an increased mutagenesis of the repaired virus. In order to approach this problem, we used Simian Virus 40 as a molecular probe. The lytic infection of SV40 in monkey kidney cells is well characterized and consists of early functions (virus absorption, uncoating, early mRNA transcription

and early protein synthesis producing the small t and large T anti-
gens), followed by initiation of viral DNA replication and then
expression of late functions (late mRNA transcription and late protein synthesis producing the VP-1, VP-2 and VP-3 coat proteins)
(see 23 for review). Due to its limited genetic content, SV40 is
extremely host dependent and therefore SV40 is an excellent tool
to investigate the induction of cellular enzymes particularly those
involved in DNA metabolism. A number of SV40 mutants have been iso-
ted and well characterized. We used thermosensitive SV40 mutants
either at the level of the early genes (tsA58) or of the late
genes (tsB201) (Fig. 1).

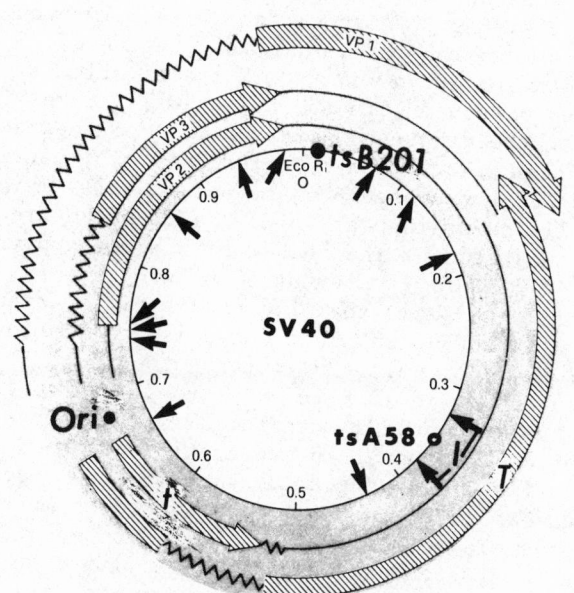

Fig. 1. Genetic map of SV40 genome (23)
     Ori represents the unique site of SV40 DNA replication.
     t is the early gene coding for the small t antigen.
     T is the early gene coding for the large T antigen where
the tsA58 mutant is mapped.
     VP_1, VP_2, VP_3 are the late genes coding for the capsid
proteins (tsB201 mutant is mapped in the VP_1 gene).
     ➡ represents the *Hind II-III* restriction enzyme sites.
The I restriction fragment contains the tsA58 mutation.
     ⟫ symbolizes the coding messenger RNA.
     ⌒⌒ symbolizes the splicing region.
     —— symbolizes the non-coding messenger RNA.
The map units start at the position of the *Eco R$_I$* restriction enzyme site.

tsA58 SV40 mutant is defective in the initiation of viral DNA replication at 41°C due to a single base-pair substitution in the large
T antigen gene (24, 25, 26). tsB201 SV40 mutant is defective in
virus production at 41°C due to a mutation in the VP-1 protein gene.
Moreover, SV40 DNA which is a small supercoiled molecule (3.5 x 10[6]
daltons) is easy to isolate and its DNA sequence is completely
known. Finally, the idea that SV40 DNA should be repaired in the
same way as cellular DNA is strengthened by the fact that SV40 mini-
chromosome shares structural similarities with mammalian chromosome
(27).

MUTAGENESIS OF UV-IRRADIATED tsB201 SV40 IN UNTREATED MONKEY CELLS

     UV-irradiation of SV40 decreases the virus survival measured
in untreated cells. Fig. 2 A shows the UV-inactivation curve of a
tsB201 SV40 mutant at the permissive temperature of 33°C. The
typical two-component curve for the UV-inactivation of this mutant
has been already observed with various animal viruses (28). In the
same experiment, the frequency of mutation from thermosensitive
growth to wild-type growth can be determined by measuring directly
the survival of UV-irradiated SV40 mutant in cells maintained at
41°C and in cells maintained at 33°C. The mutation frequency is
calculated as the ratio of the number of plaques growing at 41°C
to the number of plaques growing at 33°C. From the shape of the
curve in Fig. 2 A, it seems that the number of revertants increases
as the square of the UV-dose. This curve is similar to the one
obtained by Cleaver and Weil (29) with another ts SV40 mutant. The
slope of the curve on Fig. 2 B is equal to 2.4 which demonstrates
that UV-mutagenesis of tsB201 mutant really exhibits a dose-squared
induction kinetics. The usual interpretation of the dose-squared
relationships between UV-induced mutagenesis and UV-dose is that
two independent events are necessary to produce a mutation (30, 31,
32, 33). Two possible mechanisms have been proposed. First, two
independent lesions, such as two pyrimidine dimers on opposite
strands, might be a substrate for the mutation event. Second, the
first event might be the induction of an error-prone repair pathway
which will give rise to mutation at the level of DNA lesions. This
last hypothesis has been proposed to explain the UV-mutagenesis of
a tif-1 uvrA E. coli mutant (32) or UV-mutagenesis in yeast in the
presence of cycloheximide (33).

INDUCED MUTAGENESIS OF UV-IRRADIATED tsA58 SV40 IN UV-IRRADIATED
MONKEY CELLS

     The experimental protocol used to study UV-induced mutagenesis
of SV40 in UV-irradiated cells is schematized on Fig. 3 : virus
infection occurs for one lytic cycle at the permissive temperature
(72 h at 33°C) and then viral progeny survival is measured at 33°C
and 41°C in untreated CV1-P cells. In this assay there is no selec-

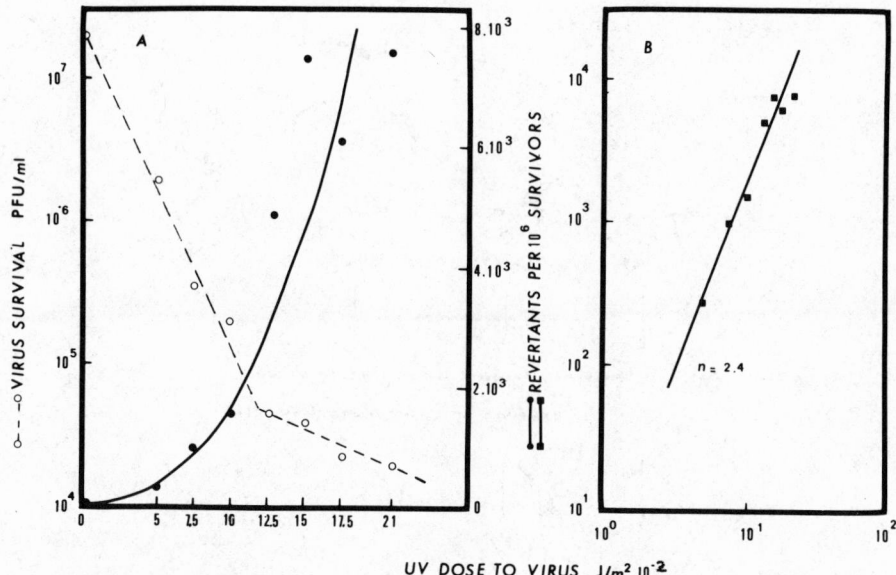

UV DOSE TO VIRUS J/m^2 10^{-2}

Fig. 2. Survival and reversion-frequency of tsB201 SV40 mutants
irradiated with UV-light and grown in unirradiated CV1-P
monkey cells.

A. Crude virus stocks of thermosensitive tsB201 late mutants
of SV40 were irradiated with a germicidal lamp at 254 nm.
The survival of UV-irradiated tsB201 was measured in
CV1-P cells maintained at 33°C for 21 days using the plaque-
assay technique (13) and the number of revertants towards
a wild-type phenotype was measured by the same assay but
in cells maintained at 41°C for 10 days. Results are the
mean of at least four determinations.

B. The slope of the curve from a log-log plotting is equal to
2.4 indicating a dose-squared relationship between UV-
induced mutations and UV-dose.

tion between the various SV40 mutants or revertants which grow
identically at 33°C (34). The mutation frequency is equal to the
ratio of virus survival at 41°C over virus survival at 33°C.

UV-irradiation of monkey kidney cells 24 h before infection
with UV-irradiated SV40 increases the virus survival compared with
untreated cells. This phenomenon -called induced virus reactivation-
is particularly efficient if one measures the kinetics of virus
production during one viral cycle. As seen in Fig. 4, the slope of
growth curves, after infection with UV-irradiated SV40 either in

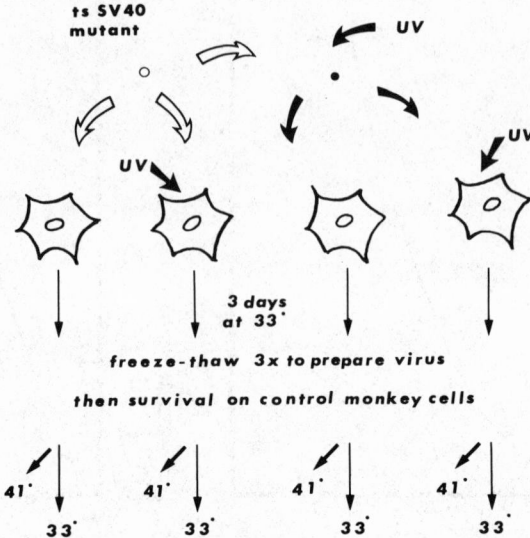

Fig. 3. Experimental protocol used to study inducible mutagenesis
of SV40 in pretreated monkey cells.
Unirradiated or UV-irradiated ts SV40 mutants (tsA58 or
tsB201) are used to infect confluent monolayers of CV1-P
monkey kidney cells treated 24 h before infection with
UV-light or chemicals. After one lytic cycle at the permissive temperature (72 h at 33°C), virus progeny is obtained
by the freeze-thawing technique and virus survival is measured by the plaque-assay technique on untreated CV1-P
cells at 33°C and 41°C.

UV-irradiated cells or in control cells, were parallel, but virus
production was much faster in UV-irradiated cells compared with
control cells. This result may indicate that enzymes from pretreated cells are able to better replicate UV-irradiated templates.
It should be pointed out that in this experiment, the multiplicity
of infection (m.o.i.) was 100 times higher for control SV40
(5×10^{-2} PFU/cell) as compared to UV-irradiated SV40 (5×10^{-4}
PFU/cell), so that the total number of viral particles per cell
was the same for the two types of virus (1 PFU is usually taken to
be equal to 100 virus particles for control SV40 and equal to about
10^4 particles for 1200 J/m^2 UV-irradiated SV40). Increased survival
was observed with unirradiated SV40 neither at that high m.o.i.
nor at lower m.o.i. (14), indicating that induced reactivation *per*

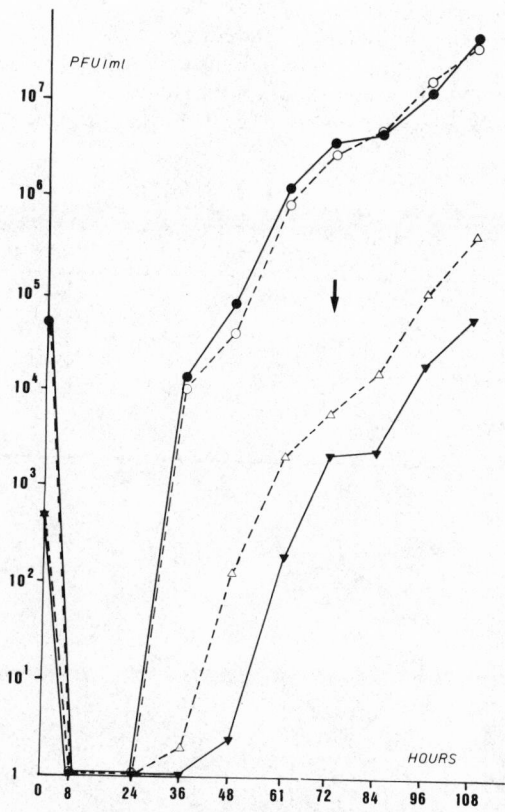

Fig. 4. Kinetics of SV40 production during one lytic cycle.
Unirradiated SV40 (circle symbols) and 1200 J/m^2 UV-
irradiated SV40 (triangle symbols) were used to infect unirradiated (closed symbols) or 10 J/m^2 UV-irradiated (open
symbols) monkey kidney CV1-P cells. Cells were infected
24 h after UV-irradiation with a multiplicity of infection
of 5 x 10^{-2} for unirradiated SV40 and 5 x 10^{-4} for UV-
irradiated SV40. After different times at 33°C, infected
cells were harvested and the virus titer was measured by
the plaque-assay method (13). The arrow indicates the end
of the first lytic cycle. Results are expressed in plaque-
forming units per ml of harvested cells (PFU/ml) and are
the mean numbers of plaques of 4-8 different values.
Reproduced form (34) with the permission of Elsevier/
North-Holland Biochemical Press

se was not due to recombination between DNA molecules inside the same cell.

The increased survival of UV-irradiated SV40 is accompanied by an increase of the reversion frequency from thermosensitive phenotype to wild type phenotype. As seen in Table I, the number of mutants is greatly increased when both cell and virus were UV-irradiated. The experiments were carried out at low multiplicity

Table I. Reversion frequency of tsA58 mutant to wild-type phenotype.

UV-irradiation (J/m^2)		Number of mutants per 10^5 survivors
SV40	Cell	
0	0	< 5
0	5	< 5
0	10	< 10
0	15	< 6
1200	0	< 10
1200	5	80
1200	10	150
1200	15	135
1500	0	< 10
1500	5	700
1500	10	570
1500	15	1130
2000	0	< 100
2000	5	3000
2000	10	450
2000	15	1500

Unirradiated or UV-irradiated tsA58 SV40 mutants were used to infect unirradiated or UV-irradiated CV1-P monkey kidney cells for one lytic cycle at 33°C (72 h) as described in Fig. 3. The number of mutants is deduced from the number of plaques growing at 41°C. An overestimated value is given (<) when no plaques appeared at 41°C. The multiplicity of infection is equal to 10^{-2} for unirradiated SV40 and 10^{-4} for UV-irradiated SV40. Each value represents the average of 4 determinations.

Table II. Wild-type phenotype of SV40 revertants isolated from plaques growing at 41°C.

Virus Survival PFU/ml	VA4554	tsA58	Plaques growing at 41°C					
			R-14-1	R-14-2	R-14-3	R-14-4	R-14-5	R-14-6
33°C	1.3×10^5	1.5×10^4	2×10^2	3×10^4	2×10^4	5×10^4	10^4	4×10^3
41°C	2.5×10^5	< 5	5×10^3	2×10^5	2×10^5	5×10^4	10^4	2×10^4

Plaques growing at 41°C during the plaque assay were isolated, freeze-thawed and virus survival was measured again at 33°C and 41°C. Plaques from the tsA58 SV40 mutant were grown at 33°C and plaques from the parental wild-type VA4554 were grown at 41°C.

of infection (m.o.i. ≃ 10^{-4} PFU/cell) to avoid any significant recombination inside the cell. We have previously shown that in the same conditions an increased mutagenesis is also observed with tsB201, an SV40 late mutant (34). In consequence, this UV-induced mutagenesis is proved with the two principal groups of SV40 mutants (early and late mutants) and therefore is independent of the cause of the thermosensitivity.

To ensure that this reversion was genetically stable, some plaques appearing at 41°C were picked up and viruses were assayed for survival at 33°C and 41°C. As seen in Table II, the viruses present in these plaques grow as well at 41°C as at 33°C or even better. This result indicates that our mutation assay really measures true genetic reversion from thermosensitive to wild type phenotype.

INDUCED MUTAGENESIS OF UV-IRRADIATED tsB201 SV40 IN CARCINOGEN-TREATED MONKEY CELLS

Treatment of monkey kidney cells with various chemical carcinogens such as UV-like compounds (Fig. 5) or X-ray like compounds (13), strongly increases the survival of UV-irradiated SV40. As in the case of UV-light, this increased survival is accompanied by a high rate of mutation of the repaired virus. Fig. 6 B shows that the frequency of reversion towards wild-type phenotype of tsB201 SV40 mutants is strongly increased when cells have been treated with N-acetoxy-acetyl-aminofluorene (AAAF) before the infection with UV-irradiated virus. A very small amount of carcinogen (less than 1 µM) is enough to induce the mutagenic process. This phenomenon is only detectable with UV-irradiated SV40 and no significant

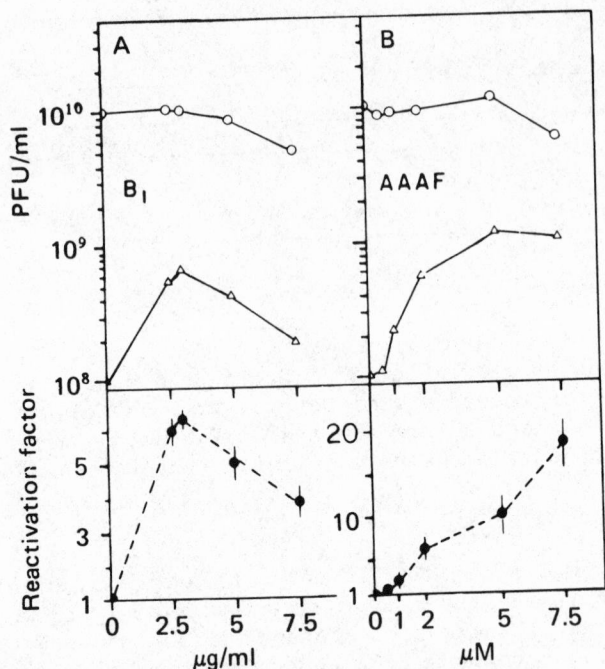

Fig. 5. Increased survival of UV-irradiated SV40 in carcinogen-
treated cells.
Top panels represent the survival of unirradiated (0) or
$1300 \ J/m^2$-UV-irradiated ($\triangle$) SV40 in monkey CV1-P cells
treated 24 h before infection by increasing amounts of
aflatoxin B_1 metabolites (A) or N-acetoxy-acetylaminofluo-
rene (B).
Bottom panels represent the reactivation factor calculated
as the ratio between virus survival in treated cells to
virus survival in untreated cells.
Reproduced from (13) with the permission of the National
Academy of Sciences of USA.

increased mutagenesis is observed with unirradiated SV40 mutant
(Fig. 6 A). A similar increased mutagenesis is observed when cells
have been treated with mitomycin C 24 h before infection with UV-
irradiated SV40 tsB201 mutant (Fig. 6 C).

As in the case of bacteria, treatment of monkey cells which
gives rise to DNA damages and to inhibition of scheduled DNA synthesis, induces an error-prone repair or replication mode responsible for the increased survival and mutagenesis of DNA-damaged
virus.

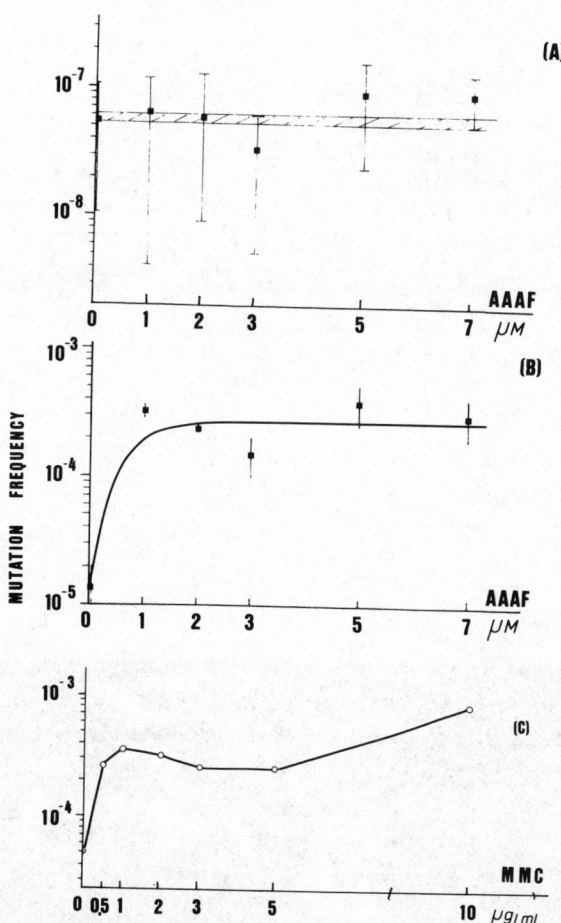

Fig. 6. Reversion-frequency towards wild type growth of tsB201
SV40 mutant in monkey cells treated with acetoxy-acetyl-
aminofluorene or mitomycin C.
  (A) Unirradiated tsB201 mutants are used to infect monkey cells
treated with increasing amounts of AAAF 24 h before infection (m.o.i. = 1 PFU/cell).
UV-irradiated tsB201 mutants (1500 J/m^2) are used to infect
monkey cells treated with increasing amounts of :
  (B) AAAF 24 h before infection (m.o.i. = 2 x 10^{-3} PFU/cell).
  (C) Mitomycin C 24 h before infection (m.o.i. = 3 x 10^{-4} PFU/
cell).
The error bars correspond to the standard deviation of the
mean.

MOLECULAR MECHANISM OF THE INDUCED MUTAGENESIS OF UV-IRRADIATED
SV40

It has been shown that UV-irradiation of bacteria or phages
could induce some deletions in their respective DNA (see 29). In
order to test this possible effect on SV40, we isolated various
revertants of tsB201 SV40 produced in UV-irradiated or AAAF-
treated cells. We isolated the supercoiled DNA of these viruses
and analyzed their genome structure by using various restriction
enzymes. The patterns of SV40 DNA isolated from several revertants
and treated by restriction enzymes *Hinf* I (Fig. 7), *Hae* III or
Hind III (data not shown) were identical with the patterns obtained
after treatment of tsB201 DNA. This result indicates that no major
deletions or additions were responsible for the reversion of SV40
tsB201 phenotype.

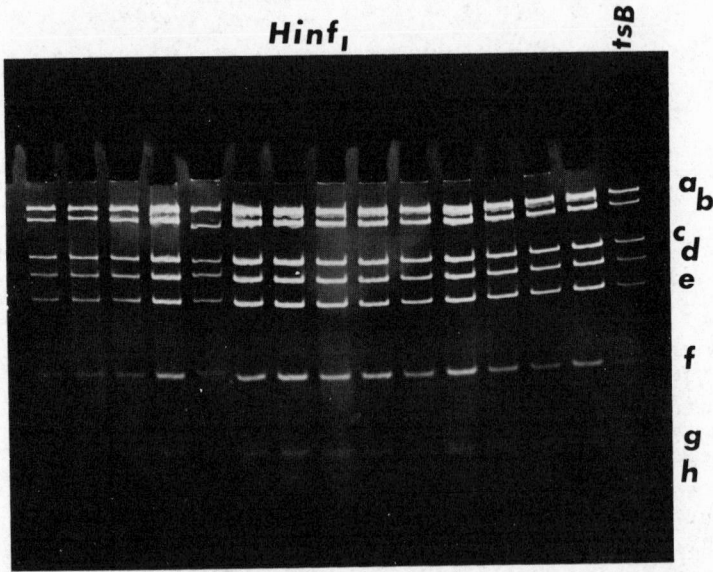

Fig. 7. Patterns of *Hinf* I digest of SV40 DNA from revertants and
        tsB201 on polyacrylamide gel.
        DNA from tsB201 SV40 and SV40 revertants were treated
        with *Hinf* I restriction enzyme and run on a 4 % poly-
        acrylamide gel. The channel labelled tsB contained DNA
        from tsB201 virus and all the other channels contained
        DNA from SV40 revertants. The letters correspond to the
        eight restriction fragments visible on this type of gel.

Table III. Nucleotide sequences of the I fragments of SV40 mutant
DNA treated with *Hind II + III* restriction enzyme.

```
              -Thr-.............        -Ala-Ala-Ala-Leu-Leu-Glu-Leu-Cys-Gly-Gly-Lys-
   Wt    5'...A.ACA.....................GCA.GCT.GCT.TTG.CTT.GAA.TTA.TGT.GGG.GGG.AAA.GCT.T...3'
(VA4554)                    217 nu
         ...T.TGT.....................CGT.CGA.CGA.AAC.GAA.CTT.AAT.ACA.CCC.CCC.TTT.CGA.A...

   Hind II                             Alu I                                    Hind III



              -Thr-.............        -Ala-|Val|-Leu-Leu-Glu-Leu
  tsA58   5'...A.ACA...................GCT.GTT.TTG.CTT.GAA.TTA....................
                             217 nu
         ...T.TGT...................CGA.CAA.AAC.GAA.CTT.AAT....................

   Hind II                           Alu I



              -Thr-.............        -Ala-|Val|-Leu-Leu-|Asp|-Leu-
  R14-10   5'...A.ACA...................GCT.GTT.TTG.CTT.GAT.TTA....................
                             217 nu
         ...T.TGT...................CGA.CAA.AAC.GAA.CTA.AAT....................

   Hind II                           Alu I
```

        DNA isolated from wild type VA4554 strain, tsA58 and
R14-10 SV40 revertant from tsA58 are digested with *Hind II-III*
restriction enzymes and the I fragments are purified (Fig. 1). DNA
sequencing is carried out according to the Maxam and Gilbert technique (35). The amino-acid sequence is deduced from the DNA sequence.
The first nucleotide on the 5' side of the I fragment is number
3476 and the last nucleotide on the 3' side is number 3734 according to the latest SV40 nucleotide sequence (23). *Hind II, Hind III*
and *Alu I* indicate the position of the restriction enzyme sites.

        In order to analyze the reversion site at the level of the
nucleotide sequence, we compared the tsA58 mutant to one of its
revertants since the tsA58 mutation is precisely mapped on the
SV40 genome (Fig. 1).

        The mutation site of the tsA58 mutant is present on the large
T antigen gene (25) in the fragment I of the *Hind II + III* restriction fragments (Fig. 1). We have sequenced the tsA58 I fragment
and found a single base-pair substitution when compared with the
parental wild-type VA4554 virus : a base-pair substitution from
C-G to T-A (Table III). When we analyzed the same I fragment from
a revertant (R14-10) obtained after infection of UV-irradiated
tsA58 SV40 in UV-irradiated monkey cells, the DNA sequence showed
another unique base-pair substitution : a transversion A-T to T-A
opposite to a possible thymine dimer site and is located nine base-
pairs apart from the original tsA mutation still present (Table

III). Since we have shown before sequencing, that the I fragment from R14-10 revertant DNA was able to complement the tsA58 DNA at 41°C by the marker rescue technique (25) (data not shown), it means that the transversion from A-T to T-A is sufficient to govern the synthesis of a thermostable T antigen protein.

It is interesting to emphasize that such an A-T to T-A transversion responsible for the R14-10 reversion has been described as mutation sites in E. coli (36) and in yeast (31) and could have been produced by replication past a pyrimidine dimer.

DISCUSSION

Our results show that treatment that damages the DNA and inhibits DNA replication of monkey cells such as UV-light, AAAF or mitomycin C, induces the expression of a cellular recovery pathway that can replicate damaged DNA faster than the constitutive enzymes but with a relaxed fidelity leading to higher mutation frequency. This error-prone repair pathway induced in monkey cells seems to be similar to the SOS repair mode described in E. coli (1, 2, 3). Although the exact molecular mechanism of the E. coli SOS repair process is still unknown, it seems that replication past lesions can be promoted by an induced polymerizing activity (7). Such a process will lead to a more efficient replication of damaged viral DNA (induced phage or virus reactivation) with a higher rate of non-complementary bases incorporated (increased mutagenesis). This interpretation is supported by the DNA sequencing data of the lacI gene of E. coli in which UV-irradiation induces single and tandem double mutations probably due to the incorporation of one or two incorrect nucleotides opposite pyrimidine dimer (36). We have obtained similar results with the system composed of SV40 infecting monkey cells for which we have previously shown that DNA replication past dimer can occur after UV-irradiation of the host cell (37). Moreover our present DNA sequence data obtained on SV40 replicated in SOS conditions indicates that erroneous incorporation of one base might have been done opposite a pyrimidine dimer (Table III).

Similar error-prone replication of viral DNA in UV-irradiated host cell has been reported by Das Gupta and Summers using UV-irradiated Herpes simplex virus infecting UV-irradiated monkey cells (38). By using the same cellular system, Lytle et al. (39) concluded that mutagenesis of UV-irradiated Herpes virus is observed only when high m.o.i. is used and that multiplicity reactivation which occurred in their experimental conditions is directly responsible for the enhanced mutagenesis. By contrast, we have shown that SV40 increased mutagenesis is only obtained at low m.o.i. ($10^{-3} - 10^{-4}$). Moreover, if we increased the m.o.i. (more than 10^{-3}) the increased mutagenesis of UV-irradiated SV40 grown in treated cells compared to untreated cells disappeared (unpublished results). We don't really explain this discrepancy although it is possible that the mutagenic enzymes are induced in small amount in the treated cells and can act on a limited number of DNA lesions. If too many damaged templates are present in the same cell the small number of lesions replicated by an erroneous process will be diluted out and a non significant increased mutagenesis will be obtained. This hypothesis might explain the absence of enhanced mutagenesis reported in some cases (40, 41). Moreover, we can postulate that the infection with UV-irradiated virus be mutagenic *per se* by giving rise to a SOS signal as it has been shown in bacteria (42). In that case, by increasing the m.o.i. we will increase the number of UV-damaged templates and then indirectly induce the SOS repair process, even in untreated cells.

The mutation assay that we have described, show a very low background, since the spontaneous reversion frequency for unirradiated ts SV40 in untreated cells is around 10^{-7}. This low spontaneous reversion rate represents an obvious advantage in studying induced mutagenesis but can be a disadvantage if we want to study mutation frequency of unirradiated virus in treated cells. To do so, we need to infect at high m.o.i. in order to produce enough virus at 33°C and to get some plaques growing at 41°C. In our experiments, we do not see a significant increase of reversion-frequency for unirradiated virus growing in treated cells compared to untreated cells that some other groups have reported (38, 39, 41). This discrepancy could be due to the lower spontaneous mutation rate we observed in our experiment.

The presence of an error-prone mode of DNA replication in mammalian cells treated with DNA-damaging agents has been demonstrated by using various viral DNA as probe. However, no evidence has been obtained that this error-prone pathway is active on cellular DNA. Since we do not think that mammalian cells would have conserved such an erroneous pathway only for viruses, this discrepancy could be due to experimental protocols. In fact, viral mutants are much more well characterized and easier to manipulate than cellular mutants. However, the recent generation of various "mutator" mutants of mammalian cells can provide in the near future a good tool for the analysis of this induced mutagenesis.

ACKNOWLEDGEMENTS

We thank very much Mrs. A. Benoit for her excellent technical assistance. We are grateful to Dr. G. Bernardi in whose laboratory the DNA sequencing has been performed and to Prof. P.C. Hanawalt and to Prof. R. Monier in whose laboratories some of these experiments were initiated. This work was supported by grant ATP-DGRST

A.650.7886 (Paris, France) and from the Coordinating Council for
Cancer Research (Villejuif, France).

REFERENCES

1. Witkin, E.M. : Ultraviolet mutagenesis and inducible DNA repair
   in *Escherichia coli*. Bacteriol. Rev., 40 : 869 (1976)

2. Radman, M. : SOS repair hypothesis : phenomenology of an induci-
   ble DNA repair which is accompanied by mutagenesis. In
   Molecular Mechanism for Repair of DNA, Ed. by P.C. Hanawalt
   and R.B. Setlow (Proc. of a workshop conference on Molecular
   Mechanism for Repair of DNA, February 1974, Squaw Valley,
   Calif.) Plenum Press, New York, p. 355 (1975)

3. Devoret, R., Goze, A., Moulé, Y. and Sarasin A. : Lysogenic
   induction and induced phage reactivation by aflatoxin B_1
   metabolites. In Mécanismes d'altération et de réparation du
   DNA : relation avec la mutagénèse et la cancérogénèse chimi-
   que, Ed. by R. Daudel, Y. Moulé and F. Zajdela (Colloques
   Internationaux du C.N.R.S., n° 256) C.N.R.S., Paris, p. 283
   (1977)

4. Hall, J.D. and Mount, D.W. : Mechanisms of DNA replication and
   mutagenesis in ultraviolet irradiated bacteria and mammalian
   cells. Prog. Nucl. Acid Res. and Mol. Biol., 25 : 53 (1981)

5. Hanawalt, P.C., Cooper, P.K., Ganesan, A.K. and Smith, C.A. :
   DNA repair in bacteria and mammalian cells. Ann. Rev. Biochem.,
   48 : 783 (1979)

6. Defais, M., Hanawalt, P.C. and Sarasin, A. : Viral probes for
   DNA repair. Adv. in Radiat. Biol., In press (1981)

7. Caillet-Fauquet, P., Defais, M. and Radman, M. : Molecular
   mechanism of induced mutagenesis, replication *in vivo* of
   bacteriophage ØX174 single-stranded, ultraviolet light-
   irradiated DNA in intact and irradiated host cells. J. Mol.
   Biol., 117 : 95 (1977)

8. McCormick, J.J. and Maher, V.M. : Mammalian cell mutagenesis
   as a biological consequence of DNA damage. In DNA Repair
   Mechanisms, Ed. by P.C. Hanawalt, E.C. Friedberg and C.F. Fox
   (Proc. ICN-UCLA Symp. on DNA Repair Mechanisms, February
   1978, Keystone, Colo.) Academic Press, New York, p. 739,
   (1978)

9. Maher, V.M., Ouelette, L.M., Curren, R.D. and McCormick, J.J. :
   Frequency of ultraviolet light-induced mutations is higher
   in xeroderma pigmentosum variant cells than in normal human
   cells. Nature, 261 : 593 (1976)

10. Chang, C.C., d'Ambrosio, S.M., Schultz, R., Trosko, J.E. and
    Setlow, R.B. : Modifications of UV-induced mutation frequen-
    cies in Chinese hamster cells by dose fractionation, cyclo-
    heximide and caffeine treatments. Mutation Res., 52 : 231
    (1978)

11. Bockstahler, L.E. and Lytle, C.D. : Ultraviolet light enhanced
    reactivation of a mammalian virus. Biochem. Biophys. Res.
    Commun., 41 : 184 (1970)

12. Bockstahler, L.E. and Lytle, C.D. : Radiation enhanced reacti-
    vation of nuclear replicating mammalian viruses. Photochem.
    Photobiol., 25 : 477 (1977)

13. Sarasin, A. and Hanawalt, P.C. : Carcinogens enhance survival
    of UV-irradiated Simian Virus 40 in treated monkey kidney
    cells : Induction of a recovery pathway ? Proc. Natl. Acad.
    Sci. USA, 75 : 346 (1978)

14. Sarasin, A. : Induced DNA repair processes in eucaryotic cells.
    Biochimie, 60 : 1141 (1978)

15. Lytle, C.D. : Radiation-enhanced virus reactivation in mamma-
    lian cells. J. Natl. Cancer Instit. Monograph., 50 : 145
    (1978)

16. Günther, M., Wicker, R., Tiravy, S. and Coppey, J. : Enhanced
    survival of ultraviolet-damaged parvovirus Lu III and Herpes
    virus in carcinogen pretreated transformed human cells. In
    Chromosome Damage and Repair, Ed. by E. Seeberg, Plenum Press,
    New York, In press (1981)

17. Rommelaere, J., Vos, J.M., Cornelis, J.J. and Ward, D.C. :
    UV-enhanced reactivation of Minute-Virus-of Mice : stimula-
    tion of a late step in the viral cycle. Submitted for pu-
    blication

18. Jeeves, W.P. and Rainbow, A.J. : γ-ray-enhanced reactivation
    of UV-irradiated adenovirus in normal human fibroblasts.
    Mutation Res., 60 : 33 (1979)

19. Jeeves, W.P. and Rainbow, A.J. : γ-ray-enhanced reactivation
    of γ-irradiated adenovirus in human cells. Biochem. Biophys.
    Res. Commun., 90 : 567 (1979)

20. Lytle, C.D., Coppey, J. and Taylor, W.D. : Enhanced survival
    of ultraviolet-irradiated Herpes simplex virus in carcinogen-
    pretreated cells. Nature, 272 : 60 (1978)

21. D'Ambrosio, S.M. and Setlow, R.B. : Enhancement of post-
    replication repair in Chinese hamster cells. Proc. Natl. Acad.
    Sci. USA, 73 : 2396 (1976)

22. D'Ambrosio, S.M. and Setlow, R.B. : Defective and enhanced
    post-replication repair in classical and variant xeroderma
    pigmentosum cells treated with N-acetoxy-2-acetylaminofluo-
    rene. Cancer Res., 38 : 1147 (1978)

23. Tooze, J. : In DNA tumor viruses, Ed. J. Tooze, Cold Spring
    Harbor Laboratory (1980)

24. Tegmeyer, P. and Ozer, H.L. : Temperature-sensitive mutants of
    Simian Virus 40 : infection of permissive cells. J. Virol.,
    8 : 516 (1971)

25. Lai, C.J. and Nathans, D. : A map of temperature-sensitive
    mutants of Simian Virus 40. Virology, 66 : 70 (1975)

26. Sarasin, A., Gaillard, C. and Benoit, A. : Molecular mechanism
    of error-prone DNA replication induced in UV-irradiated or
    acetoxy-acetyl-aminofluorene treated monkey cells. J.
    Supramol. Struct. and Cell. Biochem., 5 : 203 (1981)

27. Chambon, P. : The molecular biology of the eukaryotic genome
    is coming of age. Cold Spring Harbor Symp. Quant. Biol.,
    42 : 1209 (1977)

28. Lytle, C.D. : Host cell reactivation in mammalian cells : I.
    Survival of ultraviolet-irradiated herpes virus in different
    cell lines. Int. J. Radiat. Biol., 19 : 329 (1971)

29. Cleaver, J.E. and Weil, S. : UV-induced reversion of a tempera-
    ture-sensitive late mutant of Simian Virus 40 to a wild-type
    phenotype. J. Virol., 16 : 214 (1975)

30. Witkin, E.M. and George, D.L. : Ultraviolet mutagenesis in
    $Pol\ A$ and $UVrA\ PolA$ derivatives of $E.\ coli$ B/r : evidence
    for an inducible error-prone repair system. Genetics, 73 :
    91 (1973)

31. Lawrence, C.W., Stewart, J.W., Sherman, F. and Christensen, R. :
    Specificity and frequency of ultraviolet-induced reversion of
    iso-1-cytochrome C ochre mutant in radiation-sensitive strains
    of yeast. J. Mol. Biol., 85 : 137 (1974)

32. Witkin, E.M. : Thermal enhancement of ultraviolet mutability in a *tif-1 UVrA* derivative of *Escherichia coli* B/r : evidence that ultraviolet mutagenesis depends upon an inducible function. Proc. Natl. Acad. Sci. USA, 71 : 1930 (1974)

33. Eckardt, F., Moustacchi, E. and Haynes, R.M. : On the inducibility of error-prone repair in yeast. In DNA Repair Mechanisms, op. cit., p. 421 (1978)

34. Sarasin, A. and Benoit, A. : Induction of an error-prone mode of DNA repair in UV-irradiated monkey kidney cells. Mutation Res., 70 : 71 (1980)

35. Maxam, A.M. and Gilbert, W. : Sequencing end-labeled DNA with base-specific chemical cleavages. Methods in Enzym., 65 : 499 (1980)

36. Coulondre, C. and Miller, J.H. : Genetic studies of the *lac* repressor. IV Mutagenic specificity in the *lac I* gene of *Escherichia coli*. J. Mol. Biol., 117 : 577 (1977)

37. Sarasin, A. and Hanawalt, P.C. : Replication of ultraviolet-irradiated Simian Virus 40 in monkey kidney cells. J. Mol. Biol., 138 : 299 (1980)

38. Das Gupta, U.B. and Summers, W.C. : Ultraviolet-reactivation of herpes simplex virus is mutagenic and inducible in mammalian cells. Proc. Natl. Acad. Sci. USA, 75 : 2378 (1978)

39. Lytle, C.D., Goddard, J.G. and Lin, C.H. : Repair and mutagenesis of herpes simplex virus in UV-irradiated monkey cells. Mutation Res., 70 : 139 (1980)

40. Day, R.S. III and Ziolkowski, C. : Studies on UV-induced viral reversion, cockayne's syndrome, and MNNG damage using adenovirus 5. In DNA Repair Mechanisms, op. cit., p. 535 (1978)

41. Cornelis, J.J., Lupker, J.H. and Van der Eb, A.J. : UV-reactivation, virus production and mutagenesis of SV40 in UV-irradiated monkey kidney cells. Mutation Res., 71 : 139 (1980)

42. Devoret, R. : Characterization of a replicon-controlled function required in the initiation of the SOS pathway in *E. coli*. J. Supramol. Struct. and Cell. Bioch., 5 : 196 (1981)

DISCUSSION

DR. LAWRENCE: Even though one plaque-forming unit gets into a cell, do your virus particules have such a low viability that there's a chance inactive particles enter with plaque-forming ones, with the possibility for some kind of rescue?

DR. SARASIN: To answer your question, we have to recall that with animal viruses such as SV40, adenoviruses or herpes virus one plaque-forming unit is composed of at least a hundred virus particles which is a big difference from phages where one PFU is composed of one particle. Consequently, it is always difficult to be sure that we are not looking at some kind of recombination or rescue process. To avoid this problem, we always carry out experiments at very low multiplicity of infection, usually 10^{-2}-10^{-3}, in order to infect each cell with a small number of virus particles.

DR. LAWRENCE: Might there be some kind of recombination repair going on?

DR. SARASIN: From other experiments we carried out in my lab, we know that UV-irradiated SV40 DNA is subject to recombination in monkey cells. If such a process occurs in the experiments I described here, it could explain the increased virus survival. However, it could not explain the increased mutagenesis of progeny virus since the infecting input is only composed of one type of thermosensitive mutant.

DR. GLICKMAN: Were those experiments carried out with irradiated phage or irradiated hosts, the recombination experiments?

DR. SARASIN: We are carrying out this recombination assay right now by coinfecting the same cell with two thermosensitive SV40 mutants belonging to the same complementation group. Preliminary results showed that UV-irradiation of the cell did not significantly increase recombination between viruses.

DR. GLICKMAN: UV induces recombination in yeast very efficiently.

DR. SARASIN: In fact, UV-irradiation of viral DNA by itself, increases dramatically the recombination process, but as I said, nothing spectacular happens after UV-irradiation of cells alone.

DR. LOEB: Could you just write the change on the board. That is the one in 249 nucleotides that constituted the fragment. I assume you have a thymine dimer on one strand.

DR. SARASIN: Indeed, we have a possible thymine dimer site at the position of the reverse mutation. By using W. Haseltine's technique, we have shown that such a thymine dimer can really be induced after UV-irradiation. However, it does not prove that this dimer is responsible for mutagenesis.

DR. GLICKMAN: Is that the only one you've sequences so far?

DR. SARASIN: We have sequenced several revertants. In all of them, we found a reversion site different from the original tsA mutation and always a single base-pair substitution opposite a possible pyrimidine dimer site.

DR. LOEB: Which is the way the polymerase reads on that frame?

DR. SARASIN: We found mutations corresponding to the two possible frames of the polymerase reading.

DR. SHERMAN: Did you find a random distribution of thymine dimers on your sequencing gel or were there more at some positions?

DR. SARASIN: There was indeed a random distribution of pyrimidine dimers along the DNA fragment and no specific hot spot was found at the site of reversion. Moreover, you have to remember that we are selecting for a reversion of a ts mutant towards wild type phenotype. Consequently, all possible mutation and reversion sites are not obligatory at a position for which the probability of getting a dimer is highest.

DR. BRIDGES: Are there any adjacent thymines that you cannot dimerize?

DR. SARASIN: It is very difficult, at least in our hands, if you have something like CTT, to exclude that you are making a dimer in one place and not in the other. You see a stronger band in one position, but you can still see something in the other. In fact, we found pyrimidine dimer in every position in which it was possible to make one.

DR. GLICKMAN: I'd like to point out that you have a mutation that's not at a PydC site, just for the record. You said that SV-40 was different from the bacterial system in that UV irradiation to the virus and not to the host already gives some mutation. I think it's work pointing out that in fact many sites in M13 phage also respond without the host being irradiated. The factors are small, 4 to 6 fold, but it isn't really necessary to irradiate the host to get M13 reversion.

DR. WALKER: This is a point I was trying to make previously. You could have something that's, if you want to use the term,

error-prone repair, without it necessarily being inducible or it
could be partially inducible. In fact, putting pKM101 into E.
coli makes it look something like the system you're describing, in
that now if you UV irradiate the phage you get some mutations
showing up, although you can get more by irradiating the host.
It's as though the system is now partially constituitive.

DR. GLICKMAN: The transversion of a T to an A in a TT sequence
makes be dubious about whether or not it's caused by a dimer.
Targeted mutations are not likely to be random, and dimers may
have some coding potential. Because of the bacterial and yeast
specificity of UV it is unlikely that dimers favour transversion
and the pyrimidine-pyrimidine pathway to transversion is not a
likely one, according to the Topal and Fresco models. If I see a
transversion, I start immediately to think of a non-targeted
mutation or some lesion other than a dimer. Again, we do not know
if a dimer is involved.

DR. SARASIN: You are right, however, as I said already, we are
looking at a specific reversion which will permit a
thermosensitive enzyme to regain its wild type behavior. In these
conditions, the selection procedure is so high that all possible
mutations are probably not detected.

DR. GLICKMAN: But you have a big target.

DR. SARASIN: The target is quite small in fact, and is probably
limited to the recognition site for the initiation sequence of
SV40 DNA.

DR. THILLY: I wonder if anybody here could comment on the what I
understand to be the legion of possible photoproducts other than
pyrimidine-pyrimidine dimers. Are there reasonable frequencies of
photohydrates, deaminations of cytosine, and others? I'd just
like to know because while I'm not a UV worker myself, I see all
of us happily counting dimers and just assuming these are the
predominant causes of human cell mutations.

DR. DOUBLEDAY: We did a few experiments using poly (dC) as a
template for DNA synthesis in which we found cytosine hydrates to
be formed with approximately the same frequency as pyrimidine
dimers (Nucleic Acids Res 9:3491). When we replicated the
UV-irradiated poly (dC) with E. coli DNA polymerase I the cytosine
hydrates block synthesis in a reaction in which only dGTP is
present as substrate. This blockage is partly relieved by the
addition of dTTP, and there's incorporation of thymine which is
dependent upon the dTTP concentration. The alleviation is
complete with dCTP, so it would appear that cytosine hydrates code
for cytosine, and, to a lesser extent, thymine.

Cytosine hydrates are very labile, and can either revert back
to cytosine by dehydration, or they can deaminate; the route
depends upon the irradiation conditions, pH, buffer and so on. If
they deaminate, uracil hydrates are produced and these are really
very stable; they will be stable for a matter of days at 37º I
think. They are refractory to removal by the uracil DNA
glycosylase, and code for adenine.

DR. MAHER: Did I understand you to say that induced frequency of
the cytosine hydrates is equal to thymine dimers?

DR. DOUBLEDAY: Cytosine dimers, it's poly (dC).

DR. MAHER: What is the frequency of cytosine dimers formed when
you're using 254 nm UV?

DR. DOUBLEDAY: They're not that infrequent relative to
thymine-thymine dimers.

DR. BRIDGES: Is this single-stranded DNA?

DR. DOUBLEDAY: Yes.

DR. BRIDGES: Well, you see double-stranded is very much less.

DR. GLICKMAN: But it's still on 7-fold so it's not a marginal
lesion, it's still a real possibility.

DR. BRIDGES: You can't eliminate anything.

DR. DOUBLEDAY: We don't know if these hydrates are formed in
double-stranded DNA at all.

Which DNA polymerases are involved in the replication of this
virus? And is there any evidence for involvement of DNA
polymerase beta in the repair? Can you throw in
di-deoxynucleotides without affecting the reactivation?

DR. SARASIN: Replication of SV40 DNA is carried out by cellular
DNA polymerase α. To my knowledge, nobody has evidence of the
involvement of DNA polymerase β acting on SV40 DNA.

DR. GLICKMAN: Coming back to the dimers being blocking lesions in
SV-40. Do you know about what happens in vivo?

DR. SARASIN: In vivo, I have shown with P.C. Hanawalt that SV40
DNA replication is blocked by the first pyrimidine dimer
encountered by the leading strand. This block exists at least for
one hour at 33ºC but it is overcome after that period of time.
It is still not clear if there is a bypass of the dimer or if there

is an excision of the dimer. By electron microscopic studies, M. Mezzina has shown, in my lab, that an excision process might occur at the level of the pyrimidine dimer which blocks the replication fork.

DR. GLICKMAN: And if you've irradiated the host?

DR. SARASIN: In the type of experiments I just described, we UV-irradiated the infected cells while SV40 DNA replication is occurring. Consequently, it is difficult to know whether we are looking at an inducible process or not.

SESSION V

HUMAN SYSTEMS

AND

ENVIRONMENTAL PROTECTION

MODERATOR: CHRISTOPHER W. LAWRENCE

ANALYSIS OF CHEMICALLY INDUCED MUTATION

IN SINGLE CELL POPULATIONS

William G. Thilly

Department of Nutrition and Food Science
Massachusetts Institute of Technology
Cambridge, Massachusetts   02139

I. INTRODUCTION

     This discussion is about tools to analyze the complex relation-
ship between chemicals and mutation in a relatively simple biological
system: the exponentially growing single cell population. Three assumptions are made about this relationship: that some DNA adducts
are premutagenic lesions, that adducted DNA may be repaired, misrepaired or unrepaired at the time of DNA synthesis and that unrepaired
DNA may be replicated, misreplicated or unreplicated at the time of
mitosis. These three assumptions lead to the definition of variables
ascertainable by experimental measurement which, in turn, lead to explicit formulae which should be useful as guides in the analysis of
dose(adduct)-response(mutation) relations.

     Two technical developments have made it possible to consider
the relationship between DNA adducts and genetic change in quantitative terms.

     The first is the improvement in analytical chemistry of
chemical-DNA reaction products by a combination of high pressure
liquid chromatography, careful studies of hydrolysis conditions and
application of high resolution analytical identification techniques.
In many laboratories it is now possible to identify and measure the
DNA lesions of a variety of simple alkylating agents, base analogues,
polycyclic aromatic hydrocarbons, mycotoxins and other known
mutagens.

     The second is improvement in techniques of measuring genetic
changes in single cells of bacteria, yeasts and mammals, including
man. Much remains to be done, however, in terms of the possibilities
for systemic bias in several methods in general use. This shortcoming notwithstanding, several laboratories have carefully probed
their means of measuring mutant fractions with carefully designed
reconstruction experiments which create confidence in the accuracy
of their reports. Thus, we should now be able to learn a great
deal about the process of chemically-induced mutation by combining
the twin abilities to measure the particular DNA adducts and resulting mutant fractions in the same cell populations. A number of
specific enzyme systems which modify adducted-DNA have already been
identified and are in various stages of characterization. This
discussion, however, avoids these tempting data and is guided by
studies which have taught us to expect a large number of gene
products to intervene between formation of an adduct in DNA and
appearance of a heritable mutation. Only the ability to measure
time will be added to the abilities to measure particular adducts
and particular mutations. The spirit of this discussion is to
broadly consider the interactions of chemical mutagens with cellular processes in the form of a general model (or fable?) potentially useful in discussing mutational phenomena and, perhaps, in
planning future experiments.

II.   ASSUMPTIONS

a.   DNA Adducts are Premutagenic Lesions

Cells are grown in culture medium to which a known quantity of
a particular chemical or complex mixture is added. Sometimes a crude
tissue extract containing the microsomal and cytosolic fractions of
metabolically competent cells is also present. The chemical must
interact with the cell in some way to kill or mutate. If cell death
or mutation occurs, then we may assume that an interaction of the
chemical, a spontaneous reaction product of the chemical, or an
enzymatic metabolite of the chemical with the cell has also
occurred. Competition occurs between reaction with sites which
could lead to mutation and sites such as culture medium nucleophiles
which would not lead to mutation. The molecular identities and
quantities of these potentially mutagenic reaction products assume
primary importance, since their formation is here assumed to represent the first of a series of steps necessary for genetic mutation.
[A mutagen could conceivably be mutagenic by altering culture conditions and never interacting chemically with a cell; if we limit
ourselves to thinking about those mutagens which do act by cellular
reactions, it is implicit that we assume that an important set of
chemical mutagens act by this means.] Since some portion of the
chromatin DNA of the cell carries all of the cell's genetic information and since all known mutagenic substances studied with regard
to DNA reaction products have indeed been found to react with DNA,
we will assume that some subset of the reaction products with DNA
are the actual chemical precursors of mutation.

Two requirements exist for a chemical DNA adduct to have the
potential to cause mutation. First, it must occur in that part of
the genome in which change would be detected by the mutation assay
for phenotypic alteration. This part of the genome may be considered
the "target" for mutation having a dimension of β base pairs.
Secondly, even if an adduct forms in the mutational target, we should
consider the fact that a chemical can form more than one kind of
reaction product with DNA. That fraction of all of its reaction
products which are potentially able to cause mutation we denote f_m.
The two requirements, that the reaction take place within the mutational target and that the reaction be of the potentially mutagenic
kind are treated here as independent variables so that the joint
probability factor for both requirements being met is just βf_m.

These definitions appear to neglect nearest neighbor effects,
i.e. adjacent base pairs may influence either the formation of an
adduct or the subsequent processing of a given adduct at a particular base pair. However, these effects can be considered as part of
the target term β, or the term describing the premutagenic probability f_m, respectively, as a first approximation. DNA-adducts are

not necessarily the entire set of potentially mutagenic reaction products. As an example, direct reaction with a DNA polymerase molecule could lead to replication error and thus mutation. An implicit assumption, therefore, is that mutation at a particular gene locus has a much higher probability of resulting from a reaction in that locus than from the probability of a combination of two reactions, e.g., reaction with a fraction of polymerase molecules and then use of altered polymerase molecules in replicating the gene locus we are observing with an error frequency much greater than found in unaltered polymerase molecules.

b. Adducted DNA May Be Repaired, Misrepaired or Unrepaired

DNA repair could be defined in a number of different ways. For the purpose of this discussion, DNA repair will mean those processes which return a section of DNA containing a reaction product to a section of DNA which does not contain a reaction product and which contains a sequence of nucleic acid bases identical to the sequence existing prior to chemical reaction. If the reaction product is removed and the sequence is restored to a form which can serve as a substrate for DNA replication but is not identical to the sequence existing prior to chemical reaction (any continuous sequence of altered base sequence length or identity), we will refer to the process resulting in this change as misrepair. If no change is effected or if the process is abortive, e.g., the chemical reaction product is removed, but no substrate for DNA replication results, then the process (or lack thereof) will be called nonrepair.

The definitions of repair, misrepair and nonrepair may be useful since they are mutually exclusive in their end results. This logical construct may permit developing a rational mathematical description of what appears to be a complex mixture of molecular processes. We can first consider the effects of each on mutation.

Repair does not lead to mutation, by definition.

Misrepair requires no further enzymic activity other than DNA replication to become an altered sequence in double-standed DNA. [A single event of misrepair should give rise to a locally altered DNA structure. This, in itself, might be considered a substrate of repair, misrepair or nonrepair prior to DNA replication. In such cases the defined terms apply to the fate of the lesion at the time of attempted DNA replication.] Misrepair, once having occurred, might be imagined to have a very high intrinsic probability of giving rise to a mutation.

The mutational effects of <u>nonrepair</u> would be dependent upon the behaviour of the particular DNA reaction product in DNA replication.

c. <u>Unrepaired DNA May Be Replicated, Misreplicated or Unreplicated</u>

If replication were simply inhibited by an unrepaired lesion, one would imagine that the cell could not complete synthesis of an entire replicon and that, since other replicons without such unrepaired lesions would complete synthesis, sister chromatids would result in which one small section would be unreplicated. Possibly, chromosome breakage would result from such a hypothetical event. The probability of cell reproductive death from one or a small number of such lesions might be expected to be high.

If, however, DNA synthesis were able to proceed past a DNA reaction product, then chromosome structure need not be compromised. Any number of possible behaviours present themselves, but it is sufficient for this discussion to imagine that replication through the unrepaired site will have some finite probability of failing to duplicate the correct nucleotide sequence and thus give rise to a mutant sequence in one daughter DNA strand. The altered sequence in one DNA strand and the unrepaired lesion in the other strand of the same chromatid could be acted upon by repair, misrepair or nonrepair; however, except by a recombinogenic matching process, this chromatid does not contain the correct sequence information and may be considered as having a very high probability of giving rise to <u>at least one</u> double stranded mutant DNA sequence in the subsequent round of DNA replication.

What arises from this approach is a model in which a particular DNA reaction product is either repaired (no mutation), misrepaired (probable mutation) or not repaired (probable lethality or probable mutation). Since the processes of repair and misrepair are assumed to be dependent on sets of enzymes acting on the area of a lesion as a substrate, one can imagine that the time available between the initial chemical reaction and the initiation of attempted DNA replication represents the time available for repair or misrepair. Lesions for which repair or misrepair systems exist will have increased probabilities of repair or misrepair as the time between lesion formation and DNA replication increases. A second prediction based on the model of the adducted DNA site as substrate and the repair systems as enzymes is (a) that the probability of repair of a particular lesion will decrease as the number of total adducted sites increases, and (b) that the probability of misrepair will also decrease with increased numbers of substrates for the repair/misrepair systems. A direct consequence of these expectations is that increased substrate (DNA reaction products) is predicted to lead to a shift toward increased frequency of unrepaired reaction sites at the time of initiation of DNA synthesis.

III.   DEFINITIONS AND EXPLICIT FORMULATION OF EXAMPLES

In order to transmute the model outlined above into something at least experimentally testable, expectation must be expressed in quantitative terms relating cell mutation and survival to experimental variables.

A few definitions are required:

M,   mutant fraction, the fraction of those cells surviving treatment in which mutations at the locus(i) studied have been induced by treatment.

S,   the fraction of cells surviving treatment.

D,   the average number of chemical adducts occurring in the cellular DNA (adducts/base pair).

β,   that portion of the cellular DNA which, when mutated, will be observed as a mutation (size of mutable target, base pairs).

f_m,   the fraction of adducts (DNA) which are potentially mutagenic by virtue of their chemical identity.

q,   the probability that an unrepaired lesion will give rise to a mutation during DNA replication.

$C_R(t)$,   the capacity to repair a particular DNA lesion prior to DNA replication.

$C_M(t)$,   the capacity to misrepair a particular DNA lesion prior to DNA replication.

Purists will already have noted that the definitions skip over many obvious possibilities. However, in order to keep things from becoming unnecessarily complicated, a few simplified situations are first proposed and their behaviour considered.

The first simplification is to imagine a chemical which reacts with DNA to give only one kind of chemical adduct. The second is to imagine, in turn, that our imaginary adduct is subject to only one kind of system to remove it. Having considered some very simple possibilities, we can then combine them (multiple lesions, competition for substrates among recovery systems, etc.) and see if they make any sense in describing experimental observations.

a. Nonrepair Precedes Mutation

For mutation by an imaginary chemical which creates only one kind of DNA lesion which could be repaired or not repaired, but not misrepaired, one should observe:

$$M = \boxed{\begin{array}{c} \text{Probability of Premutagenic} \\ \text{Reaction in Target} \end{array}} \qquad [\beta fmD]$$

$$X$$

$$\boxed{\text{Probability of Nonrepair}} \qquad [e^{-C_R(t)/D}]$$

$$X$$

$$\boxed{\text{Probability of Misreplication}} \qquad [q]$$

$$M = [\beta fmD] \ [e^{-C_R(t)/D}] \cdot q$$

The probability that our genetic target is hit with regard to potential mutation at all is $\beta f_m D$. The probability that this lesion goes unrepaired is $e^{-C_R(t)}/D$, where a capacity for $C_R(t)$ repair events is pictured as randomly distributed over D randomly distributed repairable lesions. The probability of nonrepair within the target is approximated by exp-(repair events/repairable events) via application of the poisson distribution. Since the probability of the unrepaired site giving rise to mutation is q, and the three phenomena (DNA reaction, repair, DNA misreplication) are assumed to be independent events, the relationship between D and M, both directly measurable quantities, should be of the form derived. This model predicts a higher order dependence of mutation upon D than a simple linear response, as shown in Figure 1.

This higher order curve has frequently been reported as a mutation dose response in which "D" is roughly approximated by knowledge of the initial concentration of chemical in the culture medium. Examples of this kind of behaviour has been observed by us for methylnitrosourea and ultraviolet light in human lymphoblast cultures.

b. Misrepair Precedes Mutation

A second sample of this analytical process is the case in which an imaginary chemical creates only one kind of DNA lesion which can

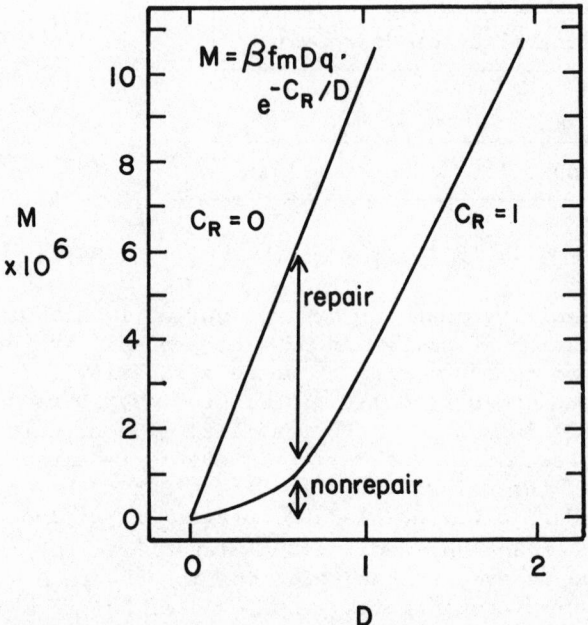

Figure 1. Mutation as a function of initial DNA reaction
products when mutation arises solely as a result of misreplication of unrepaired lesions.
The product of the constants $\beta f_m q$ has been
set at 10^{-5} for this example.

A second sample of this analytical process is the case in which an imaginary chemical creates only one kind of DNA lesion which can be misrepaired, but for which nonrepair does not lead to detectable mutation (q = o).

In such a case, one would observe the relation

$$M = \boxed{\begin{array}{c}\text{Probability of a Premutagenic}\\\text{Reaction in Target}\end{array}} \quad [\beta f_m D]$$

$$X$$

$$\boxed{\text{Probability of Misrepair}} \quad [1-e^{-C_M/D}]$$

$$M = [\beta f_m] \; D \; x \; (1-e^{-C_M(t)/D})$$

where $\beta f_m D$ is again the probability of the genetic target sustaining a potentially mutagenic event. However, the capacity to misrepair $(1-e^{-C_M(t)/D})$ is seen to be simply 1-(probability of not being misrepaired, $e^{-C_M(t)/D})$. When $C_M(t)/D$ becomes a small number, as when D increases, the function $(1-e^{-C_M(t)/D})$ approximates $C_M(t)/D$ and leads to the prediction that at high values of D

$$M = \beta f_m \; D \; x \; C_M(t)/D$$

$$= \beta f_m \; x \; C_M(t) = \text{a constant for any fixed value of t.}$$

This general form is a "saturation" type response as shown in Figure 2. There are also many examples of this form of response in the literature of mutagenesis. ICR-191 has displayed this form of behaviour in human lymphoblast experiments.

c. Nonrepair and Misrepair Precede Mutation

From our simple examples we can now make combinations of functions. For a case in which repair, misrepair and misreplication apply to the same single lesion formed by our single-adduct chemical, we can write

$$M = [\beta f_m D][[e^{-(C_M + C_R/D)} \cdot q] + [1-e^{-C_M/D}]]$$

$$= \boxed{\begin{array}{c}\text{Target}\\\text{Function}\end{array}} \; x \; \boxed{\begin{array}{c}\text{Nonrepair \&}\\\text{Misreplication}\end{array} + \text{Misrepair}}$$

This kind of combination is shown as a flow diagram in Figure 3.

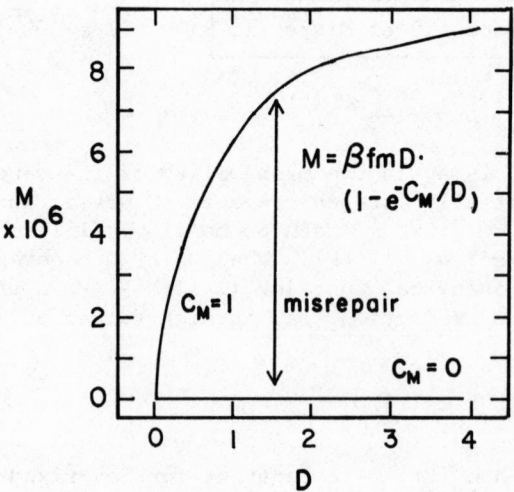

Figure 2.   Mutation as a function of initial DNA reaction products
            when mutation arises solely as a result of misrepair of
            lesions. The product of the constants βf_m has been set
            at 10^{-5} for this example.
            (Assumption for this example is that probability of mis-
            replication of an unrepaired lesion, q, is zero.)

In Figure 4 the case of repair and misrepair systems of similar capacities is considered as an example. When q is relatively small (q = 0.01) as in Figure 4, then one can discern a more rapid rise in M at low D as misrepair makes its contribution. At higher D both C_M and C_R are saturated, and mutation is seen to increase entirely as a result of nonrepair and misreplication.

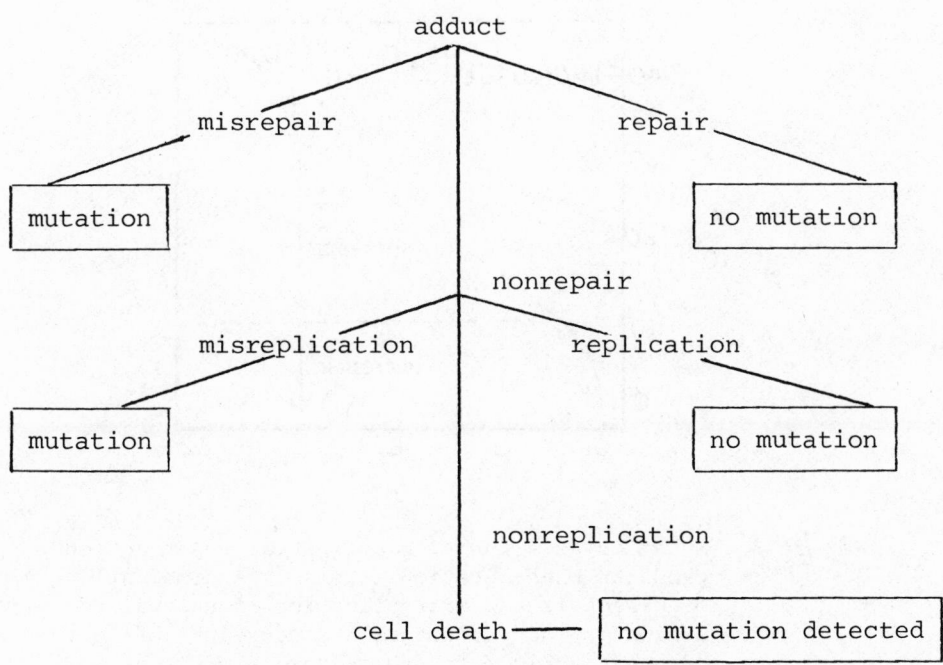

Figure 3.   Summary of the Assumptions.

IV.   APPLICATION TO EXPERIMENTAL SITUATIONS

Having considered the behaviour of these simple possibilities by way of introduction, one may now turn to real problems. In order to permit an orderly discussion, several important complications, both experimental and theoretical, will be discussed in turn,

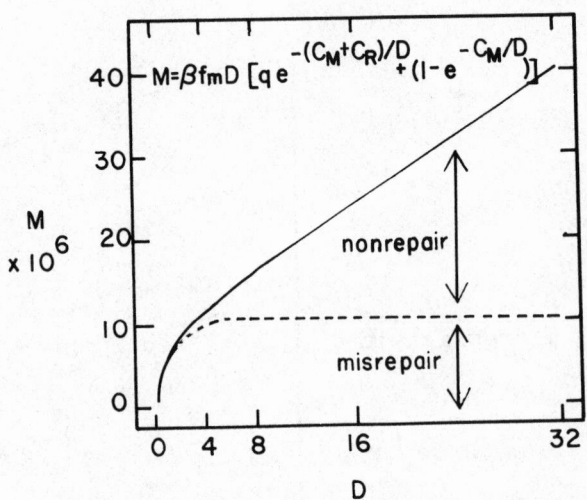

Figure 4. Mutation as a function of initial DNA reaction
           products when mutation arises as a combination
           of misrepair and misreplication of unrepaired
           lesions. The product of the constants βf_m have
           been set at 10^{-5} for this example with the
           probability of misreplication of unrepaired
           lesions, q, set at 0.10.

a. limitations in measuring mutation and DNA adducts,

b. heterogeneity of populations due to asynchrony,

c. multiple kinds of lesions from a single chemical,

d. time dependence of mutation and adduct removal,

e. inducible responses,

f. host cell reactivation,

g. behaviour of mutants in the mutational process.

a. Limitations in Measuring Mutation and DNA Adducts

   (i) Measuring Mutation. Depending on the particular cell
system chosen and the gene locus assayed for change, one has to con-
tend with a spontaneous rate of mutation which will obscure the
actual amount of induced mutation at low values of D. The parameters
chosen for Figures 1, 2 and 4 have been assigned values which cause
them to coincide roughly with our experiences with mutation induced
in human lymphoblasts for both the putative loci for hypoxanthine-
guanine phosphoribosyl transferase and thymidine kinase. These are
measured by the fractions capable of forming colonies in the presence
of 6 thioguanine and trifluorothymidine, respectively, after an appropriate period for phenotypic expression. For both of these loci
with our clonal derivative of the WI-L2 line, TK6, the spontaneous
mutant fraction cannot conveniently be lowered below 1-2 x 10^{-6} in
large cultures. Thus, the presence of a saturable repair system
can easily be obscured when it has a small capacity. (C_R is small.)
It is not really responsive to the problem to choose to study only
repair systems of apparent large capacity since it would seem that
people are exposed to somewhat lower concentrations of chemicals
than have been used thus far in experiments, and thus may only require low capacity repair systems to guard against mutation. Of
course, we have no inkling as to which, if any, chemicals are mutating people , and thus are not at liberty to rank order repair systems
in importance in terms of their apparent capacities.

   Similarly, misrepair systems of low capacity should not be
overlooked for lack of dramatic behaviour. As noted ICR-191 in human cell studies does appear to give a plateau effect of the kind
depicted in Figure 2 with the plateau occurring at an induced mutant

fraction about one hundred fold greater than spontaneous fractions. Perylene, on the other hand, demonstrates a plateau of mutation in S. typhimurium at an induced mutant fraction only five times the mean spontaneous mutant fraction and less than twice the 99% upper confidence limit on that spontaneous fraction. In both cases D is not actually known and is approximated by concentration of mutagen in medium.

Figure 5 summarizes these observations and points out the limitation imposed by spontaneous mutant fractions in such cases. A compound interacting with a low capacity misrepair system could quite conceivably double the amount of mutation in a system at extremely low concentration and yet escape detection completely.

A separate problem arises in those common cases in which the mutagen being studied is also toxic. From a theoretical standpoint, the assumption of independence of mutation and survival based on the experimental choice of a gene locus not required for survival would seem to be sufficient if one measures the survival of non-mutants and mutants under identical conditions. This latter qualification is rarely the case in practice and may have led to serious errors in the interpretation of already published data.

More common, however, is the failure to provide for sufficient cell number so that the number of surviving, induced mutants does not decrease below a minimum value, such as 100, which would yield observations of useful statistical precision, i.e. $\leq \pm 20\%$.

A final limitation may be drawn from the practical observation that estimates of mutant fractions for particular, dominant, putative, single loci, i.e., achondroplasia, are generally of the order of 10^{-5} in the human population. It would seem prudent, therefore, to direct our studies whenever possible to induced mutant fractions for forward mutation in single loci within the range of 10^{-6} to 10^{-5} bounded below by limitations of spontaneous rates and above by recognition of the need to understand systems of potential importance in human experience.

(ii) Measuring DNA Adducts. My colleagues Drs. J.M. Essigman and G.N. Wogan, working with high specific activity labelled aflatoxin B_1, are able to comfortably measure DNA adducts occurring at frequencies of less than one in 10^7 base pairs. Other laboratories have developed highly sensitive immunological fluorescent means to carry out assays with somewhat greater sensitivities. Sensitivity in the range of 10^{-7} to 10^{-8} adducts per base pair is not just a marvelous achievement of analytic chemistry, it is an absolute necessity if the relation of DNA adduction to mutation is to be understood.

The importance of this conclusion is enhanced by thinking about the number of base pairs in gene locus, such as HGPRT which, for the sake of argument, we may estimate to be about 10^3. When a

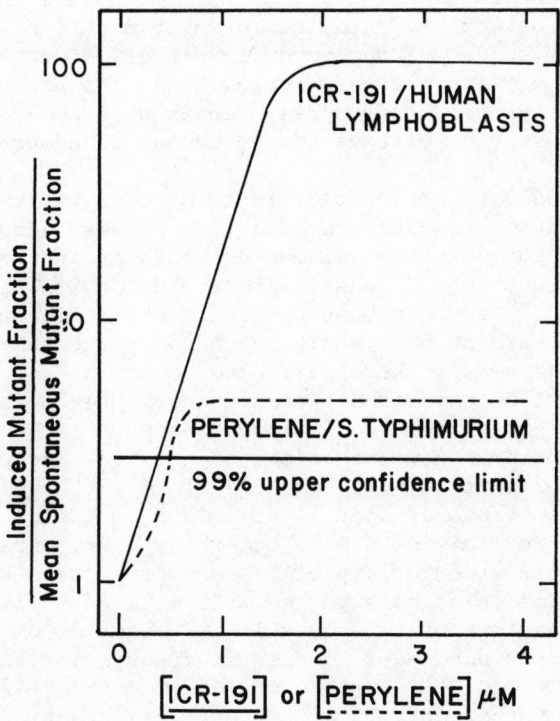

Figure 5.   Two examples of apparent saturation of mutation
as a function of chemical concentration in media.
The mutagens are equipotent at 0.5 μM, despite
twenty-fold difference in potencies at 2.0 μM.

mutant fraction of 10^{-4} is induced in human cells at the HGPRT locus, this means that a mutational event has occurred, on average, once in every $10^4 \times 10^3 = 10^7$ base pairs.

If the aflatoxin B_1-DNA adduct is a precursor of the induced mutations, we would expect that the average number of adducts per base pair would be equal to or greater than the number of mutational events per base pair. The number of adducts would, if they are the only premutagenic lesions, always be greater than the number of mutations in a gene because not all of the base pairs in a gene would give rise to a detectable mutation (cryptic mutation, β < size of continuous gene sequence), and secondly not all of the chemical reaction products within a target region need be potentially mutagenic (f_m < 1). Finally, of course, repair will reduce the number of mutations observed relative to the number of adducts formed.

We can consider this problem in terms of a molar efficiency, detectable mutations/measurable adduct. If every adduct resulted somehow in a mutation, chemical measurements at mutant fractions of 10^{-5} for targets of 10^3 base pairs would require precision in the region of 10^{-8} adducts/base pair. If the efficiency was 10%, then 10^{-7} adducts/base pair would be found at the time of treatment, if 1%, then 10^{-6} adducts/base pair, and so forth. At lower aflatoxin B_1 adduction experiments than we have previously reported, ($M \leq 10^{-5}$)   this   "molar efficiency" appears to be more than 10%, i.e., we observe fewer than one adduct/10^7 base pairs with mutant fractions near 10^{-5} at the HGPRT locus. On the other hand, similar studies of ultraviolet light induced mutation suggest that some 10^3 to 10^4 pyrimidine dimers occur for every mutation induced, indicating that a wide variety of behaviours at even this gross level of analysis is to be expected. It will do us little good to observe the behaviour of the plurality of chemical-DNA adducts which may occur at hundred-fold greater frequencies than mutation if mutations are causally related to the occurrence of a DNA-adduct accounting for only 1% of the chemical reaction products. The point of this discussion is that chemical analysis must account for those reaction products which occur at frequencies equivalent to relevant mutation frequencies, i.e., 10^{-7} to 10^{-8} events per base pair.

b. Cell Cycle and Progression Effects

It seems that those who ignore cell cycle effects in mutation studies are blessed with data that cannot be interpreted, while those who seriously set out to account for cell cycle effects get no data at all.

There are, however, many reasons why cell cycle position and effects of treatment on progression are important and thus worthy of more consideration. A few are discussed here.

   (i) Cell Cycle Time Available for Repair or Misrepair. As a
first example we can refer back to Figure 1 in which a competition
between a repair system of capacity $C_R(t)$, and the arrival of the
time to replicate the gene, repaired or not, was related to the
amount of mutation observed by the expression

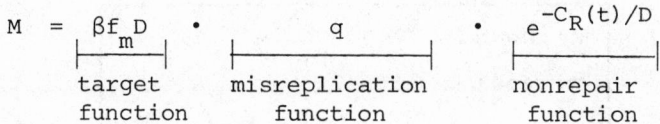

$$M = \underbrace{\beta f_m D}_{\substack{\text{target} \\ \text{function}}} \cdot \underbrace{q}_{\substack{\text{misreplication} \\ \text{function}}} \cdot \underbrace{e^{-C_R(t)/D}}_{\substack{\text{nonrepair} \\ \text{function}}}$$

   In that earlier section $C_R(t)$ was introduced as a capacity to
repair initial damage D in the time between the formation of an adduct and the time of DNA synthesis at the adduct site. We would
expect that the more time allowed to repair, the less damage remaining for misreplication and the fewer mutants to count. If we
could synchronize the growth of all of the cells, we could examine
the cell cycle dependence of mutation. We would expect that cells
having the most time to repair prior to gene replication damage
would have the lowest probability of mutation. Cells with little
time would have the highest induced mutant fraction. Figure 6 summarizes this expectation.

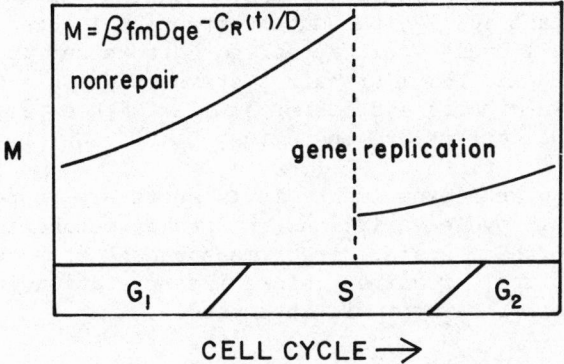

Figure 6. Example of expectation for cell cycle dependence
           of mutation when caused by nonrepair of DNA lesions
           in the absence of progression effects.

Figure 7 summarizes the expectation for the example used in Figure 2 in which mutation arose exclusively from misrepair of lesions. In this case the longer the time available for misrepair, the more frequent the mutations expected.

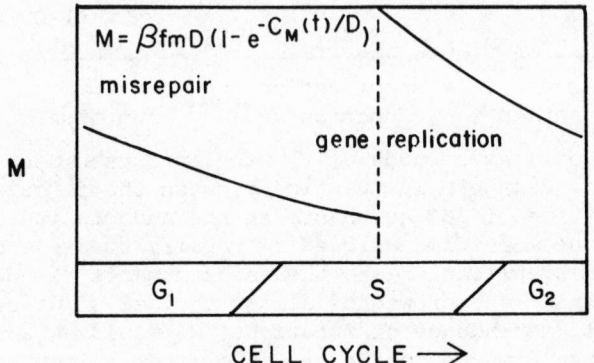

Figure 7. Example of expectation for cell cycle dependence of mutation when caused by misrepair of DNA lesions in the absence of progression effects.

Extension of the example of Figure 4 in which the cell contained similar capacities for repair and misrepair leads to the conclusion that expectation of observing significant cell cycle variation would be diminished, depending on the values of q, $C_R(t)$ and $C_M(t)$. Paradoxically, the mutant fraction induced at various times in the cell cycle would appear to be constant in cases in which neither repair nor misrepair systems exist $C_M = C_R = 0$, as well as when two or more systems of similar high capacity but opposite mutational effect are operative, $C_M(t) = C_R(t) \gg D$, $q = 1$.

In planning such experiments it is necessary to remember that there may be cell cycle variations in uptake, metabolism and target availability so that, again, direct measurement of adducts on the DNA, D, is absolutely required before interpretation in terms of repair or misrepair capacity is attempted.

Cell cycle variation may also occur in repair or misrepair rates. These could be measured by measuring the rate adduct removal, (dD/dt), as a function of cell cycle time.

   (ii) <u>Progression Delay</u>. More troublesome than cell cycle
repair or misrepair variations will be that cell cycle progression
is often effected by chemical treatment, and these progression effects vary with cell cycle position and dose. Thus, if a particular chemical lesion is repaired by a system with capacity $C_R(t)$ but
the chemical only slows progression of cells treated in the G_1 phase
of the cell cycle, the paradox of observing fewer mutants from G_1
treated cells than from G_2 treated cells could result. This phenomenon of progression delay may, in fact, be an important factor
in permitting a cell to carry out repair or misrepair in that either
would tend to permit DNA replication, and thus cell survival

   That a defect in progression delay induced by UV light may
account for the hypersensitivity of Xeroderma pigmentosum cells in
culture has been suggested by Cleaver and his colleagues.

   (iii) <u>Cell Cycle Variation in Sensitivity</u>. What would we observe if only one cell cycle stage of our treated population were
sensitive to both the cell killing and mutating effects of treatment?
Since mutant fraction is measured in terms of

$$\left[\frac{\text{colonies formed under selective conditions}}{\text{colonies formed under non-selective conditions}} \right]$$

some very bizarre observations could result.

   As an example, let us imagine a cell cycle dependence of cell
killing and mutation in which half the cells survived according to
the relation $S = e^{-D}$ and mutated by the relation $M = \beta f_m D$, and the
other half was neither killed nor mutated by treatment $S = 1, M = 0$.

What one would see for a survival curve would be

$$S = 0.5(1 + e^{-D}).$$

Mutation of only the sensitive cells in the culture would be

$$M = 0.5(0 + \beta f_m D).$$

However, what one observes in the presence of selective conditions
is not M, but M x S, since mutants counted must both be mutant and
have survived. That relation would be observed as

$$\frac{[\text{M x S}]_{\text{sensitive}} + [\text{M x S}]_{\text{insensitive}}}{\text{S sensitive} + \text{S insensitive}}$$

or:

$$\left[\frac{0.5(\beta f_m D)(e^{-D}) + 0}{0.5\, e^{-D} + 0.5} \right]$$

This observed "mutant fraction" as a function of D is shown in Figure 8 and presents the interesting possibility of observing a decreased mutant fraction as D increases. [This predicted behaviour is distinct from that discussed by Haynes for the number of mutants in a culture which increases to a maximum and decreases to zero as the negative exponential function of D (survival) (e^{-D}) overcomes the positive and essentially linear function of D (mutation) $(\beta f_m D)$.]

Behaviour similar to that depicted in Figure 8 has been observed by John DeLuca in our laboratory in his studies of UV mutation of human lymphoblasts. In this case he has suggested an S phase sensitivity to UV killing as an explanation.

The cell cycle presents many challenges to those trying to unravel mutation mechanisms. Considerable time has been invested by many laboratories in devising means to synchronize single cell populations. Considering the problems of interpreting observations from asynchronous populations, this effort certainly seems warranted.

With regard to our model involving $C_R(t)$ and/or $C_M(t)$, cell cycle and progression effects can be seen as effecting either the rate of adduct removal by cell cycle variation or the time available for adduct removal before gene replication, either by position in the cell cycle or induced progression delay.

$$C(t) = \int_0^t -(dD/dt)\, dt$$

t ————→ variable with cell cycle position and progression delay (time between damage & DNA synthesis)

↓ potential cell cycle variable

c. Multiple Kinds of DNA Adducts

Of the chemicals already examined with regard to forming DNA adducts, none has produced a unique reaction product. Aflatoxin B_1 produces more than 90% of its lesions by substitution at the N^7 position of guanine, but subsequent reactions produce substantial quanitites of depurinated positions and a stable ring opened product in addition to the remaining 10% of lesions which include at least one involving adenine.

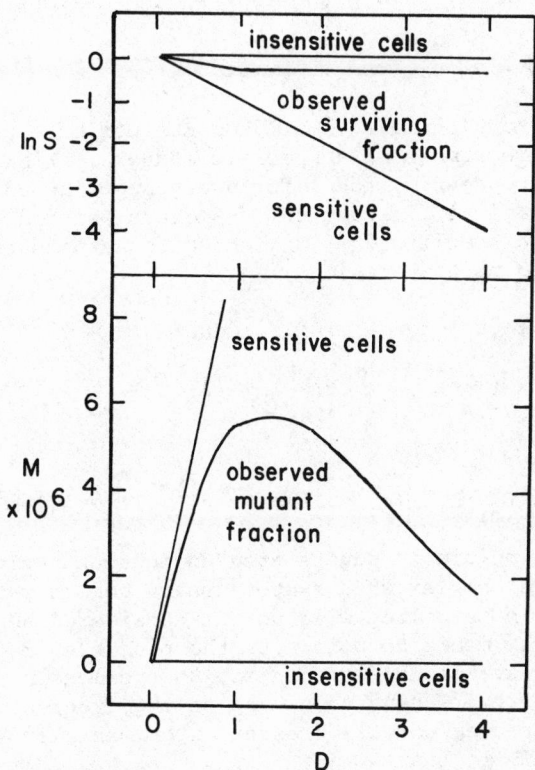

Figure 8.   Behavior of mixed population consisting of cells
            half-sensitive and half-insensitive to both the
            toxic and mutagenic effects of treatment. Examples
            of a possible problem arising from inhomogeneity in
            nonsynchronous cell cultures.

It would seem that the imaginary chemical with its single lesion which we have used so far for convenience may have been misleading. Given multiple lesions we might well expect different kinds of lesion removal systems, some independent of each other and others sharing common enzymes and substrates. Some of the lesions at low D may be subjected to misrepair, others to repair. How does one treat this complexity?

We can begin with a hypothesis that the lesions are treated independently.

$$D = \Sigma\ D_i + D_2 + D_3 + \ldots D_n$$

where D is the total number of all adducts per base pair in cellular DNA and D_i is the number of a specific adducts per base pair. If we treat them independently, then, for every specific adduct, there would be a specific f_{mi} (fraction of adducts which are chemically premutagenic), a specific C_{Ri}, C_{Mi} and q_i governing the way the cell can respond to that lesion.

The general case for a single adduct

$$M = \beta f_m D \left[(q\ e^{-(C_M + C_R)/D}) + (1 - e^{-C_M/D}) \right]$$

becomes

$$M = \beta\ \Sigma\ f_{mi}\ D_i \left[(q_i\ e^{-(C_{Mi} + C_{Ri})/D_i}) + (1 - e^{-C_{Mi}/D_i}) \right]$$

for the case of multiple adducts treated independently. Things become even more complex when systems share common rate-limiting steps, and this possibility will not be considered here. This discussion is merely meant to point out the possible problems involved. Methods are now available for identifying and measuring many adducts occurring at frequencies as low as the frequency of mutation, allowing this problem to be addressed experimentally.

d. Time Dependence of Mutation and Adduct Removal

Since we can measure time, t, specific adduct concentrations, D_i and mutation, M, can we combine our measurements in a way which will teach us something about mutation?

Adduct concentration can be measured as a function of time after treatment. If repair or misrepair systems operate on the particular adduct, it should no longer be found attached to the DNA. Unfortunately, nonrepair, in the sense of failing to restore a continuous DNA sequence, can also involve adduct removal, and this distinction should not be ignored.

For any particular adduct we may anticipate generally that it
will be removed in some determinable fashion. Some possibilities
are illustrated in Fig. 9. Rapid removal with a half-life of 10-20
minutes has been reported for 0^6 methyl guanine, biphasic removal
has been reported for thymine dimers and a number of adducts have
been reported to remain relatively constant for a period of a day
or more. In this time-dependence of specific adduct removal, we
hope to find a key to identify those lesions responsible for the
mutation caused by a chemical.

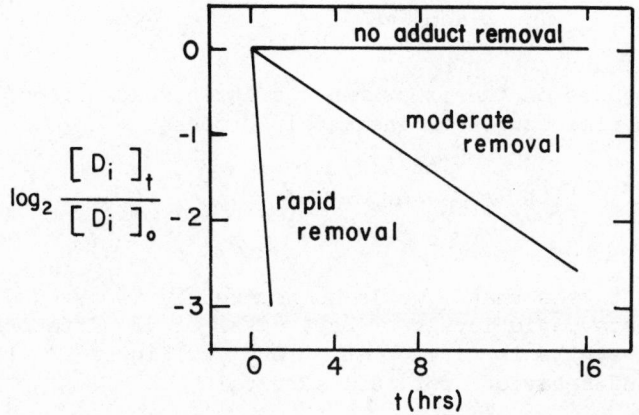

Figure 9. Examples of expectation for specific adduct removal
           from DNA as a function of time after treatment.

(i) Elkind Recovery. The key to interpreting these time dependent effects may be to use Elkind's concept of potentially lethal
damage transmuted into potentially mutagenic damages. Elkind used
split-dose experiments and measured cell survival.

The basic idea behind this approach is that a treatment which
nearly saturates the recovery system of a cell for lethal damage
will not reduce cell survival appreciably, but twice that treatment
would be very toxic. If two such identical treatments are separated
in time, the repair of the damage of the first treatment will occur,
and the second treatment will have little lethal effect because

the cell will be "ready" to repair the second round of damage as it did the first.

In explicit form, if for a single treatment

$$= e^{-\alpha D} \underbrace{}_{\substack{\text{lethal} \\ \text{target} \\ \text{term}}} \underbrace{e^{-C_R(t)/D}}_{\substack{\text{probability} \\ \text{of nonrepair} \\ \text{term}}}$$

then, two treatments separated by time Δt will produce the effect

$$S = e^{-\alpha 2De^{-C_R/2D}} \qquad\qquad \text{at } \Delta t = 0$$

and

$$S = (e^{-\alpha De^{-C_R/D}})^2 = c^{-\alpha 2De^{-C_R/D}} \qquad \text{at } \Delta t = \infty$$

The progress of the experiment as Δt increases from 0 to ∞ defines the time course of the repair process

$$C_R(t) = \int_0^{\Delta t} (dD/dt)\ dt$$

A limit on Δt less than ∞ would be imposed by removing damage to less than detectable levels or by arrival of an irreversible event such as DNA synthesis, i.e., $\Delta t = \Delta t_{max}$. Figure 10 illustrates this expected behaviour for cell survival.

(ii) <u>Application of Elkind Recovery to Mutation</u>. Expressing the expected results of split-dose experiments for mutation is not very difficult following Elkind's lead. Using the case of a single lesion acted upon by repair but not by misrepair ($C_M = 0$), we remember the expectation that

$$M = \beta f_m D\, q\, e^{-C_R(t)/D}$$

as shown in Figure 7.

If two equal treatments, D, are given but separated in time by Δt, we would expect

$$M = \beta f_m (2D)\, q\, e^{-C_R(t)/2D} \qquad \text{at } \Delta t = 0$$

and

$$M = \beta f_m (2D)\, q\, e^{-C_R(t)/D} \qquad \text{at } \Delta t = \infty.$$

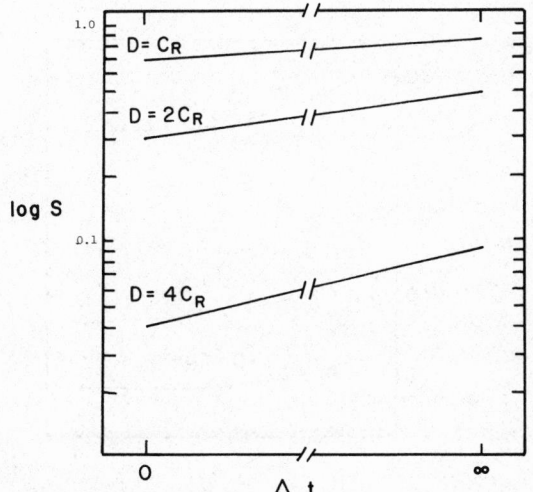

Figure 10.   Expectation for split-dose experiment in which
$$S = e^{-\alpha D e^{-C_R(t)/D}}$$

that is in which cell death arises solely as a
result of unrepaired lesions. The constant, α,
has been set equal to 1.0 for this example as has
C_R. The two treatments in combination produce the
values of D indicated.

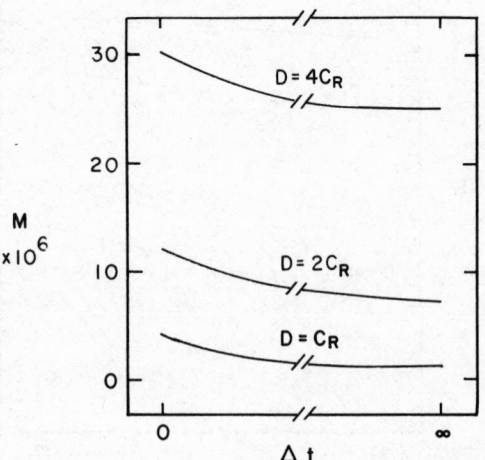

Figure 11.   Mutation as a function of time between two
             equal treatments with mutagen when mutation
             arises solely as a result of misreplication
             of unrepaired lesions. The product of the
             constants $\beta f_m q$ have been set at 10^{-5} for
             this example, and C_R has been set at 1.0.
             The two treatments in combination produce
             the values of D indicated.

(iii) <u>Combination of Adduct Removal Data with Split-Dose Removal Data</u>. Our underlying hypothesis is that some of the adducts formed in DNA by a chemical are premutagenic lesions and some are not. Chemical analysis indicates the rate and magnitude of removal of adduct as a function of time (Figure 9), while split-dose mutation experiments should show us the rate and magnitude of removal of pre-mutagenic (potentially mutagenic) lesions as a function of time (Figure 11). Assuming we have overcome problems of cell cycle variation and progression effects by synchronizing our cells, we should be able to differentiate among various DNA adducts in terms of whether or not their removal kinetics are similar to the removal kinetics of premutagenic lesions. For example, if a split-dose experiment yielded the results of Figure 12a, and three adducts were found in the DNA to be removed as shown in Figure 12b, which of the adducts would not be premutagenic?

Adduct A is not removed at all while premutagenic lesions are disappearing. It would seem that adduct A is not the precursor for these premutagenic lesions.

Adduct B is removed at a rate similar to the removal of pre-mutagenic lesions, and it would seem reasonable to postulate that it is a premutagenic precursor.

Adduct C is removed much more rapidly than the premutagenic lesions. However, adduct removal might be a rapid first step in repair, and the rate of removal of premutagenic lesions could be defined by a subsequent rate limiting step in repair.

Another level of analysis would be to ask if adducts B or C occur in sufficient quantity to account for the amount of mutation. If one did not, then the remaining lesion in this case could be identified as a strong candidate to be the primary premutagenic lesion for this chemical.

Similarly, application of this split-dose approach can be made with the case where misrepair of a particular adduct dominates the mutational response.

$$M = \beta f_m D (1 - e^{-C_M(t)/D})$$

If two equal treatments, D, are given, but separated in time by Δt, we would expect

$$M = \beta f_m (2D) (1 - e^{-C_M(t)/2D}) \qquad \text{at } \Delta t = 0$$

and

$$M = \beta f_m (2D) (1 - e^{-C_M(t)/D}) \qquad \text{at } \Delta t = \infty.$$

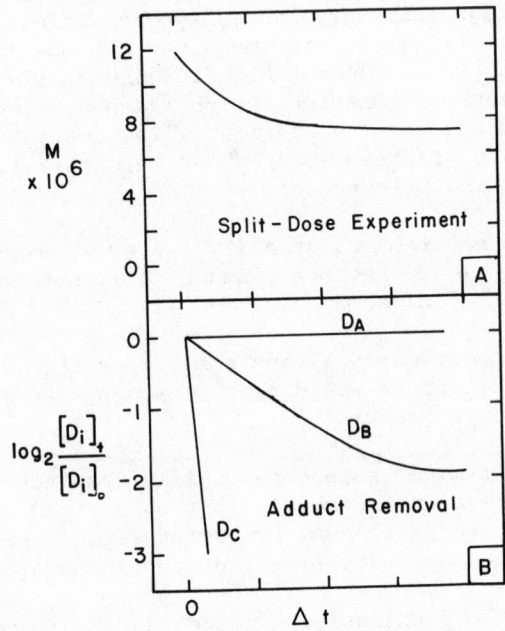

Figure 12a. Expectation of mutation as a function of time between
             two equal treatments of $D = 2C_R$, as in Figure 11.
Figure 12b. Example of expectation for removal of particular
             adducts A, B, and C, as a function of time after
             treatment.

This expectation is approximated in Figure 13. The same kinds of comparison used in Figures 12a and 12b could be made to "nominate" or eliminate hypothetical premutagenic lesions.

It should be noted that this kind of analysis is designed to find the predominant premutagenic lesion caused by a chemical among all of the adducts formed by that chemical. Extremely potent adducts (f_m = 1, q high, C_R low, C_M high, etc.) which occur in low frequencies relative to less potent but far more frequent adducts will not be detected by this approach. All of the weight of priority is placed on finding the predominant cause of the mutation observed with a particular chemical.

(iv) Molecular Proof of Premutagenic Status of an Adduct
There is a form of mutation analysis now developing in which one synthesizes specific adducts in known positions in the DNA, introduces this specifically adducted DNA into a cell, recovers the descendant DNA and determines by sequencing the mutagenic fate of that adduct in that particular cell. One can foresee a complementarity in this approach with the methods discussed above. The "nomination" of particular adducts in causing mutation by environmentally significant chemicals by the means discussed here could be followed by specific synthetic adduct studies in human cells to directly test the validity of the hypothetical designation of an adduct as a premutagenic lesion.

e. Inducible Responses

An inducible system can easily be included in our thinking by remembering that neither $C_R(t)$ or $C_M(t)$ are constants. We have already considered that the rate of specific adduct removal can very with cell cycle position so the idea of an increase of repair or misrepair capacity as a function of D is not too hard to incorporate.

Inducible repair

$$C_R(t) \text{ (noninducible)} \longrightarrow C_R(t,D) \text{ (inducible)}$$

Inducible misrepair

$$C_M(t) \text{ (noninducible)} \longrightarrow C_M(t,D) \text{ (inducible)}$$

A significant problem with interpreting data in inducible repair/nonrepair systems is that reports have noted only differences in the amount of mutation induced between two experimental conditions or between two mutants putatively different in one genetic locus in the pathway of DNA-damage metabolism. References are fre-

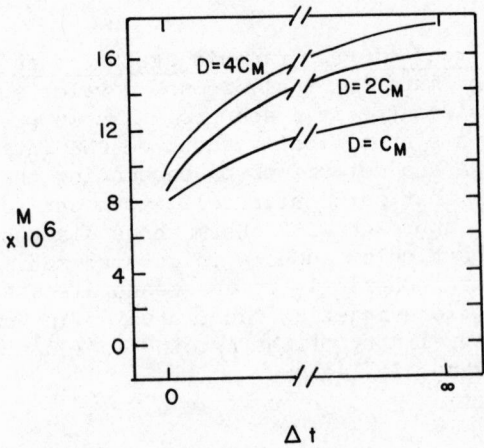

Figure 13.   Mutation as a function of time between two
             equal treatments with mutagen when mutation
             arises solely as a result of misrepair of
             lesions. The product of the constants βf_m
             have been set at 10^{-5} for this example,
             C_M at 1.0. The two treatments in combination
             produce the values of D indicated.

quently made to "error prone systems" which, in our parlance, would suggest appearance of a substantial contribution of $C_M(t,D)$ to mutagenesis. Of course, the same increase in observed mutation would be observed if a substantial decrease in a repair system occurred, $C_R(t,D) \longrightarrow 0$.

In Figure 14 we consider the kind of observation expected in an induced misrepair system. When D is less than a particular value (2), no induced response is seen. When values of D are greater than 2, the capacity of repair remains constant; but the capacity for misrepair is increased eight-fold.

This comparison (Figure 14) has used an extreme case in which the effect of the induction of misrepair is essentially instantaneous. Obviously, the time course of induction must be factored into our thinking. One easily conceived difficulty is confusion of the effect of an inducible misrepair system and the effects of inhibiting DNA synthesis which could give more time for the action of a constitutive misrepair system. Since either induction or progression effects of treatment could be functions of D itself, modeling these phenomena would require an additional term expressing $C_M(D,t)$ as an explicit function of time and DNA damage.

One may consider additional possibilities, many of which can be dealt with by combinations of $C_M(D,t)$ and $C_R(D,t)$ theoretically, but which would require combinations of time and concentration-dependent mutation studies to be dealt with experimentally.

f. Host Cell Reactivation

A number of researchers have reported experiments in which viruses are combined with host cells and mutagen treatment. These experiments involve

(1) treatment of virus/untreated cells, or

(2) untreated virus/treated cells, or

(3) treated virus/treated cells

in which the time between treatment of virus or host and infection is an experimental variable which permits study of the sequence of events. Observations in host cell reactivation experiments are generally of plaque formation on surviving virus although a number of good studies of mutation in the "rescued" viruses have appeared.

It seems that this field occasionally suffers from failure to separate variables, and one wonders if the general approach of combining a target function with a repair function as we have been doing for mutation in cells might not be helpful in thinking about these

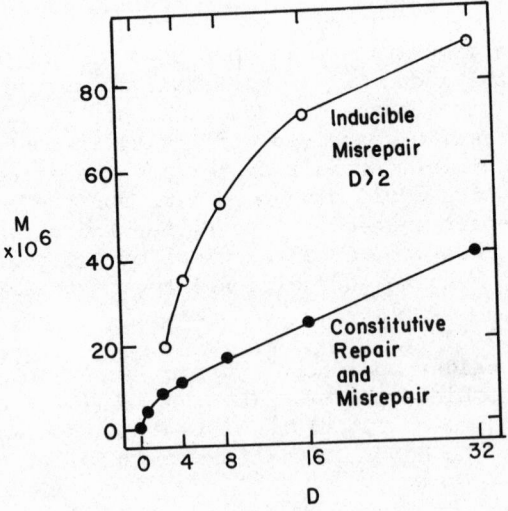

Figure 14.   Comparison of the case of constitutive repair

and misrepair

$(C_M = C_R = 1, q = 0.1, \beta f_m = 10^{-5})$ with a case
of constitutive repair $(C_R = 1, q = 0.1, \beta f_m = 10^{-5})$

and individual misrepair $(C_M = 1, D < 2; C_M = 8,$

$D > 2)$. Induction has been treated as if it were

instantaneous.

systems. Here, we will treat only the case of cell mediated virus survival although the analysis can easily be extended to mutation.

    <u>Definitions.</u>   F = fraction of virus forming plaques relative to untreated virus plated on untreated cells.

D_H = damage to host cell (adducts/base pair)

D_V = damage to virus

γ = target size for inactivating host ability to support viral growth

δ = target size for inactivating virus

fm_H = fraction of chemical adducts capable of forming lesion capable of inactivating host ability to support viral growth

fm_V = fraction of chemical adducts capable of inactivating virus

C_H = constitutive capacity of host to repair damage to its ability to support viral growth

C_V = constitutive capacity of host to repair damage to virus

For the simple case in which unrepaired damage is lethal, no inducible systems are involved and the virus does not participate in its own repair, one arrives at a suitably straightforward expression

$$F \;=\; e^{-\gamma D_H e^{-C_H/D_H}} \quad\quad x \quad\quad e^{-\delta D_V e^{-C_V/D_V}}$$

      host   virus
    target   target
         repair   host repair
       of host   of virus

This function, having separated the variables, permits greater facility in experimental design. C_H and C_V are, of course, time dependent functions and suggest means of probing the time dependence of self repair ($D_V = 0$), as opposed to repair of viruses ($D_H = 0$). Addition of an inducible function is also readily accommodated,

$$C_H(t) \;=\; C_H^{\,o}(t) \;+\; C_H(t,D)$$

where $C_H(t)$ consists of both constitutive $C_H^{\,o}(t)$ and an induced

repair ability $C_H(t,D)$. [The formulation presented here for host
cell reactivation was suggested to me by a student Robert A.Cuzick,Jr.
now deceased, who was trying to interpret a set of papers he had collected on host cell reactivation.]

Extension to mutation of viruses can be handled by treating
viral survival and survival and mutation as discussed for cell
mutation. All of the caveats about non-homogeneity of population,
cell cycle variation and combinations of repair and misrepair
systems apply.

g. Mutants in the Processes of Mutation

$$M = \beta f_m D \, [q \, e^{-(C_M + C_R)/D} + (1-e^{-C_M/D})]$$

Three terms in the general equation can be altered by mutation:
q, C_M and C_R, which represent the processes of DNA misreplication,
misrepair and repair, respectively.

For some mutagens, UV light, for instance, the general behaviour
of mutation in normal diploid human lymphoblasts suggests a dependence on nonrepair as a source of mutation of the form shown in
Figure 1. XP-A3 (complementation group C) lymphoblasts show a
simple, near-linear mutation response at lower UV fluences, which
is closely approximated by the behaviour shown in Figure 1 when
$C_R = 0$. In this case UV sensitivity to cell killing was accompanied
by an increased sensitivity to mutation.

In the search for mutants in mutation, Larry Thompson and his
colleagues have extended the observations that sensitivity to killing can be accompanied by hypermutability, as he has reported in
this meeting and elsewhere. This approach should be expected to
find mutants in the repair capacity of cells represented by a
smaller C_R in the mutants. We have reported that our lymphoblastoid
lines show survival and mutation responses to methylnitrosourea sug-
gestive of a nonrepair pathway to mutation and death. Now, we have
isolated mutants of decreased sensitivity to mutation and survival
at equivalent values of D (adducts/base pair). This may be a case
of a mutant with an enhanced repair capacity (Cuzick, R. and
Penman, B., unpublished results).

Interestingly, we have found that, at low values of D, aflatoxin B_1 mutation curves closely approximate the general form expected when misrepair was a predominant pathway, as shown in
Figure 2 (D. Kaden et al., unpublished results). Nitroquinoline-
N-oxide (NQO) also yields this form of response relative to

exogenous concentration, but we have not measured NQO adducts on
DNA. We have, however, isolated a clone apparently hypermutable by
NQO isogenic with our lymphoblastoid line TK6 (D. Kaden, unpublished
results). Such variants should be fairly easy to characterize by
the twin approaches of split-dose mutation experiments and measurement of rates of adduct removal.

Mutations effecting the misreplication term, q, may not be so
easy to find. Our search centers on looking for such "mutator"
mutants with the idea that mutants in normal DNA replication may
show heightened ability to misincorporate at damaged DNA sites.
One such "mutator" has now been found (Kaden, unpublished results)
which has an apparent ten-fold higher spontaneous mutation rate
than its isogenic parent line, TK6.

The possibilities are legion; and, despite the experimental
demonstration of dozens of independent loci effecting sensitivity
to single compounds in yeast, the study of "DNA repair" continues,
it seems to this writer, with a heavy burden of unstated or unclear-
ly stated assumptions.

In so writing, I come to the end of a discourse which has only
touched on this challenging field which, I believe, is blessed with
more powerful theoretical and experimental tools than are presently
in use. Knowledgeable persons will note that the models discussed
may not account for the presence of a mismatch repair system which
could follow behind misrepair events; this could also pertain to
misreplication happenings as long as systems can distinguish between
parent and daughter strands. Such a mismatch repair system would
be expected to be both time dependent, saturable and potentially
inducible. Students attempting to account for such behaviour in
explicit algebraic terms should be able to do so.

One telling criticism of this discussion is that my use of a
poisson expression for the probability of nonrepair

$$e^{-C_R/D}$$

and a similar expression for the probability of misrepair

$$(1-e^{-C_M/D})$$

may not be wholly justified. My reasons for doing so are not based
on any belief that such a statement is mechanistically correct, but
because it is an expression recognizable to most biologists, continuous, and thus easy to use, and expresses saturation clearly in terms
of the relative magnitudes of repair or misrepair capacities, C_M or
C_R and initial damage D.

Bruce Penman, Jan Drake and others have fairly objected to

this short cut. Penman is preparing an essay for publication in which he substitutes the parameters of an enzyme system operating on damage with first-order kinetics. It has the advantage of giving linear responses at low values of D for nonrepair mediated systems which certainly seems to fit observations as well as the poisson approximation.

Despite these shortcomings, I hope this discussion will help lay to rest the general tendency of experimentalists to interpret nonlinear increasing responses as "two hit" or higher order responses and to appreciate the potential of DNA damage to saturate the cell's systems for repair and misrepair.

A number of colleagues have spent considerable time reviewing and contributing to the way in which this discussion has developed. Bruce Penman and I worked closely in formulating the approach and restating the ideas developed by Mort Elkind in the 1950's. Dr. Elkind has been kind enough to critically review Bruce's recent Ph.D. thesis which contains several independent statements as noted above. Charles Heidelberger has also contributed helpful comments. Debra Kaden and John DeLuca have helped in deriving statements describing mixed repair/misrepair and heterogeneous population expectations, respectively. As noted, Bob Cuzick, now deceased, devised the means suggested for separating variables in host cell recovery experiments.

This work has been supported by the Department of Energy under Contract DE-AC02-77EV04267, and the National Institute of Environmental Health Sciences Grant NIH-2-P01-ES00597-11.

V. REFERENCES

DeLuca, J.G., Kaden, D.A., Krolewski, J.J., Skopek, T.R. and Thilly, W.G. (1977) Comparative mutagenicity of ICR-191 to S. typhimurium and diploid human lymphoblasts. Mutat. Res, 46: 11-18.

Elkind, M.M. and Sutton,H. (1959) X-ray camage and recovery in mammalian cells in culture. Nature, 184: 1293-1295.

Elkind, M.M. and Whitmore,G.F. (1967) Radiobiology of Cultured Mammalian Cells, Gordon and Breach, New York.

Elkind, M.M. (1977) The Initial part of the survival curve. Implications for low-dose, low-dose-rate radiation responses. Radiation Res. 71: 9-23.

Haynes, R.H. and Eckardt, F. (1979) Analysis of dose-response patterns in mutation research. Canad. J. Genetics and Cytology, 21: 277-302.

Kaufmann, W.K., Cleaver, J.E. and Painter, R.B.,(1980) Ultraviolet radiation inhibits replicon initiation in S phase human cells. Biochem. Biophys. Acta, 608: 191-195.

Park, S.D. and Cleaver, J.E. (1979) Recovery of DNA synthesis after ultraviolet irradiation of xeroderma pigmentosum cells depends on excision repair and is blocked by caffeine. Nucleic Acids Res., 6: 1151-1159.

Penman, B.W., Kaden, D.A., Liber, H.L., Skopek, T.R. and Thilly, W.G. (1980) Perylene is a more potent mutagen than benzo(α)pyrene for S. typhimurium. Mutation Res. 77: 271-277.

Thilly, W.G., DeLuca, J.G., Hoppe, H. IV and Penman, B.W. (1976) Mutation of human lymphoblasts by methylnitrosourea. Chem.-Biol. Interactions, 15: 33-50.

Thompson, L.H., Rubin, J.S., Cleaver, J.E., Whitmore, G.F. and Brookman, K. (1980) A screening method for isolating DNA repair-deficient mutants of CHO cells. Somatic Cell Genet. 6: 391-405.

Wogan, G.N., Croy, R.G., Essigman, J.D., Groopman, J.D., Thilly, W.G., Skopek, T.R. and Liber, H.L. (1979) Mechanisms of action of aflatoxin B_1 and sterigmatocystin: Relationships of macromolecular binding to carcinogenicity and mutagenicity. In Environmental Carcinogenesis, Ed. by P. Emmelot and E. Kriek. Elsevier/North Holland Biomedical Press, Amsterdam, pp. 97-121.

DISCUSSION

DR. SHAPIRO: There are these two basic classes of DNA damage; those for which repair leads to mutations and those for which non-repair leads to mutations. The kind of dose response curves you get are directly opposite for the two cases and yet environmental regulation will depend on what those response curves are. The bottom line I get to is how little progress we've made, in all the known lesions in DNA, in sorting our which is which -- which are the ones if repaired give mutations and which are the ones if left alone give mutations. In most cases we have no idea which adducts are which.

DR. THILLY: Yes, you're right. I think we must remember in both cases, and in the combination of cases, that the behavior at high concentrations at which all of us perform our experiments (in order that we shall have data to publish) do not reasonably replicate the conditions of exposure of people to nitroso pyrolidone in the bacon or other polycyclic aromatics other than benzopyrine which are present in substantial quantities in the air.

DR. GLICKMAN: I think the situation is actually much more complicated. The same lesion, depending upon where it is in the DNA, will give rise to mutation with different kinetics. I'm thinking of the secondary structure question, but I believe that the nearest neighbor is also important. Another kind of problem is that the same lesions may give rise to a frameshift, a deletion, or a base substitution. Whether you detect each event depends upon the assay. This is going to contribute to what one sees in a mutational response. I think that makes the situation considerably more complicated.

DR. THILLY: Actually, I think our algebra does address this issue. The first case, the nearest neighbor is considered in terms of what effect it would have on the two dummy variables, β and F_M. The case of the lesion producing different kinds of mutation is modeled by imagining that there are two different kinds of systems leading to those two different kinds of mutations.

DR. GLICKMAN: When one goes from the biology back to the model, it becomes difficult to discriminate.

DR. THILLY: Yes, please remember that this model is really just a first order attempt to infer what kind of responses you'd expect if a system behaves in the way we pretend these behave. I believe there are going to be a large number of gene products, dealing with each individual kind of adduct, and with such an expectation in front, it would be crazy to try and use a couple of variables to describe them. On the other hand, for the case of very low concentrations, I have tried to suggest expressions that deal with the predominant form of mutation. When I study a chemical from a toxicology point of view, I want to know what is the predominant mutation, not what is the kind of thing that happens with each individual kind of adduct. I set a priority based on the most probable cause of genetic damage.

DR. GLICKMAN: Do you care to make any comment on what pathways different agents are going through -- misrepair or misreplication, since your kinetics suggests differences in these cases?

DR. THILLY: Yes, I can for the few we've studied. For instance, aflatoxin behaves as if it were a misrepair mutagen at low concentrations, 5 nanograms/ml and less (which gives fewer than one adduct in 10^7 base pairs) down to 5% of that value. Aflatoxin is more complicated at higher concentrations, (20-30 nanograms/ml) despite an almost linear increase in DNA adduction. Above 50 nanograms/ml, mutation rises linearly to extreme mutation fractions. Gerry Wogan and I have estimated the maximum amount of aflatoxin to be expected in the liver of the unfortunate person eating aflatoxin-contaminated materials in the tropics. We came up with a number that was on the order of 5 nanograms/ml in terms of these experiments. In terms of what people are actually exposed to, I feel that aflatoxin is behaving as if it might be a misrepair mutagen. Ultraviolet light seems to be a non-repair type of mutagen, if the assumptions made here are true.

There are some systems where there is just a linear response. Now, a linear response, up to the levels that have been studied so far, can come from no repair, or a linear response can come from a complicated combination of repair systems. So a linear response is probably the most difficult to interpret, and the ones that are going to be easiest to interpret may be the non-linear responses. I guess that's my take-home lesson for mechanism studies.

DR. BRIDGES: Surely misrepair, which doesn't have a limited capacity, will give you linear response too?

DR. THILLY: Yes, if you don't go up to the saturating level.

DR. BRIDGES: With ionizing radiation and base pair substitution mutation in bacteria, that probably is the case.

DR. THILLY: I agree with Mary Lyon's interpretation of her mouse data that it is behaving as if there were a non-repair system, with non-linear mutation induction kinetics. That has been interpreted, of course, as the two-hit model or the n-hit model. I want to strongly challenge assumptions involving multiple hits. In order for a two-hit model to be a cause of targeted mutation, we must expect that there are frequencies of two lesions occurring

within a reasonable target size, equal or greater than the
observed mutation frequencies.

DR. MAHER: In the interpretation of your aflatoxin data, have you
used results from xeroderma cells, from excision repair deficient
cells, or does it all come from normally-repairing cells?

DR. THILLY: No, we have no data from excision deficient cells.
We have now, however, a little hint for the future. Debra Kaden
has isolated an NQO hypermutable cell from the normal TK6
lymphoblast line. Now, if it's also hypermutable by aflatoxin
there are two possibilities: a repair system is missing, or the
misrepair system is amplified. And we hope to discover that by
appropriate studies.

DR. MAHER: We have data from xeroderma cells using model
compounds of aflatoxin. I can't remember the nanogram doses, but
the amount bound we're talking about results in about one adduct
per 10^7 bases.

DR. THILLY: Remember, the misrepair part may possibly be only a
little part of what you see if there is a lot of repair along with
misrepair. If potential damage is being removed, and if all of a
sudden I get rid of the repair system, I could get mutation due to
non-repair and misrepair will be just whatever is added on to
that. A combination of possibilities should be considered.

DR. BRIDGES: I want to get back to your comment on the two-hit
model. Steve Sedgwick's calculations of the frequency of
overlapping daughter strand gaps that he did five or six years
ago, suggested that two pyrimidine dimers in the same thousand or
2,000 base-pair region, one on each strand, would satisfactorily
account for the mutation induction in E. coli. That's basically a
two-hit model and his sums, and I recall, worked out. It doesn't
prove that that's how it happens but the sums worked out.

DR. THILLY: It's a matter of algebra. First off, there may be
systems in which the degradation of the strand containing the
dimer is so extensive, for instance, you might take out 10,000
base pairs, where if you used reasonably high influences of UV you
might create such a condition.

DR. BRIDGES: You don't have to postulate that. If you've got one
dimer on each parental strand, then there's a probability that
when they are replicated the gaps they produce will overlap. The
gap is something like 1,000 bases long. You don't have to
postulate degradation; you just don't synthesize beyond the
dimer. And those sums work. Steve's idea was that you cannot get
recombination repair, which is error free, when daughter strand
gaps overlap so it has to go by the error-prone pathway. There is

no alternative if the cell is to survive. As I recall, the doses
in the <u>uvrA</u> WP2 strains that he was working with were 1-3 Jm^{-2}
at the most. His sums fitted, but your sums might not, though the
dimer yields should be the same.

DR. THILLY: Whenever one reaches the point where the frequency of
lesions is high enough to permit two to occur within the same
target, then of course, the objection to two-hit explanations
falls because the objection is based on the assumption that the
adduct frequencies are too low. When we deal with one adduct in
10^7 base pairs, the expectation of two adducts in a target of
10^3 base pairs would be [10^3 x 10^{-7}] = 10^{-8} or less than
the observed frequency of induced forward mutations.

DR. GLICKMAN: I think in the <u>E. coli</u> case, one has to be careful
because this is a <u>uvr</u> minus strain, as you just mentioned. The
<u>uvr</u> plus situation is not quite as straight forward in terms of
the dose-squared kinetics.

DR. BRIDGES: Bresler has postulated that you may have a two-hit
model in the excision-plus situation if you happen to have a
template dimer which is exposed by a long patch excision event.
That may have to be repaired in an error-prone way. There's no
alternative if the cell is to survive, and again there are more
than enough dimers to do that.

DR. MAHER: In your XP cells exposed to low doses, do you see a
linear relationship between mutation frequency and UV fluence as
we did?

DR. THILLY: Yes. We are unable to distinguish between a linear
response and what we observe.

DR. RIPLEY: One of the complications that hasn't been addressed
in your discussion is the chemical modification of things other
than DNA. That's the kind of thing that's addressed usually in
the phage experiments where you can treat the DNA separately from
the metabolism system which is going to work on it.

   In thinking about your cell-cycle arguments, I wondered whether
the kind of variables you set up would address the problem that
Leona mentioned earlier, about inducible systems, where the
kinetics of induction of a particular repair pathway, be it
accurate or inaccurate, might foul up your timing of when things
occur.

DR. THILLY: What I'm trying to do with this simple model is to
permit us to put these thoughts into the right place so we can
generate reasonable expectations and responses. For instance, if
I had an inducible repair system that was cell-cycle specific, I

could then think about how induced mutation would behave through the cell cycle. I could ask, if the capacity of a repair system could be tripled or increased by a factor of 10, what differences in my experiments would I observe. If by using the model I predict that a three-fold difference in repair would give only a 10% difference in the mutation through the cell cycle, I would realize that such an experiment would be fraught with difficulty in interpretation.

I do want to address your first point, that mutation may be caused by reaction with things other than a DNA. We are all very concerned with this possibility and it's addressed in a preamble of this paper. Sometimes we hear talk, for instance, of reaction with the polymerase, so that it becomes a mutator polymerase, albeit not genetically changed.

In such a case wouldn't the target be all those polymerases which could be reacted, and the altered polymerases would then be required to react with my particular gene to give rise to mutation at some higher probability than with unreacted polymerase. Even if I allowed my polymerase to be 100 times more "stupid" than a normal polymerase, I think that it just would have no measurable effect in the system; it is a variation on the two-hit hypothesis.

ADAPTABILITY OF MICROBIAL MUTAGENICITY ASSAYS TO THE STUDY OF

PROBLEM OF ENVIRONMENTAL CONCERN

Herbert S. Rosenkranz[1]*, George E. Karpinsky[1],
Monika Anders[1]*, George DeMarco[1], E. Joshua Rosenkranz[1]*,
Lynn A. Petrullo[1], Elena C. McCoy[1]*
and Robert Mermelstein[2]

[1]Department of Microbiology
New York Medical College
Valhalla, New York, 10595

and

[2]Joseph C. Wilson Center for Technology
Xerox Corporation
Rochester, N.Y. 14644
USA

INTRODUCTION

It is now well recognized that short-term microbial assays,
when run under carefully controlled conditions, have predictive
value for the detection of the potential cancer-causing properties
of most classes of chemicals. Indeed, our Laboratory has
participated both in the development and validation of such
assays. However, in an eagerness to obtain reliable data relating
to the potential genotoxic and/or carcinogenic potentials, too
much emphasis has been placed on procedural details such that
sight is lost of the fact that these assays lend themselves –
when suitably modified – to the elucidation of some basic
biological problems. There is an increasing number of these
which can be used to illustrate this point. The present report,
however, deals primarily with data obtained in our own Laboratories.

*Present address: Center for the Environmental Health Sciences,
Case Western Reserve University, Cleveland, Ohio 44106

ILLUSTRATIVE EXAMPLES

The Role of the Anaerobic Flora

Microbial assays, such as the Salmonella mutagenicity
assay developed by Ames and his associates (1), use microorganisms as test objects, hence the results obtained with such
systems reflect not only the nature of the test chemical, the
quantity as well as quality of the metabolic activation provided
by the mammalian microsomes which are usually included in the
assay, but also the unique features of the metabolic machinery
of the Salmonella and the configuration of its DNA. It must be
stressed in this respect that Salmonellae are not representative.
of all of the bacterial flora, nor for that matter of gram-
negative bacteria, nor even of the Enterobacteriaceae.

Recently much attention has been focused on the etiology
of colon cancer: genetic and diet are factors, chemicals, as well
as the bacterial flora or combinations of these have been
implicated. With respect to the bacterial flora of the colon,
there is compelling evidence to suggest that it plays a pivotal
role in colon cancer causation. Some of the evidence in support
of this comes from the following:

1.   Some chemicals with a potential of inducing colonic
cancers in rodents will not do so in germ-free animals (2-3).

2.   Members of the colonic flora can convert bile acids
and their derivatives to substances bearing a strong structural
similarity to carcinogenic polycyclic aromatic hydrocarbons
(4-10).

3.   Individuals on high-fat diets, a regimen implicated
in the development of colon cancer, will tend to be colonized
with a preponderance of anaerobic bacteria (11-17).

4.   A statistical significant relationship has been
established between patients with colon cancer and the recovery
from their stools of Streptoccus bovis (18).

In considering this colonic flora, it must be remembered
that not only is one dealing with massive amounts of these
microorganisms (it has been estimated that 40% of the weight of
human feces consists of bacteria), but that among these strict
anaerobes outnumber the usual aerobes and facultative anerobes
by a factor of 1,000:1. This anaerobic flora consists of 500
species; although several genera predominate. These anaerobic
bacteria possess a very versatile enzymic machinery capable of
converting both physiological metabolites (e.g. bile acids)

as well as xenobiotics (e.g. azo dyes) to a variety of products
which can be acted upon further by other bacteria as well as by
enzymes present in the intestinal mucosa. It must be stressed
that these anaerobic bacteria possess not only reductive enzymes
but oxidative ones as well, although they do not use molecular
oxygen as the electron acceptor.

Because cell-free extracts of strict anaerobes would be
expected to exert maximal enzyme activity only under oxygen-
free conditions, it became necessary to determine whether the
Salmonella mutagenicity assay was compatible with anaerobiosis.
In this we were fortunate that Salmonellae are facultative
anaerobes. It was thus found that a protocol of 14 hours of
anaerobic incubation in a Gas-Pak jar (BBL) {a readily obtainable
piece of equipment used in most clinical microbiology laboratories}
followed by 48 hours of aerobic incubation was optimal for the
demonstration of the mutagenicity of direct-acting mutagens
(Table I). It was further shown that some of the test chemicals,
primarily those containing nitro functions, exhibited enhanced
mutagenicity under these conditions (Table I) and some even
demonstrated mutagenicity only under these conditions. These
results presumably reflect two complementary effects: (a) the
activation of an anaerobic nitroreductase, a group of enzymes
which appears to be present in both prokaryotic and eukaryotic
cells, and (b) the stabilization of the hydroxylamine, the
presumed penultimate mutagen, which is oxygen-labile, being
reoxidized to the nitro function in the presence of oxygen.

These experiments, as well as other, established the adaptability of the Salmonella mutagenicity assay to anaerobic
conditions, a prerequisite for the use of cell-free extracts of
strict anaerobic bacteria to activate putative promutagens
which might be potential colon carcinogens. As representatives
of the human anaerobic flora, recent human isolates of
Clostridium perfringens, Bacteroides fragilis, Bacteroides
thetaiotaomicron and Bacteroides vulgatus were selected.

The presence of nitroreductases in these anaerobes could be
established as evidenced by their ability to activate nitro-
containing chemicals to mutagens (Table II). This was accomplished
by using a nitroreductase-deficient Salmonella tester strain
(TA1538NR) which does not respond to the mutagenic action of
nitrofurans, nitroimidazoles, nitronaphthalenes, nitrofluorenes
and some other nitrated polycyclic aromatic hydrocarbons (19-21).
Thus, while nitrofluorene exhibited full mutagenicity in TA1538,
its activity towards TA1538NR was abolished. However, exposure
of TA1538NR in the presence of cell-free extracts derived from
anerobes permitted expression of the mutagenic potential of
nitrofluorene (Table II). This demonstration of a nitroreductase

Table I. Effect of Anaerobic Incubation*
on Mutagenicity in Salmonella

| Chemical | Strain | Revertants per μg | |
		Aerobic	Anaerobic
Ethyl methanesulphonate	TA100	715	715
2-Bromoethanol	TA100	126	129
1,2-Epoxybutane	TA100	28	29
Propyleneimine	TA100	5000	5000
Azothioprine	TA100	0.1	1.2
6-Nitrosopurine	TA1535	< 0.06	0.5
2-Nitrofluorene	TA1538	140	244
2-Nitronaphthalene	TA1535	5	8

*Anaerobiosis was achieved by placing the petri plates into Gas
Pak jars (BBL, Cockeysville, Md.) which were incubated (37°C) in
the dark for 14 hr whereupon the plates were removed from the
jar and incubated aerobically for an additional 34 hr.

Table II. Activation of 2-Nitrofluorene to a Mutagen
          by Cell-Free Extracts from Anaerobic Bacteria

Additions	Revertants/μg	
	TA1538	TA1538NR
None	10	0.1
S9 (Hamster)	16	
B. fragilis	12	2.2
B. vulgatus	20	
B. thetaiotaoimicron	21	6.4

Table III. Activation of arylamines to Mutagens
           by Cell-Free Extracts from Anaerobic Bacteria

Chemical	Extract	Revertant/μg
2-Aminofluorene	None	0.3
	S9 (Hamster)	28
	B. fragilis	3
	B. vulgatus	3
	B. thetaiotaomicron	1
	Cl. perfringens	11
	Cl. perfringens – Heated	0
	Cl. perfringens – Aerobic	2
2-Aminoanthracene	None	0
	S9 (Rat)	74
	B. fragilis	8

Strain: TA1538

activity of anaerobic bacteria is of some import as it has been
shown recently that while the nitroreductases present in
Salmonella typhimurium possess a broad spectrum of specificities,
they are unable to activate some nitro-containing carcinogens
to mutagens, although the mutagenicity of these chemicals could
be demonstrating following their chemical reduction (22, also
see below). It would thus appear that the use of anaerobic
extracts to activate nitro-containing mutagens that might be
assimilated through the digestive tract might provide an
appropriate system for further elucidating the biotransformation
of xenobiotic and therapeutic nitro compounds.

     During the studies with nitrofluorene, it was also found
that 2-aminofluorene, which by itself is not a direct-acting
mutagen for Salmonella, {i.e. it requires activation by microsomes
in the standard Salmonella assay}: was activated by cell-free
extracts derived from anaerobic bacteria (Table III). This
phenomenon was not limited to this arylamine only but was seen
with 2-aminoanthracene as well, another promutagen. It remains
to be established that the mutagenic metabolites generated by
mammalian microsomes are indeed identical to those formed by
extracts derived from anaerobic bacteria. The mutagenic
metabolites that are formed by microsomes presumably are the
corresponding arylhydroxylamines and their esters. It is
conceivable that such oxidation products are also formed by
these extracts as anaerobes are known to undergo such oxidative
reactions (e.g. the Stickland reaction).

     In view of the multiplicity of metabolic capabilities of
anaerobic bacteria, it is not surprising that the activation of
other chemicals to mutagens has also been reported (e.g. azo
and diazo chemicals including known carcinogens) (23-26). It
has also been demonstrated that the anaerobic flora participates
in the biotransformation of bile acids including the formation
of promutagens (see below).

Cooperative Action of Anaerobes and Microsomal Enzymes in the
Activation of Promutagens

     Because of the close proximity of the anaerobic flora to
the intestinal mucosa and in view of the metabolic versatility
of anaerobes, it is likely that the multitude of reaction
products both of physiological and xenobiotic origin formed by
these anaerobes may not only be direct-acting mutagens and
carcinogens but that they may be transformed further by the
enzymes of the intestinal mucosa (e.g. microsomal oxygenases)
to ultimate carcinogens and mutagens.

It was felt that the availability of a modified mutagenicity assay using cell-free extracts derived from anaerobes would permit a determination of this possibility. Two situations were chosen to investigate the possibility of such a cooperative effect: (a) the biotransformation of arylamines and (b) the activation of bile acids to mutagens.

1. The Activation of Arylamines: The activation of arylamines to mutagens is believed to proceed through several steps, i.e. oxidation to hydroxylamines and esterification to form the ultimate DNA-reacting electrophile. Indeed, this reaction is mediated by the enzymes present in the post-mitochondrial fraction (cytosol and microsomes).

To accomplish the demonstration of a cooperative effect, cell-free extracts prepared from human strains of Bacteroides fragilis and microsomes derived from the intestinal mucosa of rats were used. When used singly, either aerobically or anaerobically, these preparation exhibited marginal levels of activity. This was, however, enhanced when mixtures of these preparations were used. Such a synergism was seen with both 2-aminofluorene and 2-aminoanthracene (Tables IV and V). That this synergism was not merely due to a the presence of co-factors in either enzymes preparations was demonstrated by a thermal denaturation experiments. Heating of either preparations resulted in a loss of enzyme activity of the mixture (Table V).

2. The Activation of Bile Acids: It has been suggested on a number of occasions both on epidemiological as well as experimental grounds that a correlation existed between colonic cancer and diets rich in animal meats. Moreover, one of the components of such a diet, the bile acids, have been implicated as promoters or as co-carcinogens (27-33) as well as metabolic precursors of polycyclic aromatic hydrocarbons, structurally resembling those that have been shown to be rodent carcinogens (4,6,7). It was felt that the procedures to investigate this further were available to us. Moreover, should a mutagenic metabolite indeed be detected, that a combination of separation procedures coupled to mutagenicity assays could result in the isolation and identification of the mutagenic chemical, in a manner similar to our previous identification of the nitropyrenes as trace contaminants in xerographic toners (34).

The first set of experiments dealt with whole human bile

Table IV.   Activations of Arylamines to Mutagens by tne
            Cooperative Action of Enzymes from Anaerobes
            and from the Intestinal Mucosa

Extract	Revertants per µg			
	2-Aminofluorene		2-Aminoanthracene	
	$-O_2$	$+O_2$	$-O_2$	$+O_2$
None	0.9	0.7	1.1	1.0
Intestinal Mucosa S9 (IM)	1.4	0	1.2	6.5
B. fragilis (B.f)	1.5	0.1	3.1	1.2
Im + B.f.	3.7	1.0	24	49

Strain: TA1538

Table V.   Effect of Heat Inactivation on the Conversion
           of 2-Aminoanthracene to a Mutagen

Enzyme	Revertants/µg
None	0.9
Rat Liver S9	48
B. fragilis Prep #2 (Bf#2)	10
Intestinal Mucosa S9 (IM)	1.2
IM + Bf#2	58
Heated IM + Bf#2	15
IM + Heated Bf#2	1.2

Strain: TA1538.
Heating: 80^0C; 5 min

obtained from non-cancer patients. Only specimens shown to be free of aerobic as well as anaerobic bacteria were used for these studies.

Initially, it was shown that such preparations were devoid of mutagenicity for the Salmonella under standard procedures both in the presence as well as in the absence of microsomal preparations derived from rat livers (Table VI). Nor did the inclusion of extracts of Bacteroides fragilis, the predominant anaerobic species in humans, result in the demonstration of mutagenic activity both in the presence as well as in the absence of microsomal preparations. To eliminate the possibility that cell-free extracts of Bacteroides fragilis were deficient in some co-factors, biles were added to growing bacteria and the culture filtrates assayed for mutagenicity both in the presence and absence of microsomal enzymes. No evidence of genetic or genotoxic activities was obtained by any of these procedures.

In view of the fact that bile acids may be considered surfactants and as such could denature proteins and inactivate enzymes, advantage was taken of the availability in the laboratory of bile-tolerant isolates of Bacteroides fragilis of human origin. Such strains are resistant to the lethal as well as inactivating activities of these chemicals. It could thus be demonstrated that "cycling" such biles through the bile-tolerant microorganisms resulted in the generation of a product endowed with intrinsic mutagenic activity (i.e. no requirement for S9; Table VI). Moreover, the addition of rat microsomes resulted in a further increase in mutagenicity (Table VI). Preparations derived from a number of patients yielded qualitatively similar results although the extent of intrinsic activity following exposure to B. fragilis alone varied, presumably reflecting the composition of the biles.

In order to determine further which component of human bile might be the promutagen, individual bile acids were investigated. It could thus been shown (Table VII) that deoxycholate, following "cycling" through bile-tolerant B. fragilis and assayed in the presence of rat microsomes exhibited mutagenic activity. Parallel cycling through other bile non-tolerant anaerobes did not result in the demonstration of a mutagenic product (Table VII). Similar results were obtained using lithocholate and moreover it could be demonstrated that intact cells were not required but that cell-free extract coupled to microsomes yielded similar results (Table VII).

Previous studies had led to the suggestion that bile acids might act as promotors or as co-carcinogens (27-33). The present

Table VI. Presence of a Mutagenic Component in Human Bile

Patient	Treatment	Revertants per µl
#1	None	0
	S9	0
	"Cycled" in B. fragilis*	41
	"Cycled" in B.f. then S9	178
#2	None	0
	S9	0
	"Cycled" in B. fragilis	0.3
	"Cycled" in B.f. then S9	9.0

Strain: TA1538
S9: Rat, Aroclor induced
*Cycled: 72 hrs. in a bile tolerant strain (G4841)

Table VII. Conversion of Deoxycholate to a Mutagen by Anaerobes

Additions		Revertants/µg
None		0
S9		0
B. fragilis	G4841	7.1
	PL3429*	0
	WH14322*	0

*Bile non-tolerant strains

preliminary findings, however, suggest that following the
cooperative action of the anaerobic flora and mammalian microsomes
on bile acids or on their metabolites mutagenic components are
generated. This is not entirely surprising in view of the
metabolic versatility of anaerobes and the demonstrated ability
of such microorganism to "aromatize" bile acids to polycyclic
aromatic hydrocarbons. However, in view of the established
relationship between mutagenicity in bacteria and carcinogenicity
in animals, the present results deserve further elucidation,
especially the identification of the genotoxic metabolite seems
warranted.

The Chemical Activation of "Non-mutagenic" Carcinogens to Mutagens

An inherit shortcoming of the microbial assay system resides
in the fact that although the tester microorganisms frequently
exhibit a broad spectrum of enzymic capabilities, occasionally
these do not mirror the situation that exists in the mammal.
Allusion to this has already been made with respect to the
nitro-containing chemicals. Thus, we have a number of extreme
situations:

1. Some chemicals (such as a nitroarenes) appear tc be
extremely potent bacterial mutagens while possessing little or
no activity towards eukaryotic cells growing in tissue culture.
This may be due to the fact that the species of origin either
lack the enzyme altogether (such as the yeast Saccharomyces
cerevisiae) or perhaps that the enzyme capability was lost upon
continued propagation in culture (e.g. murine and human?).

2. Other chemicals (such as 4-nitroquinoline-1-oxide, NQO)
appear to be strong mutagens for bacteria and yeast as well as
for cultured human cells. Presumably, the nitroreductase
recognizing NQO is of a different nature than the one that acts
upon nitroarenes and nitrofurans. In bacteria as well as in
yeast, evidence in support of this has already been obtained.

3. Finally, there are nitro-containing chemicals which,
although carcinogenic for rodents, are non-mutagenic for bacteria
and the possibility therefore exists that these chemicals (a)
are not activated by the bacteria to the ultimate mutagen,
(b) that they are activated to an ultimate mutagen which immediately
is metabolized further to the amine, possibly while still linked
to the enzyme, (c) that the chemical is indeed biotransformed to
the penultimate or ultimate mutagen but that because of conformational properties unique to the bacterial genome, reaction with DNA is
precluded (d) It is also conceivable that reaction with a DNA

Table VIII.   Activation of Lithocholate to a Mutagen

Additions	Revertants per µg
None	0
B. fragilis Extract	5.3
B. fragilis/S9 (Hamster)	8.5
Heated B. fragilis Extract	0

Lithocholate was incubated with an extract of B. fragilis G4841
(bile-tolerant) anaerobically in the presence of co-factors
(similar to S9 mix) for 18 hrs, whereupon portions were processed
for determination of mutagenicity for Salmonella TA1538 in the
presence and absence of S9 derived from Aroclor 1254-induced
hamsters. Incubation was anaerobic for 18 hrs followed by 2 days
of aerobic incubation.

Table IX.   Chemical Reduction of Nitro-Containing Chemicals

Chemical	Revertants per Nanomole	
	TA100	TA98
8-Nitroquinoline	0	0
8-Nitroquinoline + Zn/NH$_4$Cl	2.7	0.4
1,8-Dinitronaphthalene	0	0
1,8-Dinitronaphthalene + Zn/NH$_4$Cl	1.1	
1,3,6,8-Tetranitronaphthalene	2.3	1.1
1,3,6,8-Tetranitronaphthalene + Zn/NH$_4$Cl	10.5	7.5

does occur but that the product formed is "trivial", i.e. it does not induce a conformational change which leads to a mutagenic and/or genotoxic event. (e) Finally it is possible that the correct DNA adduct is formed but that the microorganism possesses unique and efficient error-free DNA repair systems which compensate for that genomic lesions.

One of the approaches for studying this lack of demonstrable bacterial mutagenicity is the inclusion of S9 (post-mitochondrial preparations) in the assay. Although, it has been demonstrated that such preparations are active in transforming niridazole, metronidazole, nitrofurans and nitrofluorene to active intermediates (19,35-37), other studies have shown that this may not be the predominant pathway as other nitro-containing chemicals are reduced all the way to the amines which, subsequently are oxidized by a different biosynthetic pathway (monooxygenases) to the penultimate hydroxylamines. This results in the demonstration of the mutagenicity of a final product which may not reflect the mutagenic potential of the initial chemical of interest. A further complication is the ability of such enzymes to oxidize the aromatic ring to mutagenic as well as non-mutagenic epoxides, the activity (or lack thereof) of which bears no direct relationship to the activity of the nitro-function that is under investigation. This appears to occur when nitropyrenes, a group of extremely mutagenic chemicals, are exposed to microsomes (20).

Another approach is to dissect the process further using non-enzymic means for activating the test chemical. Thus, if the bacteria are unable to reduce a specific chemical to the corresponding hydroxylamine and if a chemical reduction procedure were compatible with the microbial assay then this could be used to elucidate the lack of demonstrable genetic activity. Investigation revealed that the method of choice for reducing nitro groups to hydroxylamines (exposure to zinc dust and ammonium chloride) was entirely compatible with the Salmonella mutagenicity assay. Indeed, it could be shown that the non-mutagenic carcinogen 8-nitroquinoline was converted to a direct-acting chemical when Zn and NH$_4$Cl were included in the assay (Table IX). This procedure has now been extended to demonstrate or enhance the mutagenicity of dinitro- and of tetranitro- naphthalenes.

This, illustrates a further example of the flexibility of the bacterial assay procedures advantage of which can be taken to investigate and identify processes that might lead to an understanding of the basis of carcinogenicity.

Photodynamic Activation of Environmental Agents to Mutagens

The fact that visible light in the absence of added chromosomes

is mutagenic for bacteria has been described earlier (38).
This has also been extended to cultured mammalian cells and to
the demonstration of the photodynamically-induced oncogenic
transformation of cells in culture (for references see 39). In
bacteria, spectral studies have identified blue light (approximately
450 nm) as the wavelength of maximal mutagenic activity (38,39).
This presumably reflects the photodynamic activation of riboflavin,
a ubiquitous cellular component, which through the intermediacy
of singlet oxygen causes modification of the guanine residue
which is expressed as a base-substitution mutation (39).

    However, in addition to this intrinsic effect, the
photodynamic activation of chemicals of environmental concern to
mutagens has been reported (see for example ref. 40). The
chemical susceptible to this photo-induction include polycyclic
aromatic hydrocarbons usually found in mobile emissions
(e.g. diesels) (Table X). The presence of singlet oxygen,
another constituent of polluted air (see ref. 40), catalyzes the
formation of additional mutagenic principles (Table X).

    Obviously, these findings are of some import, as unlike the
short-lived singlet oxygen generated by illumination of
chromophores such as riboflavin, methylene blue, and hematopor-
phyrin, the photoactivated species formed by polycyclic aromatic
hydrocarbons and other environmental pollutants are long-lived
oxidation products (40). Their biological activity can usually
be distinguished from the singlet oxygen-mediated effects
because the latter induce mutations of the base-substitution
variety (38,39) while the former induced frameshift mutations
(40). Moreover, it should be stressed that these photoactivated
stable chemicals are direct-acting agents not requiring the
participation of microsomal systems to express their genetic
activity. Thus they might act directly at the site of contact
with the exposed individual. An added dimension that needs to
be explored further is the finding that some environmental
polycyclic aromatic hydrocarbons (e.g. phenanthrene) which
heretofore were considered neither mutagenic nor carcinogenic
can be activated by visible light and singlet oxygen to mutagens
(Table X).

    The direct-acting potential of these stable photoproducts
(i.e. they require no further metabolism by either bacterial or
mammalian enzymes) was confirmed by the demonstration that they
cause conformational changes when added to purified DNA as
evidenced by decreases in the temperature (Tm) of the helix-to-
coil transition profiles (Table XI).

    Obviously, these findings add to our perception of environmental
risks. These effects are probably not limited to polycyclic

Table X.   The Photodynamic Activation of Enviromental
           Chemicals to Mutagens

Chemical	Revertants/µg
7,12-Dimethylbenz(a)anthracene	15.4
Chrysene	0.14
3-Methylcholanthrene	5.4
Perylene	0.1
Phenanthrene + $^1O_2$	0.6
Chrysene + $^1O_2$	0.43
3-Methylcholanthrene + $^1O_2$	9.8

None of the test chemicals showed any activity in the absence of
light ($\sim$130 kJ/m^2).
Mutagenic activity is expressed as revertants per µg of starting
(unirradiated) chemical; the light source was a Dura-Test Vite
Light Singlet oxygen ($^1O_2$) was generated photodynamically by
illuminating in the presence of rose bengal.
Tester strain: TA98

Table XI.   Effect of Illumination in the Presence of Polycyclic
            Aromatic Hydrocarbons on the Configuration of DNA

Addition	µg/ml	kJ/m^2	Tm	ΔTm
None (Solvent)	0	130	70.2	0
Anthracene	50	130	63.7	-6.5
	100	130	57.1	-13.1
	100	0	71.7	+1.5
7,12-Dimethylbenz(a) anthracene	50	130	63.3	-6.9
	50	0	70.1	-0.1
3-Methylcholanthrene	50	130	57.6	-12.6
	100	130	53.1	-17.1
	50	0	68.8	-1.4
	100	0	69.8	-0.4

aromatic hydrocarbons but may be applicable to the photoproducts of widely used pesticides, some of which might also be penultimate or ultimate genotoxicants.

Studies on the Mechanism of Action of Nitroarenes

Originally, our interest in the properties of nitroarenes was awakened by the finding of potent mutagenicity in the extracts of certain xerographic copies and toners. This activity was associated with carbon black which is used as the colorant in the xerographic toners (34,41). A well-coordinated effort led to the discovery that the mutagenicity was due to nitropyrenes which were generated as contaminants in one of the steps of the carbon black manufacturing process (20,34). Recognition of this resulted in a modification of the carbon black process which greatly reduced the nitropyrene content of the final product (34).

Subsequently, nitropyrenes were recognized as almost ubiquitous by-products of mobile as well as stationary combustion processes (42-58) resulting from the facile nitration of polycyclic aromatic hydrocarbons (47,48,57). The presence of nitropyrenes in the environment raised some concern regarding the possible health implications. Accordingly, the nitropyrenes and related nitroarenes are now the subject of a number of intensive studies regarding their generation, distribution, and biological activity. In this section, we will be concerned mainly with studies in microorganisms, realizing fully that studies with eukaryotic cells and whole animals will be ultimately needed to provide information usable for risk assessement. Still, such preliminary microbial studies have proven to be most informative both by themselves and in designing experiments using eukaryotic cells.

The ultimate purpose of such studies is, of course, an understanding of the significance of the genotoxic activity of nitropyrenes; to determine whether they present a risk to higher forms of life and, if so, to remedy this situation. With respect to the latter, two approaches involving for example a modification of the diesel engine appear plausible: (a) changes in the design which will reduce the extent of formation of oxides of nitrogen and/or of acid, which are requirements for ultimate nitration of polycyclic aromatic hydrocarbons (59,60), (b) increasing the efficiency of nitration such that the polycyclic aromatic hydrocarbons are fully substituted. Mutagenicity studies have indicated that while mono- and di-nitroarenes are highly mutagenic, tetranitroarenes are not (Table XII) and (c) a modification of the combustion process which will reduce the yield of polycyclic aromatic hydrocarbons.

Some of the information acquired using nitropyrenes in

Table XII. Mutagenicity of Nitroarenes

Chemical	Revertants/nanomole
2-Nitronaphthalene	1.3
1,3,6,8-Tetranitronaphthalene	0.3
2-Nitrofluorene	14
2,7-Dinitrofluorenone	1459
2,4,5,7-Tetranitrofluorenone	860
1-Nitropyrene	453
1,8-Dinitropyrene	254,000
1,3,6,8-Tetranitropyrene	15,600

Strain: TA98, except TA100 for the nitrated naphthalenes.

Table XIII. Mutagenicity of Nitropyrenes

	Revertants/Nanomole		
	TA98	TA98NR	TA98/1,8-DNP$_6$
1-Nitropyrene	453	35	199
1,3-Dinitropyrene	144,760	24,750	2,750
1,6-Dinitropyrene	183,570	190,900	45,890
1,8-Dinitropyrene	254,000	264,160	5,840
1,3,6-Trinitropyrene	40,700	36,630	25,640
1,3,6,8-Tetranitropyrene	15,600	10,600	14,000

microbial systems is reviewed below:

1.   Nitropyrenes are direct-acting frameshift mutagens for Salmonella typhimurium (34,61).

2.   There is evidence for structural specificity, as the 1,8-dinitro-isomer is the most active chemical of the group.   As a matter of fact, 1,8-dinitropyrene is the most potent Salmonella mutagen reported thus far in the literature (34,61).

3.   The stereospecificity of the mutagenic response appears to be related to the substrate specificity of the bacterial nitroreductases.   Thus, originally rather puzzling results were obtained.   The mutagenicity of three of the six nitropyrenes studied was undiminished in Salmonella tester strains deficient in the nitroreductase which recognizes nitrofurans, nitroimidazoles, nitronaphthalenes and nitrofluorenes (Table XIII).   This originally led to the suggestion that perhaps these nitropyrenes were direct-acting mutagens, unlike the other nitro-containing chemicals which required reduction to the hydroxylamines. However, no evidence was obtained that these nitropyrenes modified purified DNA in vitro.

This led to the hypothesis that the nitropyrenes were indeed reduced to hydroxylamines but by an enzyme, other than the one acting on nitrofluorenes and other nitroarenes.   A search for such an enzyme ensued and indeed it was found that a derivative of strain TA98 resistant to 1,8-dinitropyrene (TA98/1,8 DNP$_6$) was also refractory to the mutagenic action to 1,8-dinitropyrene and related chemicals (Table XIII).   This strain retains its sensitivity to nitrofluorenes, niridazole, and others thus suggesting that different nitroreductases are involved.   Moreover, TA98/1,8 DNP$_6$, just as TA98 NR (a strain deficient in the classical nitroreductase) retains its sensitivity to 4-nitroquinoline-1-oxide, suggesting that this chemical is dependent upon still another nitroreductase for its mutagenicity. An examination of the data indicates that nitration only in the A-ring of the pyrene is recognized by the classical niroreductase while a nitro function in rings A and C depends upon the alternate nitroreductase for expression of mutagenicity.

The suggestion was made earlier that the alternate nitroreductase was different from the "classical" one which recognizes nitrofurans and related chemicals.   It could, however, also be argued that strain TA98NR and TA98/1,8 DNP$_6$ represent mutations of the same gene product, i.e. an amino acid substitution at the active site which results in changes in substrates specificity.   That this situation might exist is based upon the fact that TA98NR was originally selected as resistant to nitrofuran or to niridazole, two chemicals which are

Table XIV. Properties of Derivatives of Salmonella TA98
Deficient in Two Nitroreductases

Strain	Percent Residual Activity		
	1,8-DNP	Niridazole	NQO
TA98	100	100	100
TA98/NR	104	4	100
TA98/1,8-DNP$_6$	2	100	84
TA98/NR/1,8-DNP$_1$	1.3	0	100
TA98/NR/1,8-DNP$_2$	1.4	0	100
TA98/NR/1,8-DNP$_5$	3.9	0	100
TA98/1,8-DNP$_6$/NR$_1$	2	0	100
TA98/1,8-DNP$_6$/NR$_8$	2	0	100
TA98/1,8-DNP$_6$/NR$_{11}$	2	0	100
TA98/1,8-DNP$_6$/NR$_{12}$	2	0	100

Abbreviations: 1,8-DNP, 1,8-dinitropyrene; NQO, 4-nitroquinoline-1-oxide.
Strains designated TA98/NR/DNP were TA98/NR strains rendered resistant to 1,8-DNP;
similarly TA98/1,8-DNP$_6$/NR denotes TA98/1,8-DNP strains made resistant to
niridazole.
All strains retained plasmid pKM101 and the deep-rough character. Moreover they
all responded normally to non-nitrated mutagens.

also are endowed with base-substitution mutagenicity (62). The
nitroreductase deficient mutant might therefore have been generated
as the result of a base-substitution mutation which leads to a
gene product which is either inactive or of greatly reduced
activity.

In order to explore this possibility further, two kinds of
experiments were designed.

A. If the two mutant types (resistant to nitrofurans and
resistant to 1,8-dinitropyrene) represent mutations at the
active site of the same gene product, double mutant (i.e.
strains resistant to the mutagenicity of the nitrofurans as well
as of the nitropyrenes) should not be obtainable. In fact,
TA98NR could be made resistant to the mutagenicity of 1,8-
dinitropyrene and vise-versa TA98/1,8-DNP$_6$ was made resistant to
nitrofurans and niridazole. Both types, however, retained their
sensitivity to the mutagenicity of 4-nitroquinoline-1-oxide
(Table XIV). This, therefore is consistent with the hypothesis
of the existence of two different gene products (and possibly a
third one which is involved in the mutagenicity of 4-nitroquino-
line-1-oxide).

B. If it be argued that the two types of phenotypes (e.g.
resistance to the mutagenicity of nitrofurans and to the
mutagenicity of nitropyrenes) result from a mutation on a
single gene, then if one were to construct a mutant totally
deficient in one of these enzymes, even an inactive one,
it should also not be responsive to the mutagenicity of the
other group of chemicals. To test this we constructed a strain
of TA98 deficient in the classical nitroreductase by inducing
the enzyme deficiency through a frameshift mutation (62).   Such
a strain, due to premature change termination, should be totally
devoid of the gene product, even an inactive one or one of
greatly reduced activity. Such strains, which are insensitive
to the mutagenic action of nitrofurans, niridazole and related
chemicals, were still fully sensitive to the mutagenicity of
1,8-dinitropyrene which is a further indication that one is
indeed dealing with two separate gene products (Table XV).

Evidence for the Existence of Multiple Nitroreductases
in Eykaryotic Cells

The studies in Salmonella have indicated the probable
existence in bacteria of a multiplicity of enzymes with
nitroreductase activities. This has led to an examination for
the evidence in eukaryotic cells for a family of nitroreductases
differing in specificities. That different eukaryotic

Table XV. Properties of Salmonella Typhimurium TA98 Strains
Deficient in Classical Nitroreductase as a Result
of Frameshift Mutations

Strain	Percent Residual Activity		
	Niridazole	1,8-DNP	NQO
TA98	100	100	100
TA98NR	4.4	104	100
TA98NIR301	0.19	106	100
TA98NIR302	0.19	100	100
TA98NIR303	0.16	100	100
TA98NIR304	0.22	100	100

Abbreviations: 1,8-DNP, 1,8-dinitropyrene; NQO,
4-nitroquinoline-1-oxide.

Strain TA98NR is deficient in "classical" nitroreductase selected
as niridazole resistant. In strains TA98NIR301 - TA98NIR304 this
deficiency was introduced as the result of frameshift mutations.

Table XVI.   Effect of Microsomal Preparations
on the Mutagenicity of Nitropyrenes

Pyrene	Revertant per µg	
	-S9	+S9
1-Nitro	13,820	165
1,3-Dinitro	485,200	0
1,6-Dinitro	615,200	1230
1,8-Dinitro	631,200	0
1,3,6-Trinitro	77,600	310
1,3,6,8-Tetranitro	10,820	130

Strain TA98
The purity of the sample of 1-nitropyrene used was lower than
usual, $\sim$ 97.5%. S9 was derived from the livers of Aroclor-induced
rats.

Table XVII.   Chemical Reduction of 1,8-Dinitropyrene
to a Mutagen

Additions	Revertants per µg	
	TA98	TA98/1,8-DNP$_6$
1,8-Dinitropyrene	327,500	7,428
1,8-Dinitropyrene + Zn/NH$_4$Cl	450,800	205,600

Bacteria were incubated for 45 min. at 37^0C in the presence or absence o
zinc dust and NH$_4$Cl whereupon soft agar was added and the standard
Salmonella assay procedure was carried out. The presence of Zn and NH$_4$(
alone did not result in mutagenic activity.

nitroreductase do exist has been known for some time based upon oxygen-sensitivity, cellular localization and co-factor requirements. However, evidence regarding substrate specificity towards nitro-contraining chemicals had to be established. Studies in these and other laboratories had shown while the yeast Saccharomyces cerevisiae was completely resistant to the mutagenicity of the nitropyrenes, a group of powerful bacterial mutagens, they were exquisitely sensitive to the mutagenicity of nitrofurans (e.g. AF-2) and 4-nitroquinoline-1-oxide (20) and also upon prolonged incubation to the mutagenicity of nitrophenylenediamines (63). Similarly, while normal and xeroderma pigmentosium human cells are sensitive to 4-nitroquinoline-1-oxide they are not mutagenized by 1,8-dinitropyrene (64). Thus suggesting an analogy between bacterial and eukaryotic cells. This, in turn, set the stage for an exploration of this phenomenon in mammalian cells. Such an investigation is underway.

Metabolism of Nitropyrenes: The mutagenicity of nitropyrenes is greatly reduced when rodent S9 preparations are included in the assay mixture (Table XVI). This activity, which is present in the microsomal fraction, is dependent upon co-factors and is maximal in preparations derived from animals pretreated with Aroclor 1254. It remains to be established whether this inactivation reflects complete reduction of the nitropyrenes to the aminopyrene by-passing the hydroxylamino intermediate, possibly because the intermediate is enzyme-bound (see ref. 65), or oxidation of the aromatic moiety to a nonmutagenic product (66). It must be pointed out, however, that the first possibility appears unlikely. Thus, aminopyrene in the presence of S9 is mutagenic for Salmonella and hence some residual activity would have been expected. Current studies are designed to identified the nature of the nitropyrene metabolite generated by microsomal enzymes. The results with the nitroreductase-deficient strains suggest that the ultimate or penultimate mutagen are the hydroxylaminopyrenes.

In order to gain information on the nature of the ultimate mutagen, the adaptation of the mutagenicity assay incorporating chemical reduction with zinc dust and NH_4Cl was used. It was shown that the mutagenicity of 1,8-dinitropyrene for TA98 was increased in the presence of such a mixture, as would be expected from a penultimate intermediate (Table XVII). Indeed such a relationship has been demonstrated earlier with nitro- and hydroxylamino-naphthalenes (Table XVIII). However, more convincing was the demonstration of the potent mutagenicity of the reduction product of 1,8-dinitropyrene in strain TA98/1,8-DNP6 (Table XVII). The strain, normally is non-responsive to

Table XVIII. Comparative Mutagenicities of Nitro-
and Hydroxylamino-Naphthalenes

Naphthalene	Revertants per Nanomole			
	TA100	TA100NR	TA98	TA98NR
1-Nitro-	1.0	0	0.1	0
1-Hydroxylamino-	10.6	11.8	1.9	2.9
2-Nitro-	1.3	0.1	0.01	0.02
2-Hydroxylamino-	15.0	16.1		

Table XIX. Reduction of 1,8-Dinitropyrene to a DNA
to a DNA-Reacting Chemical

Additions	Tm (^{0}C)	ΔTm (^{0}C)
None (Solvent)	69.1	0
Zn Dust	69.1	0
NH$_4$Cl	69.1	0
ZN + NH$_4$Cl	69.1	0
1,8-DNP	69.1	0
1,8-DNP+Zn+NH$_4$Cl	67.2	-1.9
1,8-DNP+Zn+NH$_4$Cl (Anaerobic)	66.0	-3.1
Acetoxy-2-fluorenylacetamide	65.3	-3.8

Abbreviations: 1,8-Dinitropyrene, 1,8-DNP; SSC/10, 0.015M NaCl,
0.0015M sodium citrate in, 0.01M phosphate buffer, pH 7.0; AFA,
acetoxy-2-fluorenylacetamide.
    Final concentrations in the reaction mixture μg per ml;
DNA 500; 1,8-DNP 4; Zn 8; NH$_4$Cl 4; AFA 100.
Portion of the reaction mixtures were diluted with SSC/10 and
the thermal helix-to-coil transition profiles were determined.
Tm is the midpoint of the transition.

the mutagenicity of 1,8-dinitropyrene. A similar effect had
also been demonstrated earlier (55,67) with nitro- and
hydroxylamino-naphthalenes using strains deficient in the classical
nitroreductases (Table XVIII).

These experiments support the notion that hydroxylaminopyrenes
are indeed the penultimate or ultimate mutagens. If they are
the penultimate mutagens then presumably the hydroxamic acid
ester derivatives are the ultimate ones. Evidence in support of
the possibility that the hydroxylamines are ultimate genotoxicants
was the demonstration that a hydroxylamine (the reduced
1,8-dinitropyrene) was capable of modifying the configuration of
purified DNA as demonstrated (Table XIX) by decreases in the
temperature of the helix-to-coil transition profile (the $\underline{Tm}$).
The effect was maximal when reduction was carried out
anaerobically, which is consistent with the intermediacy of the
postulated oxygen-labile hydroxylamine intermediate. It should
be noted that the extent of the $\underline{Tm}$ modification induced by
1,8-dinitropyrene was of the same order of magnitude as that
caused by a 25 fold higher concentration of the direct-acting
electrophile ester of N-acetoxy-N2-fluorenylacetamide.

With respect to the nature of the chemical modification
induced by the hydroxylamine, by analogy with hydroxylamino-
naphthalenes it should be noted that esterification of the
hydroxylamino moiety is not required for DNA modification and
that the sites of adduct formation include the 0^6, C-8 or
exocyclic amino group of guanine or the exocyclic amino group
of adenine (68-70).

CONCLUSIONS

The purpose of this presentation was to illustrate that in
addition to being useful, under standard conditions, for the
detection of potential carcinogens and genotoxicants, microbial
mutagenicity assays can be used as powerful research tools to
investigate basic mechanisms related to genotoxicity and
carcinogenicity. In particular they are useful to create
experimental conditions which mimic particular ecologic niches.
Although, the results described herein reflect mainly studies
carried out in these laboratories, it must be stressed that they
represent our own interests. An understanding of the physiology
and metabolism of the indicator microorganisms should permit an
adaptation to other situations. Indeed this is being explored
in a number of laboratories.

REFERENCES

1. B.N. Ames, J. McCann, and E. Yamasaki, Methods for detecting
    carcinogens and mutagens with the Salmonella/microsome
    mutagenicity test. Mutation Res. 31:347 (1975).

2. B.S. Reddy, J.H. Weisburger, T. Narisawa, and E.L. Wynder
    Colon carcinogenesis in germ-free rats with 1,2-Dimethyl-
    hydrazine and N-Methyl-N'-nitro-N-nitrosoguanidine,
    Cancer Res. 34:2368, (1974).

3. B.S. Reddy, T. Narisawa, P. Wright, D. Vukushich, J.H.
    Weisburger, and E.L. Wynder, Colon carcinogenesis with
    azoxymethane and dimethylhydrazine in germ-free rats,
    Cancer Res. 35:287 (1975).

4. M.J. Hill, Bacteria and the etiology of colon cancer,
    Cancer 34:815 (1974).

5. M.J. Hill, Metabolic epidemiology of dietary factors in
    large bowel cancer, Cancer Res. 35:3398 (1975).

6. M.J. Hill, Role of colon anaerobes in the metabolism of
    bile acids and steroids, and its relation to colon
    cancer, Cancer 36:2387 (1975).

7. M.H. Hill, The role of unsaturated bile acids in the
    etiology of large bowel cancer, in: "Origins of
    human cancer, Book C," H.H. Hiatt, J.D. Watson,
    and J.A. Winstein, eds., Cold Spring Harbor Laboratory,
    (1977).

8. W.E.C. Moore and L.V. Holdeman, Discussion of current
    bacteriological investigations of the relationship
    between intestinal flora, diet and colon cancer,
    Cancer Res. 35:3418 (1975).

9. S.M. Finegold, V.L. Sutter, P.T. Sugihara, H.A. Elder,
    S.M. Lehmann, and R.L. Phillips, Fecal microbial
    flora in Seventh-day Adventist population and control
    subjects, Am. J. Clin. Nutr, 30:1781 (1977).

10. J. Slemrova and R. Edenharden, Die Bedeutung des bakteriellen
    Steroidabbau für die Atiologie des Dickdarmkrebses.
    VII. Zur Methodik der Identifizierung von Gallensäure-
    abbauprodukten, Zentralbl. Bakteriol. (Orig. B) 164:236
    (1977).

11. B. Goldin and S.L. Gorbach, Alteration in fecal microflora
    enzymes related to diet, age, Lactobacillus supplements
    and dimethylhydrazine, Cancer 40:2421.

12. E.L. Wynder, Nutrition and cancer, Fed. Proc. 36:1309 (1976).

13. B.S. Reddy, T. Narisawa, R. Maronpot, J.H. Weisburger and
    E.L. Wynder, Animal models for the study of dietary
    factors and cancer of the large bowel, Cancer Res. 35:
    3421 (1975).

14. B.S. Reddy, A. Mastromarino and E.L. Wynder, Further leads
    on metabolic epidemiology of large bowel cancer,
    Cancer Res. 35:3403 (1975).

15. B.S. Reddy, J.H. Weisburger and E.L. Wynder, Effect of high
    and low risk diets for colon carcinogenesis on fecal
    microflora and steroids in man, J. Nutr. 105:878 (1975).

16. B.S. Reddy, S. Mangat, A. Sheinfil, J.H. Weisburger and
    E.L. Wynder, Effect of type and amount of diety fat
    and 1,2-dimethylhydrazine on biliary bile acids, fecal
    bile acids, and neutral sterols in rats, Cancer Res.
    37:2132 (1977).

17. A.B. Lowenfels and M.E. Anderson, Diet and cancer, Cancer
    39:1809 (1977).

18. R.S. Klein, M.T. Catalano, S.C. Edberg, J.I. Casey, and
    N.H. Steigbigel, Streptococcus bovis septicemia and
    carcinoma of the colon, Annals Intern. Med. 91:560
    (1979).

19. H.S. Rosenkranz and W.T. Speck, Mutagenicity of metronidazole:
    activation by mammalian liver microsomes, Biochem.
    Biophys. Res. Commun. 66:520 (1975).

20. R. Mermelstein, E. McCoy and H.S. Rosenkranz, The microbial
    mutagenicity of nitroarenes, in: "The genotoxic effects
    of airborne agents," Brookhaven National Laboratory
    Symposium, in press.

21. H.S. Rosenkranz and R. Mermelstein, The Salmonella
    mutagenicity and the E. coli Pol A$^+$/Pol A$_1$- Repair
    assays: Evaluation of relevance to carcinogenesis, in:
    "The Predictive Value of Short-Term Screening Tests
    in the Evaluation of Carcinogenicity," G.M. Williams,
    R. Kores, H.W. Waaijers, and K.W. van de Poll, eds.,
    Elsevier/North Holland (1980).

22. G.E. Karpinsky, E.C. McCoy, H.S. Rosenkranz and R. Mermelstein,
     The chemical activation of non-mutagenic nitrated poly-
     cyclic aromatic hydrocarbons to mutagens, Mut. Res.
     (1982).

23. R.B. Haveland-Smith and R.D. Combes, Genotoxicity of the
     food colours Red 2G and Brown FK in bacterial systems;
     use of structurally-related dyes and azo-reduction,
     Fd. Cosmet. Toxicol. 18:223 (1980).

24. J.P. Brown and P.S. Dietrich, Mutagenicity of anthraquinone
     and benzanthrone derivatives in the Salmonella/microsome
     test: Activation of anthraquinone glycosides by enzymic
     extracts of rat cecal bacteria, Mutation Res. 66:9
     (1979).

25. J.P. Brown, A review of the genetic effects of naturally
     occurring flavonoids, anthraquinones and related
     compounds, Mutation Res. 75:243 (1980).

26. C.P. Hartman, G.E. Fulk and A.W. Andrews, Azo reduction of
     trypan blue to a known carcinogen by a cell-free
     extract of a human intestinal anaerobe, Mutation Res.
     58:125 (1978).

27. T. Narisawa, N.E. Magadia, J.H. Weisburger, and E.L. Wynder,
     Promoting effect of bile acids on colon carcinogenesis
     after intrarectal instillation of N-methyl-N'-nitro-N-
     nitrosoquanidine in rats. J. Natl. Cancer Inst. 55:
     1093 (1974).

28. B.S. Reddy, Role of bile metabolites in colon carcinogenesis.
     Cancer 36:2401 (1975).

29. B.S. Reddy, T. Narasawa, J.H. Weisburger, and E.L. Wynder,
     Promoting effect of sodium deoxycholate on colon
     adenocarcinoma in germfree rats, J. Natl. Cancer
     Inst. 56:441 (1976).

30. A.B. Lowenfels, Does bile promote extra-colonic cancer?
     Lancet 7:239 (1978).

31. B.S. Reddy, K. Watanabe, J.H. Weisburger and E.L. Wynder,
     Promoting effect of bile acids in colon carcinogenesis
     in germ-free and conventional F33 rats, Cancer Res. 37:
     3238 (1977).

32. S.J. Silverman and A.W. Andrews, Bile acids, co-mutagenic
     activity in the Salmonella-mammalian microsome
     mutagenicity test, J. Natl. Cancer Inst. 59:1557 (1977).

33. B.S. Reddy and K. Watanabe, Effect of cholesterol metabolites
     and promoting effect of lithocholic acid in colon
     carcinogenesis in germ-free and conventional F344
     rats, Cancer Res. 39:1521 (1979).

34. H.S. Rosenkranz, E.C. McCoy, D.R. Sanders, M. Butler, D.K.
     Kiriazides and R. Mermelstein, Nitropyrenes: isolation,
     identification and reduction of mutagenic impurities
     in a carbon black and toners, Science 209:1039 (1980).

35. H.S. Rosenkranz and W.T. Speck, Activation of nitrofurantoin
     to a mutagen by rat liver nitroreductase, Biochem.
     Pharmacol. 25:1555 (1976).

36. H.S. Rosenkranz and L.A. Poirier, An evaluation of the
     mutagenicity and DNA modifying activity in microbial
     systems of carcinogens and non-carcinogens, J. Natl.
     Cancer Inst. 62:873 (1979).

37. J.L. Blumer, A. Friedman, L.W. Meyer, E. Fairchild, L.T.
     Webster, Jr. and W.T. Speck, Relative importance of
     bacterial and mammalian nitroreductases for niridazole
     mutagenesis, Cancer Res. in press (1981).

38. W.T. Speck and H.S. Rosenkranz, Base substitution mutations
     induced in Salmonella strains by visible light (450 nm),
     Photochem. Photobiol. 21:369 (1975).

39. W.T. Speck and H.S. Rosenkranz, Phototherapy for neonatal
     hyperbilirubinemia - A potential environmental health
     hazard to newborn infants: A review, Environm.
     Mutagenesis 1:321 (1979).

40. E.C. McCoy, J. Hyman and H.S. Rosenkranz, Conversion of
     environmental pollutants to mutagens by visible light,
     Biochem. Biophys. Res. Comm. 89:729 (1979).

41. G. Löfroth, E. Hefner, I. Alfheim and M. Moller, Mutagenic
     activity in photocopies, Science 209:1037 (1980).

42. T.C. Pederson and J.-S. Siak, The role of nitroaromatic
     compounds in the direct-acting mutagenicity of Diesel
     particle extracts, J. Appl. Toxicol. in press.

43. D. Schuetzle, F.S.-C. Lee, T.J. Prater and S.B. Tejada, The
    identification of polynuclear aromatic hydrocarbon
    derivatives in mutagenic fractions of Diesel particulate
    extracts, Intern. J. Environm. Analyt. Chem. 9:1 (1981).

44. G. Löfroth, Comparison of the mutagenic activity from Diesel
    and gasoline powered motor vehicles to carbon particulate
    matter, in: "Second symposium on application of complex
    environmental mixtures," Plenum Press, in press (1981).

45. T.C. Pederson and J. Siak, Characterization of direct-acting
    mutagens in Diesel exhaust particulates by thin-layer
    chromatography, Ann. Meeting Amer. Chem. Soc. Abstracts,
    Div. Environm. Chem. (1980).

46. D. Schuetzle, T.J. Prater, T. Riley, A. Durisin and I.
    Salmeen, Analysis of nitrated derivatives of PAH and
    determination of their contribution to Ames assay
    mutagenicity for Diesel particulate extracts, Fifth
    Intern. Symp. Polynuclear Aromatic Hydrocarbons,
    Columbus, Ohio Abstracts.

47. J.N. Pitts, Jr., K.A. Van Cauwenberghe, D. Grosjean, J.P.
    Schmid, D.R. Fitz, W.L. Belser, Jr., G.B. Knudson and
    P.M. Hynds, Atmospheric reactions of polycyclic
    aromatic hydrocarbons: Facile formation of mutagenic
    nitro derivatives, Science 202:515 (1978).

48. J.N. Pitts, Jr., Photochemical and biological implications
    of the atmospheric reactions of amines and benzo(a)
    pyrene, Phil. Trans. Royal Soc. London A, 290:551
    (1979).

49. J. Jäger, Detection and characterization of nitro derivatives
    of some polycyclic aromatic hydrocarbons by fluorescence
    quenching after thin-layer chromatography: Application
    to air pollution analysis, J. Chromatog. 152:575 (1978).

50. Y.Y. Wang, S.M. Rappaport, R.F. Sawyer, R.E. Talcott and
    E.T. Wei, Direct-acting mutagens in automobile exhaust,
    Cancer Letters 5:39 (1979).

51. H. Tokiwa, R. Nakagawa, K. Morita and Kamachi, Analysis of
    mutagenic nitro compounds in environmental samples
    (in Japanese), Environmental Mutagen Soc. of Japan,
    Abstracts, p. 18 (1979).

52. C.M. King, C.Y. Wang and P.O. Warner, Evidence for the presence of nitro aromatics in airborne particulates, Proc. Amer. Assoc. Cancer. Res. p. 83 (1980).

53. H.E. Kubitscheck, and D.M. Williams, Mutagenicity of fly ash from fluidized - bed combuster during start-up and steady state operating conditions, Mutation Res. 77:287 (1980).

54. E.T. Wei, Y.Y. Wang and S.M. Rappaport, Diesel emissions and the Ames test: A commentary, J. Air Poll. Control. Assoc. 30:267 (1980).

55. C.Y. Wang, M.S. Lee, C.M. King and P.O. Warner, Evidence for nitroaromatics as directing-acting mutagens of airborne-particulates, Chemosphere 9:83 (1980).

56. National Academy of Sciences, U.S., Health Effects of Exposure to Diesel Exhaust. The Report of the Health Effects Panel of the Diesel Impact Study Committee, National Research Council - National Academy of Sciences, Washington, D.C.

57. H. Tokiwa, R. Nakagawa, K. Morita, and Y. Ohnishi, Mutagenicity of nitro-derivatives induced by exposure of aromatic compounds to nitrogen dioxide, Mutation Res. in press.

58. J.N. Pitts, Jr., A.M. Winer, D.M. Lokensgard, S.D. Shaffer, E.C. Tuazon and G.W. Harris, Interactions between diesel emissions and gaseous co-pollutants in photochemical air pollution: Some health implications, Proc. EPA Intern. Symp. "Health Effects of Diesel Engine Emissions", in press.

59. Anonymous, New burner reduces nitrogen oxide emissions, Chem. Eng. News. March 30, p. 19 (1981).

60. R.L. Bradow, Diesel particle emissions, Bull. N.Y. Acad. Med. 56:797 (1980).

61. R. Mermelstein, D.K. Kiriazides, M. Butler, E.C. McCoy, and H.S. Rosenkranz, The extraordinary mutagenicity of nitropyrenes in bacteria, Mutation Res. in press.

62. W.T. Speck, S.L. Blumer, E.J. Rosenkranz, and H.S. Rosenkranz, Nirizadole nitroreductase deficiency in bacteria: An effect of genotype on mutagenicity of niridazole, Cancer Res. in press (1981).

63. V.W. Mayer, and C.J. Goin, Induction of mitotic recombination
     by certain hair-dye chemicals in Saccharomyces
     cerevisiae, Mutation Res. 78:243 (1980).

64. C.F. Arlett, J. Cole, B.C. Broughton, J. Lowe, and
     B.A. Bridges, Mutagenic effects in human and mouse cells
     by a nitropyrene in The Genotoxic Effects of Airborne
     Agents, Brookhaven Laboratory Symposium in press (1981).

65. E.C. McCoy, L.A. Petrullo, H.S. Rosenkranz, and R. Mermelstein
     4-Nitroquinoline-1-oxide: Factors determining its
     mutagenicity in bacteria, Mutation Res. in press (1981).

66. P. Sims, Qualitative and quantitative studies on the metabolism
     of a series of aromatic hydrocarbons by rat-liver
     preparations, Biochem. Pharmacol. 19:795 (1970).

67. E.C. McCoy, E.J. Rosenkranz, L.A. Petrullo, H.S. Rosenkranz
     and R. Merrmelstein, Structural basis of the mutagenicity
     in bacteria of nitrated naphthalene and derivatives,
     Environm. Mut. 3: in press (1981).

68. F.F. Kadlubar, J.A. Miller, and E.C. Miller, Hepatic
     microsomal N-glucuronidation and nucleic acid binding
     of N-hydroxy arylamines in relation to urinary bladder
     carcinogenesis, Cancer Res. 37:805 (1977).

69. F.F. Kadlubar, J.J. Miller, and E.C. Miller, Guanyl O^6-
     arylamidation and O^6-arylation of DNA by the carcinogen
     N-hydroxyl-1-naphthylamine, Cancer Res. 38:3628 (1978).

70. F.F. Kadlubar, L.E. Unruch, F.A. Beland, K.M. Straub, and
     F.E. Evans, In vitro reaction of the carcinogen,
     N-hydroxy-2-naphthylamine, with DNA at the C-8 and N^2
     atoms of guanosine and at the N^6 atom of adenine,
     Carcinogenesis 1:139 (1980).

DISCUSSION

DR. GLICKMAN: Your assumption or belief that the penultimate mutagen is hydroxylamine seems odd because this mutagen is suspected to induce GC to AT transition and I believe you were looking at a frame shift reversion in the Salmonella system.

DR. ROSENKRANZ: It's an arylhydroxylamine, or its hydroxamic acid ester adduct. I did not mean that free hydroxylamine was the mutagen.

DR. THILLY: The importance of transposition effects in environmental mutagenesis in people is an interesting issue. Of course, Graham Walker, among other people, have mentioned it as a fine way of mutating genes, and probably the best way you could wipe out a gene is to insert a virus in the middle of it. I believe Shapiro showed a dozen years ago that SV-40 was a fair mutagen for HGPRT in V79 cells.

It occurs to me, in thinking about human mutagenesis, that by virtue of our assays, we look essentially at base pair substitution and not at potential transposable elements or viral insertions. We don't really have a feeling, when we talk about the mutational load in people, whether or not we're dealing with a spontaneous error in the hereditary process, whether it's chemically induced, or whether it is the result of viral fragment insertion. Are we really at this state of ignorance or do I simply misunderstand the problem?

DR. ROSENKRANZ: I don't know.

DR. ARLETT: I'm not going to take up what is mutation in man, but I think it's appropriate to say here that 1,8-dinitropyrene is a very effective mutagen in mouse lymphoma cells, a fact which helps to bridge the studies in bacteria and in mammalian cells. One of the interesting things about this mutagen, which the bacterial studies might not have predicted, is that although clearly a very potent frame-shift mutagen in bacteria, it may not be so in mouse cells. It's become something of a truism that oubain resistance can't be induced by frame-shift mutagens, though we don't really know whether this is true. If we assume this to be correct, however, the fact that we do get oubain-resistant mutants in our mouse lymphoma cells suggests that this mutagen is behaving differently in mouse cells than it is in bacteria. Our tester strains are giving us some indication of mutation but they're also giving us indication of a different type of mutation that we see in mouse cells. I'm not yet sure what we really see in human cells.

A final comment about 1,8-dinitropyrene; you have to incubate the cells with it for a minimum of at least 24 hours, with 48 hours being the optimum time, in order to get mutation at all. If you do a C times T plot of mutation, whichever end point we take, oubain resistance, TG resistance and so on, we get more mutations in 48 hour treatments than our 24 hour treatments. The question is, have we got an inductive response?

DR. MAHER: May I ask what are the conditions of the cells during that time, thinking about our Salmonella metabolism studies, where you're incubating for 24 and 48 hours.

DR. ARLETT: Division time in the presence of maximum concentrations is absolutely normal.

DR. THILLY: We've observed mutation for the human lymphoblasts using 1,8- and 1,3-dinitropyrine but at much higher concentrations than needed for bacterial mutations.

DR. MAHER: It is our experience with similar experiments, but not using the nitropyrenes, that if surviving cells are allowed to divide two to three times during the 48 hr incubation period, an initial cell killing is obscured unless the survival has been decreased to less than 37 percent.

DR. MERMELSTEIN: Could you indicate how the level of mutagenicity of these compounds in mouse lymphoma cells compares to positive controls in the Salmonella test?

DR. ARLETT: I don't think it's a particularly difficult job to determine whether compound X or Y is mutagenic or not in any particular system, but I do think it's extraordinarily difficult to answer the question of how mutagenic it is. I will address this problem in the final discussion.

DR. ROSENKRANZ Let me make just one statement with respect to Salmonella. To demonstrate the mutagenicity of nitroarenes in Salmonella you'd have to have conditions allowing enzyme (nitroreductase) induction; following enzyme induction, mutagenicity of nitropyrenes in Salmonella is observed at concentrations that are about 100 times lower than those causing any toxicity.

FINAL DISCUSSION:

CONCLUSIONS AND SYNTHESIS

MODERATOR: CHRISTOPHER W. LAWRENCE

DR. LAWRENCE: We have information about the way mutagens interact with DNA, and attempts have been made to systematize this into a rational scheme which would perhaps allow predictions to be made about the mutagenicity of untested compounds. What is the present status of this approach?

DR. SHAPIRO: I've predicted one mutagen in my lifetime so perhaps its a mild credential. I think what chemists can to some extent predict is DNA reactivity. You can extrapolate from known classes of DNA reactivity to other related chemicals and say this shows some possibility of reacting with DNA. For example, alkylating agents give a good correlation, reasonable as these things go, between SN-1 reactivity and a tendency to mutagenesis and carcinogenesis. Other chemicals that have this type of reactivity are potential candidates for mutagenicity, though other unpredictable things may intervene, such as metabolism, detoxification mechanisms and repair. These aren't inherent in the chemistry: you start with the reactivity and then get the rest of the information before you get to mutagenicity.

DR. GLICKMAN: I think an important point is that you have to distinguish between two classes of compound. One class includes those chemicals that react directly with DNA; the chemist can say yes or no and you can test it; it reacts or doesn't react. The other class includes those chemicals metabolized perhaps to different products in different parts of the body. This class presents the real problem. It is much more difficult to ask the question what these chemicals are going to produce. Yet it is the pertinent question. I don't feel very much confidence that prediction is really going to be achieved within the next 10 or 20 years with this class of mutagen.

DR. SHAPIRO: So the chemist has his best chance with direct-acting mutagens.

DR. GLICKMAN: That is likely to be correct, but even the "simple" ones like sodium bisulfate are not well understood. It might, however, be possible just to identify them.

DR. BRIDGES: It seems to me the chemist can help an awful lot but he's clearly not to be allowed to sit by himself and tell you what's going to happen. When trying to evaluate genotoxicity I think the chemist must be an important part of the team that does it, however, and particularly, I think, the chemist can help you in designing the right sort of mutagenicity test, can tell you how to do your test. I would like to build on a point that Herb Rosenkranz spoke to this morning; that if you confine yourself to a standard protocol, in a considerable minority of cases you're

going to get results which are really not very informative,
because you didn't think about the chemistry of the substance when
you designed the standard protocol. I view with utter horror the
increasing tendency to lay down a protocol whereby you do
such-and-such a test in such-and-such a way, and if you have done
it and it's negative, you're safe. To me this is not science;
this is something else. I would like to see chemists sitting down
with biologists and thinking about tests and how substances might
be metabolized.

DR. LAWRENCE: Following up on an earlier point, what do we know
about the potential for enzymes to activate compounds? Are there
likely to be many enzymes which can convert relatively unreactive
substances into active ones? What do we know about the repertoire
of enzymatic activities that can do these things?

DR. ROSENKRANZ: Once you start looking, you may find them. We
just haven't looked.

DR. ARLETT: We always prefix any remarks we make about results
from mouse lymphoma cells with the statement that we don't really
know what metabolism this cell has.

DR. LAWRENCE: so that's a big area of ignorance at the present⌐

DR. SHAPIRO: May I point out, while we're looking at areas of
ignorance, there is a great area of ignorance even in chemistry.
After a chemical does get to DNA, we do not always know what
particular property of DNA it has changed that leads to its
biological relevance. For example, with the polycyclic aromatic
hydrocarbon, one can point to the amino group of guanine as a site
of reaction, yet there is a great ignorance as to how benzopyrene
diolepoxide, for example, works; it doesn't seem to perturb DNA
very much, interfere with it, unwind it or do anything terribly
dramatic. There's a great ignorance as to how it produces a
biologically significant event. It is much more carcinogenic than
other hydrocarbons are, while others of similar size are not
carcinogenic. Its adduct just lies in the minor groove as an
inert blocking bulk, saying to an enzyme, thou shall not pass:
that doesn't give any explanation for its potent properties
compared to other polycyclics of equal bulk, which could lie there
just as well and which can also react with DNA. There is a great
area of mystery here, where more knowledge is needed.

DR. ROSENKRANZ: I mentioned earlier the nitroarenes, in looking
at the rather simple ones, the nitronaphthalenes and the
hydroxylaminonaphthalenes, we find that 2-nitronaphthalene is a
"good" mutagen while the 1-nitro isomer is a "poor" one.
2-Aminonaphthalene is carcinogenic while 1-aminonaphthalene is
not. The penultimate metabolites, the corresponding

hydroxylamines, both react mainly with the same base in the DNA
but they form different adducts. In one case (1-hydroxylamino-)
it is with the O^6 of guanine, in the other case
(2-hydroxylaminoaphthalene) it is with the C-8 of guanine and the
exocyclic amino group of guanine and adenine.

DR. SHAPIRO: Different carcinogens react around the entire
periphery of bases at different sites in a way that leaves you
dizzy. There's no unifying principle of geometry or anything.

DR. ROSENKRANZ: Probably the conformation of the DNA adduct in
one case is different from the other.

DR. THILLY: In order for something to be dangerous as a chemical
in causing cancer or genotoxicity, somebody has to make it and
distribute it so people get exposed to it and people have to take
it up. Once they take it up, the physical chemical properties of
the particular compound will have effects on distribution, the
metabolism, the various reactions for getting the stuff excreted.
Only a very small amount of the total chemical taken into an
animal is going to have any reaction with cellular macromolecules,
most of which are not DNA, though because of the high
nucleophilicity of DNA, we're going to have a disproportionate
cross-section for DNA. Then we have a wide variety of places in
DNA that can react: we have all of those that have been discussed
here. We then have a plethora of repair systems which as far as
we can now tell are relatively specific each representing a wide
variety of gene products in which we may expect that the specific
chemical identity of the adduct is going to determine if and what
kind of mutation is going to take place. If mutation is someway
related to the appearance of a carcinogenically potent cell -- and
that's a big if -- we then have to get the darn thing expressed;
we are exposing ourselves to chemicals which cause tissue-level
changes. The hypermutability of XP cells has been thought to be
important in the appearance of skin cancer, but in fact, it's just
as reasonable on an arithmetical basis to propose that the
hypersensitivity to killing, will cause tissue-level changes and
increase the probability of expression of previously-existing
mutants in the skin. In other words, we could have both an
initiation and a promotional effect caused by sunlight in persons
who are both hypersensitive and hypermutable by UV.

This is to point out that even after the DNA is repaired and
any change that has been taken place is fixed, we still have the
opportunity for chemicals in the environment to alter the
probability of expression of unfortunate health events. By the
time you get done with this series of reactions, the poor chemist
is rightfully thinking of clamming up and wishing the whole thing
hadn't been brought up in terms of making predictions from
structures.

Persons of widely different training and outlook have got to work together as a community to try and unravel these very complex problems. The analytical chemist isn't going to be able to work it out because some of these biological phenomena are very specialized; persons trained in P-450 studies are generally not trained in genetics; people trained to work in genetics are generally not trained in tissue pathology.

DR. LAWRENCE: Let us move on to another question. Obviously, because of their greater sophistication and power of analysis, prokaryote systems have told us most of what we know about basic mutagenic mechanisms. Yet, on the other hand, it's natural enough that we should have a great interest in eukaryotes, and particularly ourselves. To what extent does present knowledge about eukaryote mechanisms tell us that they are reasonably similar or substantially different from prokaryotes?

DR. RIPLEY: I think that that's not necessarily the most useful way of thinking about it. It's certainly not the way a person who is interested in mechanism of mutation necessarily views it. I think that, particularly starting from the point of view of chemical change in the DNA, that what we are really talking about is how do enzyme systems interact with these changes in DNA. There can be large differences among prokaryotes. There are probably large differences among different kinds of eukaryotes. It's not necessarily the most relevant question to ask whether there is a difference between prokaryotes and eukaryotes in terms of mechanism; maybe the most relevant question is how does the enzyme system interact with the altered DNA.

DR. LAWRENCE: A possible difference however might be that eukaryotes, because they have far more DNA to replicate under normal circumstances, have evolved mechanisms to do this much more accurately. They might have accessory factors associated with their polymerases which recognize much more subtle distortions of a template then exist in either viruses, which generally have rather small genomes, or bacteria which have not much larger ones.

DR. BRIDGES: I think you generalize about bacteria. You mustn't forget the point that Graham Walker made earlier; that is, that we tend to assume that E. coli and Salmonella are typical. They are in no way typical. You can go out into nature and pick up various methanobacter, methanomonas, proteus species, and they don't mutate with mutagens because they haven't got error-prone repair. If we used a testing system employing Proteus mirabilus, we wouldn't be as worried today about environmental mutagens as we are. I think DNA damage is something a cell has to deal with, you can't just leave it there and do nothing. There are so many different ways in which cells can handle it, and we're just beginning to understand some of them, that you really cannot predict exactly how the cells are going to handle them even from one species to another within the same genus, and certainly not from one class to another.

DR. SAMSON: Of course, that applies to the whole animal as well. Different organs handle DNA damage differently, and that's very important in terms of testing, too. You're going to get different dose responses according to which organ you choose to make the test with.

DR. LAWRENCE: So what is the implication for prediction and testing?

DR. SAMSON: Take the worst view, I suppose.

DR. BRIDGES: It's easy to say that, but it's very hard to put it into practice because you end up without an industrial society, because you could end up with no chemicals you can use; all of them, somewhere, do something to something. If you make worse case assumptions, you could end up by banning everything, so you can't make worse case assumptions. Somehow the scientific community has got to really knuckle down to the problem of making some sort of quantitative predictions, taking into account all of the various problems. We just cannot go on worse case analyses all the time.

DR. GLICKMAN: An example of the kind of problem one can get into is the question of ionizing radiation. In E. coli, using the laci system, we estimated $2 \times 2 \ 10^{-13}$ nucleotide changes per rad. It was delightful to see that this rate is the same in yeast.

But one is not measuring point mutation in eukaryotes their frequencies are about 3 per 10^{-11} nucleotides per rad, and this is not significant compared to a 5 order of magnitude higher deletion frequency. One is really looking at a very different class of mutation and estimating risks and the like is not really very simple.

DR. SHAPIRO: I'd like to go just on from that point. When you come out of a conference like this you're always overwhelmed by how complicated everything is, and there are mechanisms within mechanisms. Yet it's a big jump to go from saying there's something complicated to saying that you can't therefore draw any conclusions at all. Logically, that would mean the regulatory agencies should draw up an alphabetical lists of all the chemicals in their environment, and ban say every 10th or 50th, depending on their resources. The point is, as information accumulates one does move from a position of total ignorance to the capability, with greater and greater probability, of correctly selecting things that are harmful. To just say it's complicated, so we can't make any statement, is to deny the responsibility that the information you have gives you.

DR. LAWRENCE: We know that there are mechanisms which can be saturated or are inducible and others we know less about. To what extent does our knowledge of these mechanisms, either in prokaryotes or different kinds or eukaryotes, allow us to define the problem which has been central to much of risk-assessment; is the linear extrapolation from high acute doses down to low doses, or from acute doses to chronic doses, a good one? What does our information about basic mechanisms tell us about that, if anything at all.

DR. BRIDGES: What do people think about Bill Thilly's point about the low capacity systems? I was much taken up by this because the cross-link repair pathway we uncovered in E. coli, which is rep dependent, is a low capacity system. It seems to be able to cope only with somewhere between 2 and 5 cross links in the chromosomes. It seems to be error prone, and we wouldn't have seen it unless we had eliminated most of the other repair in the organism and been lucky enough to have found a mutant which seemed to block it. Adaptation is also really quite a low capacity system, as is class 3 single strand break repair in E. coli. Some of these are error free, some are error prone in these systems, but they're there. We tend to ignore them in favor of the massive excision repair and the massive daughter strand gap repair, which are easily detected. These others are often difficult to assay biochemically and they're often difficult even to identify in the first place. How important do we think they are?

DR. THILLY: The whole question of saturable low capacity repair system is really a variation on the more practical issue in "testology" of insisting that compounds be evaluated at the lowest detectable limit. Returning to the earlier phrasing of the question, I think we know enough to be able to say that linear extrapolation from high concentrations does not make biological sense; we know too many systems which are clearly non-linear. Chemicals should be tested at a concentration to which human cells, whether they be gonadal or other cells, are really exposed. If concentrations in that range cause a significant change in the rate of genetic change, then I think it's an important response. The mechanisms principally responsible for genetic change at relevant concentration levels are the ones that should be studied in order to understand the chemical/biological reaction leading to genetic disease in people.

Those low capacity systems are important. At high concentrations they're all saturated. When experiments are performed at high mutagen concentrations, the effects of saturable low capacity systems would be masked. The low capacity systems

should be of highest priority in the study of how people will
respond.

DR. LAWRENCE: You seem to be implying that we might well expect
to see a lot of low capacity systems perhaps because we're exposed
to most damaging agents at rather low concentrations, and so
during the course of evolution such systems are probably highly
efficient. Should we be looking for more of such low capacity
system perhaps by knocking out all the other systems?

DR. L. PRAKASH: The only thing that bothers me about that is why
is it so difficult to demonstrate an adaptive response.

DR. SAMSON: They're very easily saturable.

DR. GLICKMAN: If you're looking at mutagenesis after very low
doses of mutagen, it is, of course, difficult to see an effect
that you can really measure, particularly with a high background
level. However, if you are in a position to say this agent will
cause a specific kind of damage and chose a mutation which reverts
by that specific event, one might be in a position to do a great
deal better than we can do now. Specificity is here the key. We
don't, however, yet have the capacity to do this in mammalian
systems. We don't yet even have agents of known specificity.

DR. L. PRAKASH: But in terms of mutagenicity for higher
organisms, people, why does it really matter what the specificity
of the mutagen is. All you really care about is that it's causing
mutation.

DR. GLICKMAN: It matters for two reasons: The first reason is
that in order to detect the mutagen which is only causing a
restricted kind of change among the large background, one has to
look at a system which specifically detects that change. So, in
order to detect a low level event, mutagenic events, one needs to
know the specificity.

DR. L. PRAKASH: You can use a forward mutational system, then you
no longer need specificity.

DR. GLICKMAN: The point is they don't exist at the moment and I
feel they're needed; specific systems in mammalian cells which
allow one to look for a specific mutational event. Supposedly
simple cases, such as alkylating agents, are suppose to cause
mutation by GC to AT, but, as Margaret Fox has shown, if one uses
a forward mutation system it is not alway possible to find
mutations.

DR. THILLY: Alkylating agents also cause frame shift mutations in

mammalian cells. People suffer from many kinds of mutations, they
don't suffer from base pair substitutions only.

DR. GLICKMAN: The point is that I don't believe that systems
exist which allow low level mutagenesis to be estimated
effectively. This has largely to do with our inability to make
use of specificity.

   A second point is, "What does it matter?" It matters greatly
in that the genetic effect of a given agent will, at least to some
extent, depend on whether it's causing base substitution or frame
shifts. The genetic potential of a given event is going to depend
very much on whether it knocks out a gene or whether it causes a
single base substitution which rarely produces a genetic
deficiency; it will alter the gene sequence, it may even alter the
protein, but it's not necessarily going to result in a change in
the phenotype.

DR. MAHER: But, if you measure changed phenotype, what difference
does it make? Who cares whether it was due to a frame shift or a
decision or a base pair change.

DR. GLICKMAN: You care if the mutation frequency is one in a
million, and is composed of a thousand different potential
events. The chemical that you're looking at is causing 10 of
those. You want a system which detects that given event and not
other events. Then you can ask whether the presence or absence of
that repair system is changing the mutagenicity of that chemical.
It will work only under conditions where you're detecting those 10
events and not as 10 events in the background of a thousand
events. I think it's that kind of specificity which is absent in
any of the testing systems in mammalian work.

DR. SARASIN: One way to study this problem could be the use of
plasmid which contains an eucaryotic and a procaryotic origins of
replication and various gene targets. Then you can treat in vitro
and in vivo the plasmid DNA with carcinogens and then give it to
human cells in order to be repaired or mutagenized. After that
step, one could rapidly analyze mutations in E. coli host and even
determine the genetic change.

DR. MAHER: If I understand it rightly, your virus is treated in
vitro like isolated DNA. If you treat isolated DNA with the
well-studied mutagen N-acetoxy AAF you get a fluorene adduct with
the acetyl group still on, bound mostly at the C-8 position of
guanine, but with some at the N-2 position. But our recent
studies show that when you treat the human cell itself with this
mutagen you do not get that acetylated adduct, although the adduct
you obtain is mutagenic in these cells. This is because human

cells have a whole series of de-acetylases. Therefore, results obtained with the vrius will differ from results in the human cell.

DR. SARASIN: You can treat the plasmid DNA in vitro or you can transfect cells with the plasmid and then treat the plasmid DNA inside the cell. In this condition, plasmid DNA will receive exactly the same carcinogen adducts as cellular DNA.

DR. SHERMAN: I'd like to point out another approach that I think is being considered, involving the new procedures of recombinant DNA. One could actually clone DNA from human cells and sequence it. If you look at a long sequence, and if the fidelity of your sequencing is strong enough, you can detect alterations. You could look at the same homologous sequence in a somatic cell or even in a germ line cell, for example, in a patient before and after chemotherapy. You could detect very rare changes, in some cases you would have a complete knowledge of the dose that was given, and you also could measure the adducts directly in the DNA. This information could be related directly to results from short-term microorganism testing systems.

In the very near future, you might be able to tell base pair changes within a human associated with a specific dose, and you could normalize that to your short-term testing system. You could also do this in a population where you could have people exposed to industrial hazards or pollutants. It would take into account all the complexity of repair systems and just give you a final answer. Of course, you'd have to relate this to some functional targets; there are some parts of DNA which when altered do not produce phenotypes, although that knowledge is also almost available now.

DR. GLICKMAN: I think that is certainly true, and for many of us around the table quite clearly the direction we have to go. I've described some results in E. coli and in terms of specificity that can now be better done by DNA sequencing. To a certain extent, however, one has great difficulty in terms of mutagenic specificity using DNA sequencing because of the large target. One has to compromise between large target and little information and a little target and no information, or if you chose well, useful information. But in the future, many of us around the table are clearly going to be sharing experiences in the question of specificity in higher organisms, using these techniques.

DR. SHERMAN: There is a large army of trained people who can sequence DNA, and believe they will run out of things to sequence pretty soon.

DR. THOMPSON: Does this apply only to germ line mutations?

DR. SHERMAN: It could be from a germ line cells.

DR. THOMPSON: How do you distinguish between spontaneous mutations and induced ones?

DR. SHERMAN: You do not look at the same change at a site. In other words, you can sequence several kilobase-pairs, and find changes at different sites. It is not like a tester system where you are looking at reversion or forward mutation at a certain site. One can either examine a large population and look for polymorphism or one can examine an individual before and after a specific exposure, which would be the best way. You could do it both in a somatic cell and a germ cell line.

DR. BRIDGES: The problem is that such methods are essentially going to be workable only following high dose treatments, I think. That gets us back to the problem -- and this is true with all the whole animal data that you ever get from mutagenicity, or indeed carcinogenicity, studies -- that with very rare exceptions that you're looking so far up the dose response curve that the results don't necessarily bear much relation to what's happening at very low doses.

DR.SHERMAN: You are not just using the frequencies of DNA changes; you also have the potential of measuring the amount of DNA adducts as well as using short-term testing systems. If the DNA adducts and the tester systems can be measured at low doses and if simple relationships exist, then the values can be normalized and extrapolated to lower doses.

DR. MAHER: In theory, in what cells of the patient's body would you think you could find the change after chemotherapy?

DR. SHERMAN: People are doing this from blood.

DR. SARASIN: It has been claimed that one ml of a blood sample from people receiving high chemotherapy treatment is sufficient, when you use the technique I described previously. You can check for the presence of platinum adducts for example, in just one ml of a blood sample. This result can be useful to determine the maximum dose of a chemotherapeutic drug to use.

DR. GLICKMAN: You can detect the lesion, but that doesn't give you any genetic information. Unless you have a specific selective system, one cannot sequence the mutation.

DR. SHERMAN: I understand that. I am saying this procedure will give you the DNA changes, and then you need some theoretical extrapolation. This method could also be used to deduce the

frequencies of gross aberrations, such as chromosomal
rearrangements, the so-called transpositional type changes.

DR. THILLY: I think Fred's on the right track here. It's just a
matter, if you will, of coming up with the plumbing to finish it
off. There are plenty of cells in a ml of blood. Dick Albertini
has already shown that he can measure gene locus mutations in
chemotherapy-treated patients in far less than a ml of blood.
There's no reason to imagine that the technology can't be worked
out, if we, as a community, decide that that will give, as you
suggest, a final answer.

DR. SHERMAN: Well, at least, in my mind, it looks potentially as
if it would be a good approach, and should certainly be considered
seriously at the present time. The technology is really
advancing, and the sequencing procedures are constantly being
improved.

DR. ARLETT: I'd like to address the problem of how we measure
potency. I think we can answer the question of whether compound
X is a mutagen or not, but that isn't sufficient for most people.
They don't simply want to know is it or is it not a mutagen; they
want to know how mutagenic it is. I find this a very difficult
problem; the question is, what do we mean by potency?

Now, in our example, with 1,8-dinitropyrine, we found that we
got very clear and unequivocal induction of mutation to oubain
resistance, thioguanine resistance, ara-C resistance, and
methotrexate resistance after 48 hour treatments, in a
dose-dependent fashion. There was zero toxicity with these
treatments, and this gives a problem because potency is often
normalized against survival.

Such a procedure works for MMS, EMS, 4NQO or ionizing radiation
and UV, and you can pull together a whole ragbag of things in this
way, but all our dinitropyrine data really goes up the vertical
axis. This implies nitropyrine is infinitely mutagenic and that
is absolutely nonsense. The next thing which we can do is to plot
induced mutation frequency against dose, or more properly against
concentration x exposure time, because the nitropyrine exposure
were 48 hour treatments with about 1-2 mg/ml, the 4NQO very much
lower doses for 2 hours and so on. Using this type of plot, we
possibly have something like a 1,000-fold difference between the
two mutagens, and that, I think, is much more meaningful, although
I'm not sure whether it's the whole answer. It doesn't mean I
think that nitropyrine is necessarily a very safe sort of
chemical, compared to 4NQO. I don't think this exercise actually
tells anybody how potent nitropyrine is, either as mutagen or
carcinogen.

DR. GLICKMAN: When talking about potency of a chemical, one really has to decide why you're interested in it, and from what point of view -- from a mechanism point of view or legislative point of view? When one talks about dose are you talking in terms of exposure, i.e., teaspoons of mutagen, or about actual dose, i.e., adducts per base pair? I want to make a comment on potency of chemicals in terms of adducts per nucleotide. The observation is that whether one is talking about O^6-ethyl, O^6-methyl, aflatoxin, benzo(a)pyrene, or any bulky adduct -- the actual mutation frequency induced per adduct does not differ enormously in the mammalian cell. That's a very big surprise because these lesions are chemically very different.

DR. MAHER: We have chosen to compare the mutagenic potency of various chemical carcinogens in human cells on the basis of how many mutants are induced per DNA adduct. But when we did so, we found that the number was much higher if the target population consisted of excision repair minus human cells, rather than excision repair-proficient normal cells. This is to be expected since I presented data indicating the frequency of mutations depends upon the average number of DNA adducts remaining unexcised when DNA replication occurs. Therefore, it really isn't correct to compare the frequency of mutations induced as a function of the initial number of total adducts bound to DNA if you are working with cells which are capable of removing these adducts. Working with the excision-deficient XP12BE cell strain, and a series of structurally related aromatic amines, we find a striking similarity between the number of mutants induced per DNA adduct. However, we also recognize from our work with ENU that the relationship cannot be expected to hold true if the predominant lesion such as the phosphotriester formed by ENU (56 percent of the bound material) is not active in producing mutations.

DR. GLICKMAN: We observed this when we were comparing our results in Chinese hamsters for O^6-ethyl G with Wogan's result with aflatoxin.

DR. THILLY: About this teaspoon business. It is sometimes difficult for a person who has worked in a genetics community and sees the doses in the real world of public health in terms of adducts that teaspoons help. The points that Colin has raised are extremely relevant because they are filters through which the chemical must go before it gets to the DNA. If one of those filters is absolute, that would make the chemical lack potency in terms of human genetics, even though I might create a condition where I put a lot of adducts from it on the DNA.

DR. GLICKMAN: The chemist wants to predict potency. That is in terms of adducts. In each case we need to define in whose terms one is measuring.

DR. THILLY: I'm pretty sure the public wants it in terms of, "will these chemicals hurt us?"

DR. ARLETT: We have to make use of the data and the potency issue is an important one in any attempt to make use of the information we have.

DR. ROSENKRANZ: May I ask a question? The duration of the nitropyrene exposures was 48 hours. What would be the risk if I were to be exposed 24 times for 2 hour periods?

DR. ARLETT: If we plot concentration times exposure time, we see fewer mutants with 24 hours exposure than we do with 48 hours, raising the possibility of an inductive response. Not only is there this question of the public wanting to know whether different substances will harm them, but we have an enormous problem in the possible variability in response by different members of the public. For example, in normal individuals the range of radiation response measured in cells is about 130 rads plus or minus 15.

We don't yet know the range of response of human cells to ENU, MMU, and so on, but I suspect we shall find quite a wide range of response from individual to individual. It's also conceivable that individuals themselves may change.

Participants

Front Row: Thompson, Maher, Walker, Lawrence, Sherman, Shapiro,
              Ripley, Glickman
Second Row: Thilly, Bridges, Rosenkranz, Arlett, Eberle, Loeb,
              Samson, Doubleday
Third Row: Ohlsson-Wilhelm, Jachymczyk, Haladus, Polakowska,
              Sarasin